STRUCTURAL
INTERMETALLICS
and Intermetallic
Matrix Composites

IIT Kharagpur Research Monograph Series

Published Titles:

Structural Intermetallics and Intermetallic Matrix Composites, *Rahul Mitra*

Digital Geometry in Image Processing, *Jayanta Mukhopadhyay, Partha Pratim Das, Samiran Chattopadhyay, Partha Bhowmick, and Biswa Nath Chatterji*

Mathematical Techniques for Wave Interaction with Flexible Structures, *Trilochan Sahoo*

Microfluidics and Microscale Transport Processes, *edited by Suman Chakraborty*

Modeling of Responsive Supply Chain, *M.K. Tiwari, B. Mahanty, S. P. Sarmah, and M. Jenamani*

Micellar Enhanced Ultrafiltration: Fundamentals & Applications, *Sirshendu De and Sourav Mondal*

STRUCTURAL INTERMETALLICS
and Intermetallic Matrix Composites

RAHUL MITRA

CRC Press
Taylor & Francis Group
Boca Raton London New York

CRC Press is an imprint of the
Taylor & Francis Group, an **informa** business

CRC Press
Taylor & Francis Group
6000 Broken Sound Parkway NW, Suite 300
Boca Raton, FL 33487-2742

First issued in paperback 2019

ISBN-13: 978-1-4665-1186-6 (hbk)
ISBN-13: 978-0-367-37769-4 (pbk)

Library of Congress Cataloging-in-Publication Data

Mitra, Rahul, 1966-
   Structural intermetallics and intermetallic matrix composites / Rahul Mitra.
      pages cm. -- (IIT Kharagpur research monograph series ; 6)
   Includes bibliographical references and index.
   ISBN 978-1-4665-1186-6
   1. Intermetallic compounds. 2. Metallic composites. 3. Composite materials. 4.
Alloys. I. Title.

TA481.M58 2015
669'.9--dc23   2014045019

Visit the Taylor & Francis Web site at
http://www.taylorandfrancis.com

and the CRC Press Web site at
http://www.crcpress.com

Contents

List of Figures

Preface

Significant research has been carried out on structural intermetallics for several decades, involving both experimental and theoretical approaches. As a result, the structure–property relations of these materials are reasonably well understood, which has led to a road map for further research to develop high-performance materials for several diverse engineering applications. Work is in progress in many parts of the world to develop selected multicomponent intermetallic alloys based on silicides and aluminides for specific applications, particularly at elevated temperatures and in different types of extreme environments. Of course, there are excellent reviews and book chapters on many of these intermetallics. This monograph has been drafted as a part of the Diamond Jubilee Series of the Indian Institute of Technology Kharagpur. The author has been working on silicides for two decades and has also taught topics related to intermetallic alloys for a postgraduate course on advanced materials. A student learner often finds it difficult to grasp the complexities of the structure of intermetallics and their effect on various physical and mechanical properties. Keeping the requirement of students in mind, the first four chapters of this monograph are devoted to necessary fundamental aspects including thermodynamic principles, phase diagrams and crystal structures, processing methods, deformation and fracture mechanisms of ordered intermetallics, and oxidation behavior with mechanisms for protection against environmental degradation. The fifth chapter focuses on possible applications on the basis of the attractive properties of aluminides and silicides. The last four chapters contain exhaustive reviews of the existing literature on selected structural silicides and aluminides. The contents of this monograph are expected to be helpful to students interested in learning about intermetallics, as well as professionals beginning their research in this area.

The author would like to thank Professor K. K. Ray and Professor S. K. Roy, senior colleagues of his department, for their encouragement to write this monograph. The assistance received from Dr. Monali Ray, a postdoctoral fellow in my research group, in preparing the reference lists for different chapters in a very short time is gratefully acknowledged. The author owes a lot to all his students and collaborators for their contributions in extending my understanding of the subject. The author would also like to thank Dr. Gagandeep Singh and Ms. Marsha Pronin, editors at Taylor and Francis, for their valuable guidance during preparation of the manuscript. This monograph would not have been possible without the constant support and encouragement received from his wife, Mrs. Barnali Mitra, and daughter, Miss Rituparna Mitra, as well as the blessings of his parents, Mr. Paritosh Kumar Mitra and Mrs. Smrity Rani Mitra.

Rahul Mitra
Kharagpur, India

Abbreviations

APB	Antiphase boundary
APD	Antiphase domain
BDTT	Brittle-to-ductile transition temperature
CRSS	Critical resolved shear stress
CSF	Complex stacking fault
CSL	Coincidence site lattice
CT	Compact tension
CTE	Coefficient of thermal expansion
DS	Directionally solidified
EAM	Embedded atom method
HIP	Hot isostatically pressed
HP	Hot pressed
KW	Kear–Wilsdorf
LPPS	Low-pressure plasma spraying
MA	Mechanical alloying/mechanically alloyed
PM	Powder metallurgy
Poly	Polycrystalline
RT	Room temperature
SC	Single crystal
SEM	Scanning electron microscope
SENB	Single-edge notch bend
SHS	Self-propagating high-temperature synthesis
SISF	Superlattice intrinsic stacking fault
TEM	Transmission electron microscope
UHV	Ultrahigh vacuum
WB TEM	Weak-beam transmission electron microscopy

1

Phase Equilibria and Structure

1.1 Introduction

In many of the binary equilibrium phase diagrams for alloys, new phases are found at intermediate concentrations and their range of existence does not extend to pure components. Either these phases are line compounds or they are characterized by their nonstoichiometric composition and extended range of compositions. The line compounds with a fixed ratio of metallic components are often called *intermetallic compounds*. This terminology is appropriate only for stoichiometric compositions. It is not suitable for alloys with nonstoichiometric or extended range of compositions, and therefore such materials are referred to as intermetallic phases or alloys.

For substantial or complete solid solubility, the Hume-Rothery rules need to be satisfied: (i) the difference of atomic radii should not exceed 15%; (ii) the difference of electronegativity (chemical affinity) should be small; (iii) the crystal structures of solute and solvent must match; and (iv) the number of valence electrons should not be very different. The formation of intermetallic phases is preferred when the aforementioned rules are not satisfied. For example, both gold and copper have a face-centered cubic (fcc) structure, but the difference between their lattice constants is ≈12.8%, which promotes the formation of intermetallics in the Cu–Au system.

The formation and microstructural evolution of intermetallics depend on their thermodynamic stability. Very often, metastable phases with inhomogeneous compositions are formed through solidification, and suitable heat treatment is required for the evolution of equilibrium phases. For desirable mechanical properties or for carrying out forming operations, it may be necessary to stabilize desirable metastable phases through the addition of suitable alloying elements. Furthermore, the mechanical properties of the intermetallics are strongly dependent on their crystal structures. Hence, knowledge of phase equilibria along with crystal structures is necessary to understand the processing–structure–property relations of various binary and multicomponent intermetallic alloys.

1.2 Stability of Intermetallic Phases

A reduction of the Gibbs free energy of the system provides the driving force for the formation of intermetallics. The stability of the intermetallic phase depends not only on the reduction of free energy due to its formation but also on the free energies of the phases in equilibrium with the intermetallic phase. An example of a two-component system (A–B) is shown in Figure 1.1a.[1] In this system, α and β are solid-solution phases along with a stable intermetallic, I, and a metastable intermetallic, I′. In Figure 1.1b, the Gibbs free energies of the phases present at temperature T are plotted as a function of atomic fraction, $c = c_B$.[1] For the phases coexisting in equilibrium, the first derivatives of the Gibbs free energy (dG/dc) are equal, such that the chemical potentials or partial molal free energies are equal. Thus, common tangents can be drawn to G–c curves for the phases in equilibrium, as shown in Figure 1.1b. The stable intermetallic phase exists over the homogeneity range

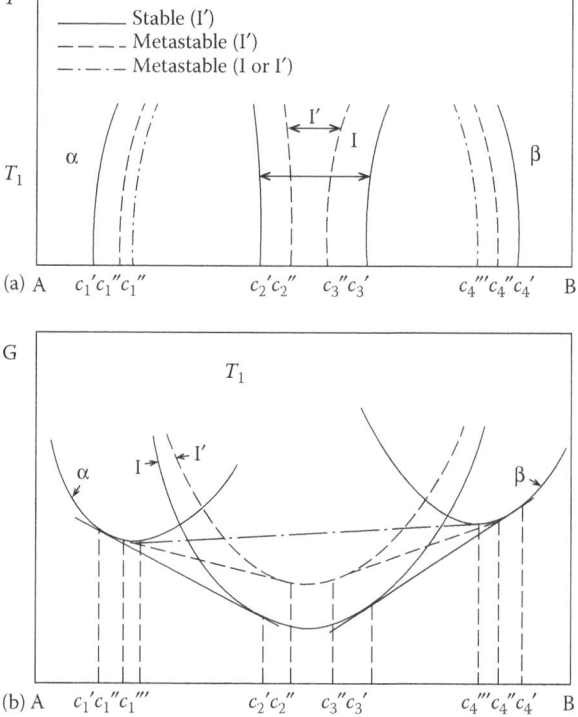

FIGURE 1.1
Schematic illustration of the thermodynamic stability of intermetallic phases (I=stable phase, I′=metastable phase, α and β are solid-solution phases): (a) phase equilibria in the temperature (T)–composition (c) diagram; and (b) the corresponding free energy (G)–composition (c) diagram for true and metastable equilibria at absolute temperature, T_1.

between c_2' and c_3'. The atomic fractions of the phases α and β coexisting with I have compositions of c_1' and c_4', respectively. The metastable intermetallic phase, I', exists over the narrower homogeneity range c_2'' to c_3''. The atomic fractions of the phases coexisting with I' are the phases α and β, with compositions of c_1'' and c_4'', respectively.

The formation of ordered solid solutions is preferred in a binary alloy system if the bonding between unlike constituents is stronger than that between like atoms. In such cases, each atom tries to have the maximum number of unlike nearest neighbors. This is the example of a regular solution with large negative exchange energy[2]:

$$H_0 = H_{AB} - \frac{\left(H_{AA} + H_{BB}\right)}{2} \ll 0 \tag{1.1}$$

where H_{AB}, H_{AA}, and H_{BB} are the heats of formation of A–B, A–A, and B–B bonds, respectively. The heat of formation in the case of a binary intermetallic alloy system (such as Ni–Al or Fe–Al) varies with its composition, increasing to a maximum value and then decreasing. The variation of enthalpy with concentration for binary Ni-, Ti-, and Fe-aluminides is plotted in Figure 1.2 on the basis of the experimental data.[3–7] The heat of formation of intermetallics is usually determined using experiments based on solution calorimetry with the help of high-temperature calorimeters specially designed for such experiments.[8] First, the intermetallic alloy is dissolved in a liquid metal used

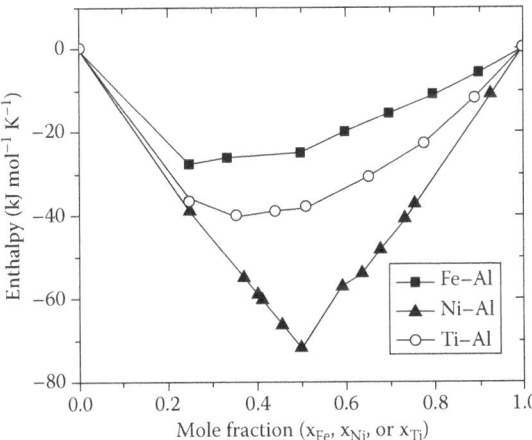

FIGURE 1.2
Plots depicting the variation of enthalpy of formation of binary intermetallic phases with temperature in Ni–Al, Ti–Al, and Fe–Al systems. The data have been taken from the literature. (From Desai, P. D., *J Phys. Chem. Ref. Data*, 16, 1, 109–124, 1987; Kubashewski, O. and W. A. Dench, *Acta Metall.*, 3, 339–346, 1955; Nash, P. and O. Kleppa, *J. Alloys Compd.*, 321, 228–231, 2001; Huang, W. and Y. A. Chang, *Intermetallics*, 6, 487–498, 1998; Samokhval, V. V., et al., *Russ. J. Phys. Chem.*, 145, 1174, 1971.)

as a solvent, and the heat of solution is determined. Subsequently, the heat of solution of the unreacted elemental mixture is measured. The difference between these two heats of solution provides the heat of formation of the intermetallic phase.

For stoichiometric intermetallic compositions, strictly periodic arrangements of atoms are attained. In the case of an intermetallic phase, AB, with the composition $c_A = 0.5$ and an ordered body-centered cubic (bcc) structure, the body-centered and corner positions of a unit cell are occupied by atoms of A and B, respectively, or vice versa. In other words, the ordered bcc unit cell comprises two simple cubic sublattices. If each sublattice site is occupied by only one type of atom, that is, either A or B, each A atom will have a B atom as its nearest neighbor.

1.3 Nomenclature of Crystal Structures

Two types of notation, Strukturbericht and Pearson's symbols, are normally used for the nomenclature of crystal structures of different intermetallic phases. Strukturbericht symbols are a partly systematic method for specifying the structure of a crystal. Here, the structures named A are monatomic (either X or Y, e.g., Al and Fe), Bs are diatomic with equal numbers of atoms of each type (XY, e.g., NaCl, NiAl, and FeAl), Cs have a 2:1 atomic ratio (X_2Y or XY_2, e.g., $MoSi_2$ and $NbSi_2$), D0s are 3:1 (XY_3 or X_3Y, e.g., Al_3Ti and Fe_3Al), E and H are used for perovskite and spinel structures, respectively, and Ls represent ordered cubic structures. It is customary to write the aformentioned structure notations with examples of real materials. A1 (fcc), A2 (bcc), A3 (hexagonal close-packed [hcp]), A4 (diamond), and A9 (graphite) are some examples of monatomic phases. The only example of A with a diatomic composition is the A15 structure, and the examples of intermetallic phases having this structure are Cr_3Si and Mo_3Si. Examples of diatomic phases are the B1 (NaCl), B2 (CsCl), B3 (zinc blende), and B11 (CuTi) structures. Similarly, typical C-type structures are $C11_b$ ($MoSi_2$), C14 (Laves—$MgZn_2$), C49 ($ZrSi_2$), etc., whereas typical D-type structures are $D0_{11}$ (Fe_3C, cementite), $D0_{22}$ (Al_3Ti), $D0_{23}$ (Al_3Zr), etc. Some of the L-type structures are $L1_0$ (AuCu), $L1_1$ (CuPt), $L1_2$ (Cu_3Au), etc.

The crystal structures of the intermetallics can be any of the seven Bravais lattices: cubic (c), hexagonal and rhombohedral (h), tetragonal (t), orthorhombic (o), monoclinic (m), and triclinic (a). The unit cells with each of these crystal structures can further be classified as primitive (P), body centered (I), face centered (F), side-face centered or base centered (S), and rhombohedral (R). Fourteen possible Bravais lattices are represented by the following notations: primitive cubic (cP), face-centered cubic (cF), body-centered cubic (cI), rhombohedral hexagonal (hR), primitive hexagonal (hP), primitive tetragonal (tP), body-centered tetragonal (tI), primitive orthorhombic (oP), body-centered orthorhombic (oI), face-centered orthorhombic (oF), side-centered orthorhombic (oS), primitive monoclinic (mP), side-centered monoclinic (mS), and primitive triclinic (aP). Besides the nature of the atomic arrangement, the number of atoms per unit cell is also included in the notation for a complete description of the unit cell. For example, the notation for the structure of any ordered fcc alloy can be written as cP4, as there are four atoms in its unit cell. In a similar manner, a body-centered, tetragonal-structured unit cell with eight atoms can be referred to as tI8.

1.4 Crystal Structures and Phase Diagrams of Silicides

The major silicide phases of interest for high-temperature structural applications are drawn from the following binary phase equilibrium systems: Mo–Si, W–Si, Ti–Si, Nb–Si, and Cr–Si. The crystal structures and lattice constants of different silicide-based intermetallics are shown in Table 1.1.

1.4.1 Molybdenum Silicides

The binary Mo–Si phase diagram shows the presence of stoichiometric compounds with compositions Mo_3Si and $MoSi_2$.[9] On the other hand, Mo_5Si_3 has a homogeneity range of 3 at.% Si. While $MoSi_2$ has a body-centered tetragonal (bct) structure ($C11_b$, tI8) with eight atoms in the unit cell (Figure 1.3a), the tetragonal unit cell of Mo_5Si_3 has 32 atoms (20 Mo atoms and 12 atoms of Si, $D8_m$, tI32) (Figure 1.3b).[10] Mo_3Si has a cubic structure (A15, cP8) comprising eight atoms in its unit cell, with six atoms of Mo and eight atoms of Si (Figure 1.3c). The bct structure of $MoSi_2$ has a fixed c/a ratio of 2.452 and appears similar to three bcc unit cells, stacked one on top of another with the body-centered site occupied by the atom of Mo or Si, alternately. It has been shown by Francwicz[11] that the c/a ratio of approximately 2.45 remains unchanged with minor alloying of tetragonal-structured $MoSi_2$, and is essential for the stability of $C11_b$ crystal structure. Alloying with transition-metal elements such as Nb, Ti, and Cr, having an atomic radius and an electronic structure close to those of Mo, substitutes Mo sites, while alloying elements such as Al with atomic radius comparable to Si occupy the Si sublattice sites. Alloying $MoSi_2$ with other elements to an extent that exceeds the limit of 3 at.% has been observed to affect the stability of the bct structure.

$MoSi_2$ has a hexagonal structure (C40, hP9) (Figure 1.3d) at high temperature (1900°C). The lattice vectors of $Mo(Si,Al)_2$ formed on alloying with Al in excess of 3 at.% also possess C40 structure.[12–14] Interestingly, the c/a ratio for the perfect hexagonal arrangement is $6^{1/2} = 0.2449$ nm,[15] which is very close to that of the $C11_b$ structure ($c/a = 0.2452$ nm). The [001], ½[111], ½[331], and [110] directions in the $C11_b$ (110) plane are equivalent to [01$\bar{1}$0], 1/3[11$\bar{2}$0], [01$\bar{1}$0],

TABLE 1.1

Crystal Structure and Lattice Constants of Silicides

Silicides	Crystal Structure	Structure and Space Group	Lattice Parameters (nm)
$MoSi_2$	Body-centered tetragonal	$C11_b$ (tI6), I4/mmm	$a = 0.3202$ $c = 0.7845$
Mo_5Si_3	Body-centered tetragonal	$D8_m$ (tI32), I4/mcm	$a = 0.959$ $c = 0.487$
Mo_3Si	Cubic	A15 (cP8) Pm3n	$a = 0.4892$
Mo_5SiB_2	Body-centered tetragonal	$D8_1$ (tI32), I4/mcm	$a = 0.6013$ $c = 1.103$
$Mo(Si,Al)_2$	Hexagonal	C40 (hP9), $P6_222$	$a = 4.644$ $c = 6.548$
WSi_2	Body-centered tetragonal	$C11_b$ (tI6), I4/mmm	$a = 0.3211$ $c = 0.7868$
Ti_5Si_3	Hexagonal	$D8_8$ (hP16), I4/mcm	$a = 0.7444$ $c = 0.5143$
$NbSi_2$	Hexagonal	C40 (hP9) $P6_222$	$a = 4.7971$ $c = 6.592$
Nb_5Si_3	Body-centered tetragonal	α: $D8_1$ (tI32)	α phase: $a = 0.656$ $b = 1.187$
		β: $D8_m$ I4/mcm	β phase: $a = 1.0$ $b = 0.507$
$CrSi_2$	Hexagonal	C40 (hP9) $P6_222$	$a = 0.4428$ $c = 0.6363$

Source: Mitra, R., *Inter. Mater. Rev.* 51, 1, 13–64, 2006.

and $1/3[\bar{2}110]$, respectively, in the C40 (0001) plane. While the $C11_b$ structure is characterized by ABAB... type stacking along the c axis, the C40 lattice shows ABCABC... type stacking. Hence, a stacking fault in the (110) plane of the $C11_b$ lattice would lead to the creation of localized C40-type structure.

The tetragonal structure of Mo_5Si_3 is quite different from that of $MoSi_2$ (compare Figure 1.3a and b), as the former material exhibits the following characteristics:[16] (i) the value of a (lattice parameter) is greater than c such that $a/c \approx 2$; (ii) close-packed planes are absent; and (iii) the –Si–Mo–Si– chains are along the [100] and [010] directions, while the –Mo–Mo– and –Si–Si– chains are along the [001] direction. In $MoSi_2$, the close-packed planes and directions are distinct, and the –Si–Mo–Si– chains in $MoSi_2$ are along the [001] direction, while the –Mo–Mo and –Si–Si chains are along the [100] and [010] directions. The interatomic bond along the [001] direction containing the –Si–Mo–Si– chain is believed to be stronger and more directional compared with either Mo–Mo or Si–Si bonds.

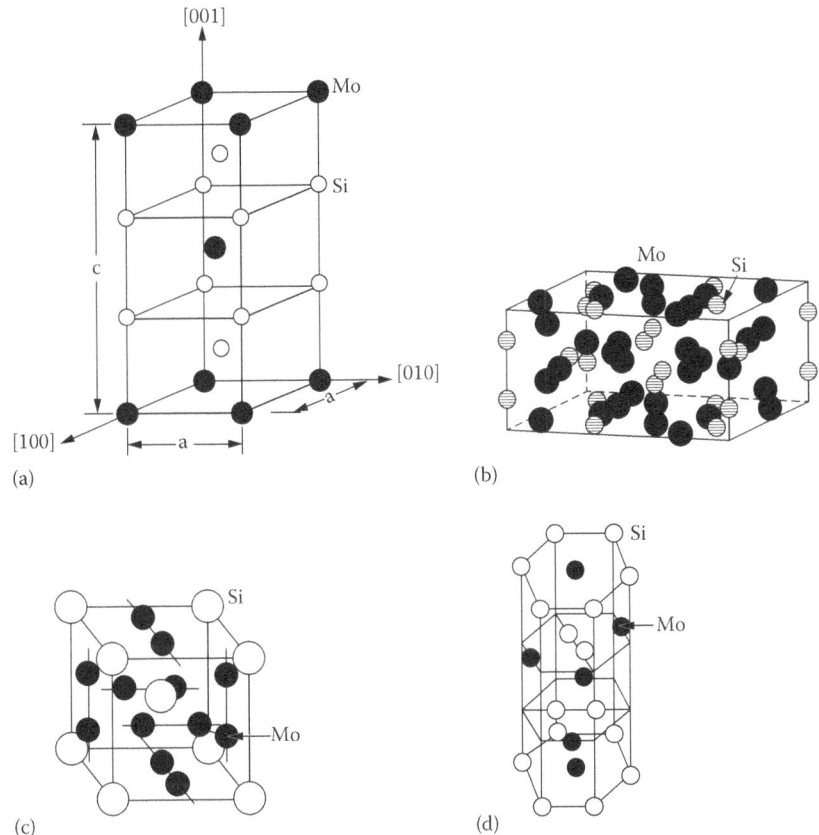

FIGURE 1.3
Schematic illustrations depicting the unit cells of (a) MoSi$_2$ (C11$_b$, tP8); (b) Mo$_5$Si$_3$ (D8$_m$, tI32); (c) Mo$_3$Si (A15, cP8); and (d) MoSi$_2$ (C40, hP9).

The Mo-rich section of the ternary isothermal phase diagram[13] of the Mo–Si–B system corresponding to 1600°C is shown in Figure 1.4. Mo–Si–B ternary alloys can be designed to have the optimum volume fractions of α-Mo, Mo$_3$Si, and Mo$_5$SiB$_2$ phases. All three phases have a nearly fixed composition with a limited solubility for other elements and hence provide microstructural stability at high temperatures. The α-Mo phase has a bcc structure with the solubility for Si and B atoms being 3 and <1 at.%, respectively, while the Mo$_3$Si possesses a single-phase composition close to 76Mo–24Si (at.%).[17] On the other hand, Mo$_5$SiB$_2$ possesses a bct structure (D8$_1$, tI32) with 32 atoms in the unit cell, comprising 20 atoms of Mo, 4 atoms of Si, and 8 atoms of B (Figure 1.5). In the unit cell of Mo$_5$SiB$_2$, three layers can be identified, the first comprising only Mo atoms, the second having only Si atoms, and the third having a mixture of Mo and Si atoms. It is interesting to note that the Mo-nearest neighbors

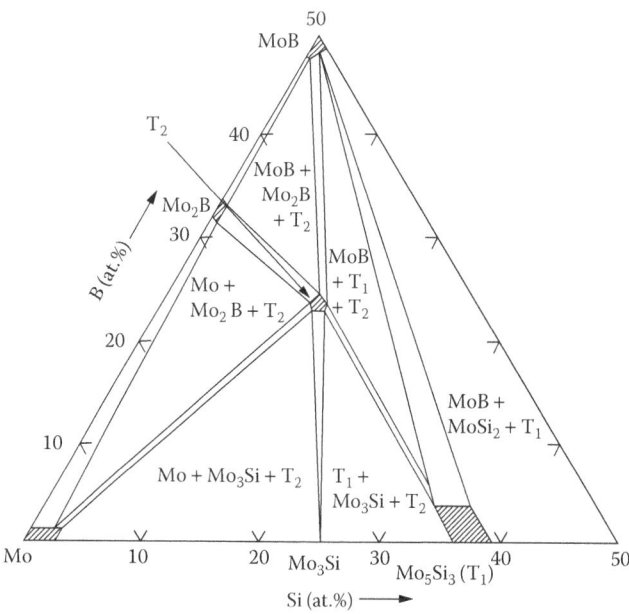

FIGURE 1.4

The Mo-rich section of the ternary isothermal phase diagram of the Mo–Si–B system corresponding to 1600°C. The position of Mo_5SiB_2 in this phase diagram is shown as T_2.

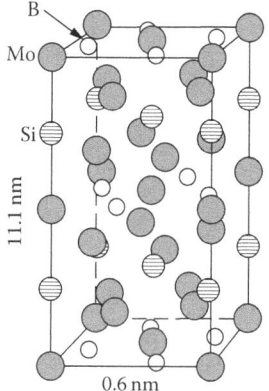

FIGURE 1.5

Schematic illustration of the unit cell of Mo_5SiB_2 ($D8_1$, tI32).

of the Mo sites in the unit cell are in bcc arrangement, which implies that the solubility of transition-metal atoms in Mo_5SiB_2 is similar to that in the bcc-Mo. The coefficient of thermal expansion anisotropy (α_c/α_a) of Mo_5SiB_2 has been found to be 1.4 at 500°C,[18] which is significantly lower than that ($\approx$2.2) of Mo_5Si_3.[16]

1.4.2 Tungsten Silicides

WSi_2 has a crystal structure, lattice constants (Table 1.1), and an electronic structure[14] similar to those of $MoSi_2$, with which it forms an isomorphous solid solution. Even the melting temperatures and coefficients of thermal expansion of $MoSi_2$ and WSi_2 are similar. Therefore, W has been considered as a suitable alloying element for strengthening $MoSi_2$.[19] However, the deformation behavior and brittle-to-ductile transition temperature (BDTT)[20,21] of $MoSi_2$ and WSi_2 single crystals are significantly different, even though both have the same $C11_b$ crystal structure, which is probably due to the more directional nature of bonds in the latter material.[22]

1.4.3 Titanium Silicides

Examination of the binary Ti–Si phase diagram[9] indicates that Ti_5Si_3 has a homogeneity range of 4 at.%, whereas other intermetallics (Ti_3Si, Ti_5Si_4, TiSi, and $TiSi_2$) in this phase diagram are line compounds. It has been reported that Ti_5Si_3 has high solubility for C, N, and O, which play a role in stabilizing the hexagonal structure.[23,24] Ti_5Si_3 has a hexagonal Mn_5Si_3-type structure (D8_8, hP16) having 16 atoms with 10 atoms of Ti and 6 atoms of Si (Figure 1.6).[10] The unit cell of Ti_5Si_3 contains two sublattices of Ti and one of Si. The Ti sublattice (1) contains atoms at positions with coordinates, (1/3, 2/3, 0), (2/3, 1/3, 0), (2/3, 1/3, 1/2), and (1/3, 2/3, 1/2). In a similar manner, the atoms of the Ti sublattice (2) are located at (0, 0.2358, 1/4), (0.7642, 0.7642, 1/4), (0.2358, 0, 1/4), (0.2358, 0.2358, 3/4), (0.7642, 0, 3/4), and (0, 0.7642, 3/4). Furthermore, the atomic locations of Si are (0, 0.5992, 1/4), (0.4008, 0.4008, 1/4), (0.5992, 0, 1/4), (0.5992, 0.5992, 3/4), (0.4008, 0, 3/4), and

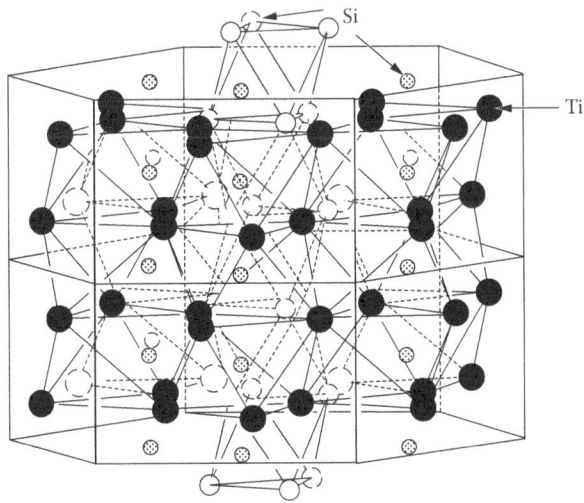

FIGURE 1.6
Schematic illustration of the Ti_5Si_3 unit cell (D8_8, hP9).

(0, 0.4008, 3/4). The [0001] or c-direction contains the chains of Ti and Si atoms.[25] The inherent brittleness of Ti_5Si_3 is due to its low crystal symmetry and strong covalent bonding, which increases the Peierls stress. Therefore, the dislocation density in Ti_5Si_3 is found to be very low.[26,27]

1.4.4 Niobium Silicides

The intermetallic phases in the binary equilibrium Nb–Si phase diagram are $NbSi_2$ and Nb_5Si_3.[28] $NbSi_2$ is a stoichiometric compound having a hexagonal, C40 crystal structure (Figure 1.3d) similar to that of $Mo(Si,Al)_2$. However, the deformation mechanisms of $NbSi_2$ have been reported[29] to differ from those of $Mo(Si,Al)_2$, and it possesses a much lower BDTT because of the intrinsic differences in the directionality of its bonding.[12,22,29–31] The alloying[32,33] of $NbSi_2$ with Mo or W by the substitution of Nb sites stabilizes the $C11_b$ structure in the regions corresponding to superlattice intrinsic stacking faults (SISF). On the other hand, the substitution of Si sites by Al atoms or Nb sites by Ti atoms stabilizes the C54 structure.

The Nb–Si binary phase diagram[28,34,35] shows a eutectic reaction point at 1880°C for the composition of 18.2 at.% Si, where the reaction taking place is $L \rightarrow$ bcc-Nb solid solution $+ Nb_3Si$. A eutectoid reaction at around 1700°C leads to the decomposition of the Nb_3Si phase into an Nb solid solution (Nb_{ss}) and Nb_5Si_3. While Nb_{ss} has a bcc structure, Nb_5Si_3 has a tetragonal structure. Based on the initial composition, the Nb_{ss} or Nb_5Si_3 present could be primary or secondary phases, the latter being the product of a eutectoid reaction. In the binary Nb–10Si and Nb–16Si alloys, the microstructure is comprised of large primary Nb_{ss} particles dispersed in a matrix of eutectic mixture of Nb_{ss} and Nb_5Si_3 phases. In another study, Mendiratta et al.[36] have reported that the primary Nb_{ss} particles become supersaturated with Si as the temperature decreases along the solvus, which causes the precipitation of silicides. The morphology of primary Nb_{ss} depends to a great extent on the method of processing. The presence of metastable Nb_3Si has been reported particularly in cast or directionally solidified composites, because of the sluggish nature of the eutectoid reaction. Annealing for a long period at 1500°C may complete the eutectoid reaction, forming Nb_{ss} and Nb_5Si_3 phases. It has been reported that on alloying with Mo, the eutectic reaction forming Nb_3Si and Nb_{ss} is inhibited, and that leading to the occurrence of Nb_5Si_3 and Nb_{ss} phases is preferred,[27] which suggests an alteration of the phase diagram to be investigated further.

The high-temperature yield and creep strengths of the Nb_{ss}–Nb_5Si_3 composite may be improved by alloying[34,35,37–47] with Ti, Hf, Cr, Mo, and Al. It has been observed that alloying of the Nb–Si system with Mo (5–25 at.%) suppresses the formation of Nb_3Si and that the microstructure in the as-solidified alloy contains Nb_{ss} and $(Nb,Mo)_5Si_3$ phases.[48] As Nb and Mo form an isomorphous solid solution, the Mo concentration of Nb_{ss} increases more significantly than that of $(Nb,Mo)_5Si_3$ with an increase in the bulk Mo content

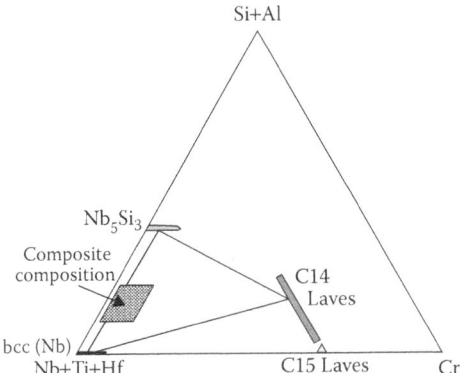

FIGURE 1.7
Schematic representation of the Nb–Si–X ternary phase diagram. (From Balsone, S. J., et al., *Materials beyond superalloys: Exploiting high-temperature composites*. GE Research and Development Report No. 2001CRD098, 2001; Bewlay, B. P., et al., *A review of very-high-temperature Nb-silicide based composites*. GE Research and Development Report No. 2002GRC172, 2002; Zhao, J.-C., et al., In *Proceedings of the 2001 Intermetallics Symposium on Structural Intermetallics*, eds. K. J. Hemker, et al. TMS, Warrendale, PA, 483–491, 2001.)

of the alloy. The microstructure represents an equilibrium between the three phases, Nb_{ss} and Nb_5Si_3 with alloying elements in solid solution, as well as Cr_2Nb-based Laves phases, having C14 or C15 structure, which is represented in the schematic ternary phase diagram[42,44,47] of Nb–Si–X (X is the alloying element) in Figure 1.7. Depending on the composition, the constituents may be 40–55 vol.% Nb_{ss}, 45–55 vol.% Nb_5Si_3, or 0–15 vol.% C14 Laves phases. While the atoms of Ti and Hf substitute the Nb sites to form a solid solution, Al atoms occupy the sites of the Si sublattice. The alloying additions need to be limited to prevent the formation of undesired phases. For example, Ti addition of less than 25 at.% does not form Ti_5Si_3 but is sufficient to destabilize the NbCrSi phase. Hence, Ti addition is kept low. Hf addition is also kept within 10 at.% to prevent the formation of Hf_5Si_3 and Hf_2Si, while Al addition is restricted to disallow the formation of Nb_3Al. Again, the maximum solubility of an alloying element in Nb_{ss} or Nb_5Si_3 while in equilibrium with one another is different from (normally, less than) that when the individual phases are in isolation.

1.4.5 Chromium Silicides

The Cr–Si binary phase diagram indicates the presence of four different intermetallics with the following compositions: Cr_3Si, Cr_5Si_3, CrSi, and $CrSi_2$.[9] $CrSi_2$ is almost a line compound, with a homogeneity range of less than 1 at.%, and has a hexagonal C40 structure similar to $Mo(Si,Al)_2$ and $NbSi_2$ (Figure 1.3d). Interestingly, the BDTT and deformation behavior of $CrSi_2$ closely resemble those of $Mo(Si,Al)_2$, and differ significantly from $NbSi_2$.[49]

On the other hand, Cr_3Si has a much higher melting point than $CrSi_2$, with a solid solubility of about 3 at.%, and a cubic A15 crystal structure similar to that of Mo_3Si (Figure 1.3c). The scope for alloying Cr_3Si provides the possibility of enhancing the creep and yield strengths at elevated temperatures.[50]

1.5 Crystal Structure and Phase Diagram of Aluminides

The major aluminides of interest to the scientific community are from the following binary phase equilibrium diagrams: Ni–Al, Fe–Al, and Ti–Al. The intermetallics NiAl and Ni_3Al from the Ni–Al system, FeAl and Fe_3Al from the Fe–Al system, and Ti_3Al, TiAl, and Al_3Ti from the Ti–Al system have been investigated extensively. The data on crystal structures and lattice constants of different aluminides are shown in Table 1.2.[51–58]

TABLE 1.2

Crystal Structures and Lattice Constants of Different Aluminides

Silicides	Crystal Structure	Structure and Space Group	Lattice Parameters (nm)
Ni_3Al	Face-centered cubic	$L1_2$, cP4 Pm3m	0.357
NiAl	Body-centered cubic	B2, cP2 Pm-3m	0.2887
Al_3Ti	Tetragonal	DO_{22}, tI8 I4/mmm	$a=0.3809$ $c=0.847$
TiAl	Tetragonal	$L1_0$, tP4 P4/mmm	$a=0.3997$ $c=0.4081$
Ti_3Al	Hexagonal	DO_{19}, hP8 $P6_3$/mmc	$a=0.5729, c=0.4574$
Fe_3Al	Face-centered cubic	DO_3, cF16 Fm3m	0.5655
FeAl	Body-centered cubic	B2, cP2 Pm-3m	0.291

Source: Villars, P. and L. D. Calvert, *Pearson's Handbook of Crystallographic Data for Intermetallic Phases,* ASM, Metals Park, OH, 1985; Dey, G. K., *Sadhana* 28(1/2), 247–262, 2003; Stoloff, N. S. and C. T. Liu, The physical and mechanical metallurgy of Ni_3Al and its alloys. In *Physical Metallurgy and Processing of Intermetallic Compounds,* eds. N. S. Stoloff and V. K. Sikka, CBS, New Delhi, India, 1997, 15–211; Hughes, T., et al., *J. Appl. Phys.,* 42, 3705–3716, 1971; Norby, P. and A. N. Christensen, *Acta Chem. Scand. A,* 40, 157–159, 1986; Menon, E. S. K., et al., *J. Mater. Sci. Lett.,* 15, 1231–1233, 1996; Gehlen, P. C., *Metall. Trans. A,* 2, 4, 1249–1250, 1971; Shu, X., et al., *J. Mater. Sci. Tech.,* 17, 6, 601–604, 2001.

1.5.1 Nickel Aluminides

The binary equilibrium phase diagram of the Ni–Al system contains the following intermetallic phases starting from the Al-rich side: Al_3Ni, Al_3Ni_2, AlNi (or NiAl), Al_3Ni_5, and $AlNi_3$ (or Ni_3Al).[59] Among these intermetallics, NiAl has the highest congruent melting point of 1638°C at the stoichiometric composition. The homogeneous phase field of NiAl is quite wide ranging, from 45 to 59 at.% Ni. It should be noted that Ni_3Al is the first intermetallic formed by a peritectic reaction (liquid $+$ NiAl $\rightarrow$ Ni_3Al) at temperatures $\leq$1395°C, as aluminum is added to nickel. Ni_3Al has a homogeneity range of $\approx$4.5 at.% around its stoichiometric composition. Due to their high melting/liquidus temperatures, NiAl and Ni_3Al are the intermetallics of interest in the Ni–Al binary system.

The crystal structures of NiAl and Ni_3Al are shown in Figure 1.8a and b, respectively. NiAl possesses ordered cubic B2 (cP2, CsCl structure). This can be visualized in the form of two interpenetrating primitive cubic cells, where Ni and Al atoms occupy the cube corners of the first and second sublattices, respectively (Figure 1.8a). NiAl is known to be strongly ordered even above $0.65\,T_m$ with an intrinsic disorder parameter $\leq 5 \times 10^{-3}$. The calculation of the electronic band structure of NiAl has shown evidence for strong Ni d–Al p hybridization along $\langle 111 \rangle$ directions between the nearest-neighbor Ni–Al atom pairs.[60] This suggests that a strong covalent bond exists between the nearest-neighbor Ni and Al atom pairs along the $\langle 111 \rangle$ direction, whereas there is weak ionic repulsion between second-nearest-neighbor atoms along $\langle 100 \rangle$. These directional bonds are superimposed on the nondirectional metallic bonds between unlike and similar metallic atoms. The strong and weak atomic bonds along $\langle 111 \rangle$ and $\langle 100 \rangle$, respectively, are responsible for the elastic anisotropy of NiAl. The strong Ni–Al covalent bond can also be used to explain the high negative heat of formation ($\approx$–72 kJ mol^{-1}, as shown in Figure 1.2),[61,62] the high congruent melting temperature (as mentioned in Section 1.2) at the stoichiometric composition, and the low degree of intrinsic disorder or high ordering energy.[55] Interestingly, the order in NiAl is preserved until temperatures above $0.65\,T_m$, with the intrinsic disorder parameter $<5 \times 10^{-3}$.[63] An order

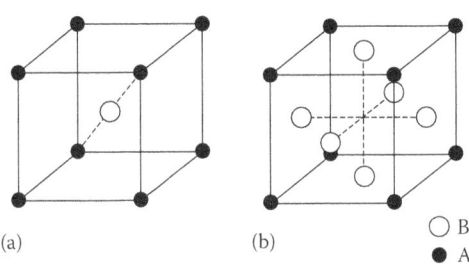

(a) (b)

○ B
● A

FIGURE 1.8
Schematic illustrations of the unit cells of (a) NiAl (B2, cP2) and (b) Ni_3Al (L1$_2$, cP4).

to disorder transformation is reported to occur in NiAl at 600°C.[64] The ordered bcc (B2) structure of NiAl is reported to be stable for large deviations from stoichiometry, with the persistence of long-range order for both Ni- and Al-rich compositions. Deviations for stoichiometry are accommodated by the presence of antisite defects, where excess Ni atoms occupy Al sites in the Ni-rich compositions. However, the Al-rich compositions of NiAl have been found to have vacant Ni sites. The lattice constant of NiAl depends on Ni:Al ratio within the permitted range of compositions, and its maximum value is observed for the stoichiometric composition.

A martensitic transformation has been reported on quenching Ni-rich NiAl.[65] The martensitic phase has either an ordered face-centered tetragonal $L1_0$ (tP4) structure or a rhombohedral unit cell with 7R stacking sequence. The M_s temperature rises sharply with increasing Ni content, and has been determined as ≈ -273°C (0 K) for Ni–40% Al and ≈ 727°C (1000 K) for Ni–32% Al.[66–68] The reversible shape memory effect as found in NiAl has the following characteristics: (i) martensitic transformation is thermoelastic in nature[58,59]; (ii) parent and product phases have ordered structure[68]; and (iii) twins are present inside the martensite.[69,70]

Ni_3Al has an ordered fcc structure ($L1_2$, cP4), with its unit cell containing three atoms of Ni and a single atom of Al, as shown in Figure 1.8b. This suggests that the Ni atoms occupy the face-centered positions, whereas Al is located at the unit cell corners. The $L1_2$ structure of Ni_3Al can be visualized as stacking of three close-packed planes in the sequence ABC. In off-stoichiometric compositions of Ni_3Al, antisite substitutions are formed on both Ni-rich and Al-rich sides by excess atoms of the other constituent. It has been shown, however, by Gao and Bacon that the presence of a Ni atom at the Al site is energetically more favorable compared with that of an Al atom at the Ni site.[71] Moreover, the formation energy of antistructure (e.g., the location of the Ni atom at the Al-site) has been reported to be lower than that of vacancy in Ni_3Al. The major difference in the structure of off-stoichiometric Ni_3Al from that of NiAl is that the antisite defect formation is preferred only in the Ni-rich side, whereas vacancies are stable in the Al-rich compositions.

1.5.2 Titanium Aluminides

The equilibrium phase diagram of the binary Ti–Al system shows the presence of three intermetallic phases: Al_3Ti, TiAl, and Ti_3Al.[9] According to this binary phase diagram, the formation of Al_3Ti occurs during cooling from the Al-rich liquid phase by a peritectic reaction:

$$\text{Liquid} + \alpha\text{-Al (s)} \rightarrow Al_3Ti \text{ (s)} \tag{1.2}$$

Al_3Ti is found to possess two different types of crystal structures: $L1_2$ (cP4) in metastable form and $D0_{22}$ (tI8) in the equilibrium state, as shown in Figure 1.9a and b, respectively. The $D0_{22}$ structure is derived from $L1_2$ by

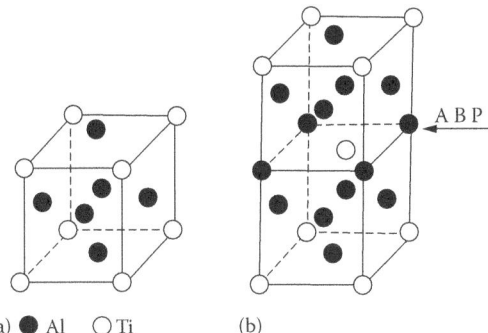

(a) ● Al ○ Ti   (b)

FIGURE 1.9
Schematic illustration of Al_3Ti unit cell: (a) $L1_2$ and (b) $D0_{22}$.

introducing an antiphase boundary (APB) with a displacement vector of $\frac{1}{2}\langle 110\rangle$ on every (001) plane. If a series of $L1_2$ unit cells are vertically stacked one on top of another, the formation of the APB is equivalent to the translation of the upper $L1_2$ unit cell with respect to that below along the $\langle 110\rangle$ direction by half the diagonal length.[72]

The Ti–Al phase diagram proposed by Murray[9] near the γ-TiAl phase field has been modified significantly by several researchers.[73,74] The location of various intermetallic phases in the modified phase diagram is schematically shown in Figure 1.10. The modified part of the phase diagram shows the γ-TiAl phase to be in equilibrium with α-Ti_3Al (A3, hP2) until the melting temperature of 1445°C. For alloys having a composition in the range of 40–49 at.% Al, the β phase forms as the primary phase in equilibrium with liquid, and α forms through a peritectic reaction[73]:

$$\text{Liquid} + \beta \rightarrow \alpha \tag{1.3}$$

For the composition range of 49–55 at.% Al, the primary phase is found to be α. The γ phase forms during solidification by the following peritectic reaction[73]:

$$\text{Liquid} + \alpha \rightarrow \gamma \tag{1.4}$$

Whereas the α phase exists until temperatures up to the melting point for alloys having 46–50 at.% Al, a β phase field is found for lower Al content (<44 at.%).[73] As expected, this peritectic solidification leads to the formation of a nonuniform microstructure with chemical segregation. The equilibrium microstructure obtained through suitable heat treatment contains either single-phase α or an $\alpha+\gamma$-phase field, depending on the Al concentration. It is obvious from Figure 1.10 that the α to $\alpha+\gamma$ transformation temperature decreases with decreasing Al content. At the concentration of 45 at.% Al, a single-phase α structure can be obtained by heat treatment above 1300°C.

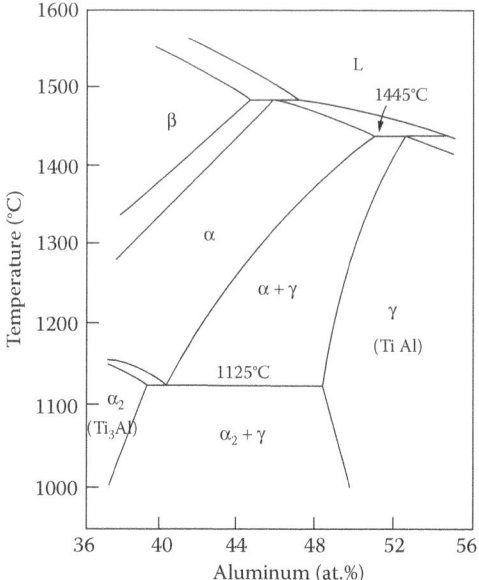

FIGURE 1.10
Modified Ti-rich part of the Ti–Al binary phase diagram showing the positions of various intermetallic phases. (From McCullough, C., et al., *Acta Metall.*, 37, 5, 1321–1336, 1989; Huang, S. C. and P. A. Siemers, *Metall. Trans. A*, 20, 10, 1899–1906, 1989.)

The modified phase diagram shown in Figure 1.10 indicates that at about 1118°C, the α phase containing $\approx$40 at.% Al transforms into $\alpha_2+\gamma$ through a eutectoid reaction. The overall transformation starting from the β phase occurs through the following steps:

$$\beta \rightarrow \alpha \rightarrow \alpha_2 + \gamma \tag{1.5}$$

The α_2 phase with a composition of Ti_3Al has a DO_{19} (hP8) structure (hexagonal with eight atoms). In the γ-based alloys containing the hypereutectoid α, the transformed structures containing eutectoid $\alpha_2+\gamma$ possess lamellar morphology with interlamellar spacing in the range of 0.1–1 µm. The intermetallic Ti_3Al (α_2) is known to become disordered at 1180°C to form the α phase (A3, hP2).[75] The close-packed (0001) planes in Ti_3Al contain Al and Ti atoms sharing nearest-neighbor bonds. The structure of α_2 can be described in terms of four interpenetrating primitive sublattices, three of which contain Ti atoms and one is made of Al atoms. The following orientation relationship is maintained in the microstructure in the course of transformation: $\{110\}_\beta//(0001)\alpha_2//\{111\}_\gamma$ and $\langle 111\rangle_\beta//\langle 1120\rangle\alpha_2//\langle 110\rangle_\gamma$.

On ternary or quaternary alloying of DO_{19}-structured α_2-Ti_3Al, Nb, Mo, Zr, and Hf substitute in the Ti sites,[76,77] whereas Ga, Sn, and Si occupy the Al sites.[78] The nature of bonding in γ-TiAl and α_2-Ti_3Al is primarily metallic, due to Ti d–Ti d interactions, with a weak covalent contribution because of hybridization between Al p and Ti d orbitals.[79] The addition of suitable alloying elements is expected to enhance the metallic character of bonding. Furthermore, stress-induced $\alpha_2 \rightarrow \gamma$ transformation has been reported to occur during deformation of γ-TiAl-based alloys with duplex microstructures comprising $\alpha_2 + \gamma$ phases.[80] Recently, significant attention has been focused on the development of β-solidifying TiAl-based alloys, where the β phase is stabilized by the addition of Mo and Nb.[74] It has also been reported that the β-stabilizing effect of Mo is four times stronger than that of Nb. During solidification, these alloys follow the following pathway[81–83]: $L \rightarrow L + \beta \rightarrow \beta \rightarrow \beta + \alpha \rightarrow \alpha + \beta + \gamma \rightarrow \alpha + \beta + \beta_0 + \gamma \rightarrow \alpha + \beta_0 + \gamma \rightarrow \alpha + \alpha_2 + \beta_0 + \gamma \rightarrow \alpha_2 + \beta_0 + \gamma$, where β_0 is the secondary precipitate phase. These alloys can be hot worked in the temperature range for $\alpha + \beta$-phase stability[84] or suitably heat treated to obtain the desirable multiphase microstructure.[81]

The ordered B2 (cP2) phase is derived from the high-temperature bcc-structured β phase of titanium. The composition of the B2 phase is usually Ti_2AlX,[85] in which the constituent X is the β-stabilizer alloying element. In this phase, Ti and Al atoms occupy two separate primitive sublattice sites. Alloying elements (X) such as Mo and V substitute in the Al site, whereas Hf and Zr substitute in the Ti site.[77]

Besides α_2 and B2, the O phase having the composition Ti_2AlX and possessing an orthorhombic crystal structure (oC16) is also of interest.[86] The O phase can be considered a distorted form of the α_2 phase (Figure 1.11) in which the Nb atoms distinctly occupy one of the sublattices originally occupied by Ti in the α_2 phase. It has been shown that the formation of the O phase occurs from α_2 supersaturated with Nb by the following transformation[87]: $\alpha_2 \rightarrow \alpha_2$ (Nb lean) + O (Nb rich). This transformation involves the diffusion

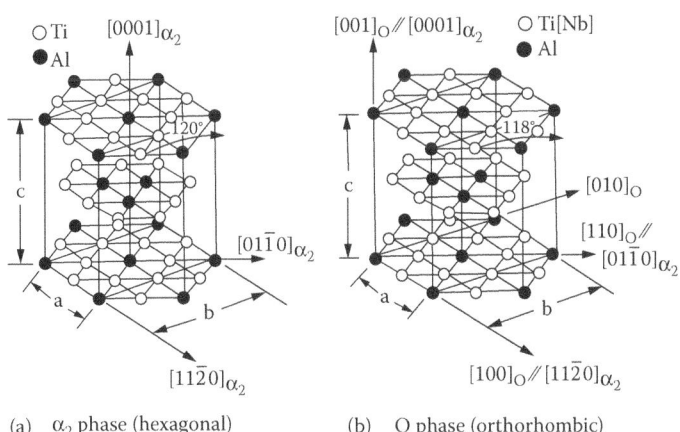

(a) α_2 phase (hexagonal)   (b) O phase (orthorhombic)

FIGURE 1.11
Schematic representation of the crystal structures of: (a) α_2 and (b) O phases.

and partitioning of Nb. The orientation relationships between the α_2 and O phases are as follows: $[0001]\alpha_2//[001]_O$, $[11\bar{2}0]\alpha_2//[100]_O$, and $[01\bar{1}0]\alpha_2//[110]_O$.

The ternary Ti–Al–Nb system contains the phase fields comprising the single-phase field, O, and the ternary-phase field of O+B2+α_2, which separates the α_2+B2, α_2+O, and O+B2-phase fields. The O phase has been found to possess significantly high solubility for Nb, and is stable until 1000°C. The ordered O phase is present as either O1 or O2.[88,89] Nb and Ti atoms randomly occupy the same sublattice in the O1 phase with the former element being present as the substitutional solute, whereas the Nb atoms comprise a separate sublattice in the O2 phase.

1.5.3 Iron Aluminides

The Fe–Al phase diagram from the ASM Metals Handbook[90] has been significantly modified on the basis of experimental results. The intermetallics in the Fe–Al phase diagram[91] have the compositions Fe_3Al, $FeAl$, $FeAl_2$, Fe_2Al_3, Fe_2Al_5, and $FeAl_3$. Out of these, only FeAl and Fe_3Al are of interest for structural applications, as the rest of the phases are known to be very brittle. The unit cells of FeAl (B2, cP2) and Fe_3Al (D0$_{22}$, cF16) are depicted schematically in Figure 1.12.

FeAl has ordered bcc (B2, cP2) structure. The concentration of Al in FeAl varies from 35 to 50 at.% Al. The melting temperature decreases with Al concentration, till it reaches 1250°C at 52 at.% Al. At higher Al content, FeAl remains B2 up to the melting point. The Fe_3Al transforms from an ordered cubic crystal structure (D0$_3$, cF16) to a defective ordered bcc (B2, cP2) structure above 541°C, which is the critical temperature for ordering. The Fe-rich portion of the Fe–Al phase diagram shows the existence of three bcc phases, a disordered solution (α-bcc), an ordered bcc phase (B2-structured FeAl), an

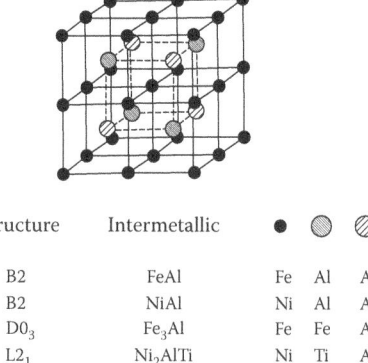

Structure	Intermetallic	●	◍	◩
B2	FeAl	Fe	Al	Al
B2	NiAl	Ni	Al	Al
D0$_3$	Fe$_3$Al	Fe	Fe	Al
L2$_1$	Ni$_2$AlTi	Ni	Ti	Al

FIGURE 1.12
Schematic illustration of the crystal structures of FeAl (B2) and Fe$_3$Al (D0$_3$).

ordered cubic phase (D0$_3$ Fe$_3$Al), and two two-phase regions (α + D0$_3$ and α + B2 in the composition range of 18–25 at.% Al above 550°C).

The defect structure of B2 FeAl is complex, with the presence of substitutional atoms, vacancies, or both.[92] Due to deviations from stoichiometry in the B2 alloys, antisite defects are commonly observed in the Fe-rich alloys. Here, the Fe atoms occupy the vacant Al sites. However, in the Al-rich combinations, Fe vacancies are observed, whereas excess Al atoms do not occupy the vacant Fe sites. Fe-aluminides can contain up to 10 at.% vacancies. At 800°C (homologous temperature = 0.71), the vacancy concentration is found to be ≈40 times higher than that in normal metals at their melting temperatures.

1.6 Summary

Intermetallics such as aluminides and silicides can be either line compounds with a single composition (e.g., Al$_3$Ti, MoSi$_2$, and Mo$_3$Si) or alloys with a limited compositional homogeneity range (Fe$_3$Al and FeAl) at any temperature. The negative enthalpy of formation of the intermetallics is significantly high, with NiAl showing the highest value among the aluminides. The crystal structures of the ordered intermetallics are found to be quite varied: cubic (Mo$_3$Si, Ni$_3$Al, NiAl, FeAl, etc.), tetragonal (Al$_3$Ti, MoSi$_2$, Mo$_5$Si$_3$, etc.), hexagonal (Ti$_5$Si$_3$, Ti$_3$Al, etc.), or orthorhombic (Ti$_2$AlNb). The nature of bonding in these intermetallics is usually metallic between the atoms of the same metal and partly covalent between the atoms of metals with different valencies or electronegativities. For example, hybridization of transition-metal (e.g., Ni) d-shell and Al p-shell electrons leads to a strongly directional nature of bonding. During the last few decades, significant research has been focused on understanding the phase transitions in binary and ternary intermetallic alloy systems with emphasis on conditions and mechanisms. This has helped in developing suitable processing schedules to obtain desirable combinations of phases with appropriate morphology in the microstructures. The defect structure of the intermetallics is found to vary significantly with their composition, with some of the alloys showing a greater concentration of vacancies and antisite substitution, which also influence the mechanical properties.

References

1. Ellner, M. and B. Predel. 1994. Bond characterization from thermodynamic properties. In *Intermetallic Compounds: Vol. 2, Principles*, 91–125, eds. J. H. Westbrook and K. L. Fleischer. Chichester, UK: Wiley.

2. Gottstein, G. 2006. *Physical Foundations of Materials Science*, Chapter 4, 107. Noida, India: Springer.
3. Desai, P. D. 1987. Thermodynamic properties of selected binary aluminum alloy systems. *J Phys. Chem. Ref. Data* 16(1):109–124.
4. Kubashewski, O. and W. A. Dench. 1955. The heats of formation in the systems titanium-aluminium and titanium-iron. *Acta Metall.* 3:339–346.
5. Nash, P. and O. Kleppa. 2001. Composition dependence of the enthalpies of formation of NiAl. *J. Alloys Compd.* 321:228–231.
6. Huang, W. and Y. A. Chang. 1998. A thermodynamic analysis of the Ni-Al system. *Intermetallics* 6:487–498.
7. Samokhval, V. V., P. A. Poleshchuk, and A. A. Vecher. 1971. Thermodynamic properties of aluminum-titanium and aluminum-vanadium alloys. *Russ. J. Phys. Chem.* 145:1174.
8. Oehme, G. and B. Predel. 1978. Hochtemperatur-kalorimeter zur genauen Bestimmung von Mischungsenthalpien. *Thermochim. Acta* 22:267–273.
9. Murray, J. L. 1987. *Binary Alloy Phase Diagrams*, eds. T. B. Massalski, H. Okamato, P. R. Subramanian, and L. Kacprazak, 1:173, 2:1333, 1631, 2054. Materials Park, OH: ASM.
10. Mitra, R. 2006. Mechanical behavior and oxidation resistance of structural silicides. *Inter. Mater. Rev.* 51(1):13–64.
11. Frankwicz, P. S., J. H. Perepezko, and D. L. Anton. 1993. Phase stability of MoSi$_2$ with Cr additions. In *High-Temperature Ordered Intermetallic Alloys V*, eds. I. Baker, J.D. Whittenberger, R. Darolia, M.H. Yoo, 288:159.
12. Inui, H. and M. Yamaguchi. 2001. Deformation mechanisms of transition-metal disilicides with the hexagonal C40 structure. *Intermetallics* 9:857–862.
13. Nowotny, H., E. Dimakopoulou, and H. Kudielka. 1957. Investigations in the ternary systems: Molybdenum-silicon-boron, tungsten-silicon-boron and system: VSi$_2$-TaSi$_2$ (in German). *Monatsh. Chem.* 88:180–192.
14. Bhattacharyya, B. K., D. M. Bylander, and L. Kleinman. 1985. Comparison of fully relativistic energy-bands and cohesive energies of MoSi$_2$ and WSi$_2$. *Phys. Rev. B* 32(12):7973–7978.
15. Inui, H., M. Moriwaki, K. Ito, and M. Yamaguchi. 1998. Plastic deformation of single crystals of Mo(Si, Al)$_2$ with the C40 structure. *Philos. Mag. A* 77(2):375–394.
16. Chu, F., D. J. Thoma, K. McClellan, P. Peralta, and Y. He. 1999. Synthesis and properties of Mo$_5$Si$_3$ single crystals. *Intermetallics* 7(5):611–620.
17. Rosales, I. and J. H. Schneibel. 2000. Stoichiometry and mechanical properties of Mo$_3$Si. *Intermetallics* 8(8):885–889.
18. Choe, H., D. Chen, J. H. Schneibel, and R. O. Ritchie. 2001. Ambient to high temperature fracture toughness and fatigue-crack propagation behavior in a Mo–12 Si–8.5B (at.%) intermetallic. *Intermetallics* 9:319–329.
19. Petrovic, J. J. and R. E. Honnell. 1990. SiC reinforced-MoSi$_2$/WSi$_2$ alloy matrix composites. *Ceram. Eng. Sci. Proc.* 11:734–744.
20. Ito, K., H. Inui, Y. Shirai, and M. Yamaguchi. 1995. Plastic deformation of MoSi$_2$ single crystals. *Philos. Mag. A* 72(4):1075–1097.
21. Ito, K., T. Yano, T. Nakamoto, H. Inui, and M. Yamaguchi. 1999. Plastic deformation of single crystals of WSi$_2$ with the C11$_b$ structure. *Acta Mater.* 47(3):937–949.
22. Tanaka, K., H. Inui, M. Yamaguchi, and M. Koiwa. 1999. Directional atomic bonds in MoSi$_2$ and other transition-metal disilicides with the C11$_b$, C40 and C54 structures. *Mater. Sci. Eng. A* 261:158–164.

23. Nowotny, H. 1963. Alloy chemistry of transition element borides. In *Electronic Structure and Alloy Chemistry of the Transition Elements*, ed. P. A. Beck, 179–220. New York: Interscience.

24. Sambasivan, S. and W. T. Petuskey. 1994. Phase chemistry in the Ti–Si–N system: thermochemical review with phase stability diagrams. *J. Mater. Res.* 9(09):2362–2369.

25. Thom, A. J., M. K. Meyer, Y. Kim, and M. Akinc. 1994. Evaluation of A5 Si3Zx intermetallics for use as high temperature structural materials. In *Processing and Fabrication of Advanced Materials III*, eds. V. A. Ravi, T. S. Srivatsan, and J. J. Moore, 413–438. Warrendale, PA: TMS.

26. Rosenkranz, R., G. Frommeyer, and W. Smarsly. 1992. Microstructures and properties of high melting point intermetallic Ti_5Si_3 and $TiSi_2$ compounds. *Mater. Sci. Eng. A* 152 (1–2):288–294.

27. Mitra, R. 1998. Microstructure and mechanical behavior of reaction hot-pressed titanium silicide and titanium silicide-based alloys and composites. *Metall. Mater. Trans. A* 29(6):1629–1641.

28. Schlesinger, M. E., H. Okamoto, A. B. Gokhale, and R. Abbaschian. 1993. The Nb-Si (niobium-silicon) system. *J. Phase Equil.* 14(4):502–509.

29. Moriwaki, M., K. Ito, H. Inui, and M. Yamaguchi. 1997. Plastic deformation of single crystals of $NbSi_2$ with the C40 structure. *Mater. Sci. Eng. A* 239–240:69–74.

30. Inui, H., M. Moriwaki, K. Ito, and M. Yamaguchi. 1998. Plastic deformation of single crystals of $Mo(Si,Al)_2$ with the C40 structure. *Philos. Mag. A* 77(2):375–394.

31. Nakamura, M. 1994. Elastic constants of some transition-metal-disilicide single crystals. *Metall. Mater. Trans. A* 25(2):331–340.

32. Nakano, T., M. Kishimoto, D. Furuta, and Y. Umakoshi. 2000. Effect of substitutational elements on plastic deformation behaviour of $NbSi_2$-based silicide single crystals with C40 structure. *Acta Mater.* 48(13):3465–3475.

33. Kim, W. Y., H. Tanaka, A. Kasama, and S. Hanada. 2001. Microstructure and room temperature fracture toughness of Nb_{ss}/Nb_5Si_3 *in situ* composites. *Intermetallics* 9(9):827–834.

34. Bewlay, B. P., M. R. Jackson, J.-C. Zhao, and P. R. Subramanian. 2003. A review of very-high-temperature Nb-silicide-based composites. *Metall. Mater. Trans. A* 34(10):2043–2052.

35. Zhao, J.-C., B. P. Bewlay, M. R. Jackson, and L. A. Peluso. 2001. Alloying and phase stability in niobium silicide *in situ* composites. GE Research and Development Report No. 2001CRD090.

36. Mendiratta, M. G., J. J. Lewandowski, and D. M. Dimiduk. 1991. Strength and ductile-phase toughening in the two-phase Nb/Nb_5Si_3 alloys. *Metall. Trans. A* 22(7):1573–1583.

37. Rigney, J. D., P. M. Singh, and J. J. Lewandowski. 1992. Environmental effects on ductile-phase toughening in Nb_5Si_3–Nb composites. *JOM* 44(8):36–41.

38. Subramanian, P. R., M. G. Mendiratta, and D. M. Dimiduk. 1996. The development of Nb-based advanced intermetallic alloys for structural applications. *JOM* 48(1):33–38.

39. Subramanian, P. R., M. G. Mendiratta, D. M. Dimiduk, and M. A. Stucke. 1997. Advanced intermetallic alloys—Beyond gamma titanium aluminides. *Mater. Sci. Eng. A* 239–240(0):1–13.

40. Bewlay, B. P., J. J. Lewandowski, and M. R. Jackson. 1997. Refractory metal-intermetallic *in situ* composites for aircraft engines. *JOM* 49(8):44–45.

41. Bewlay, B. P., M. R. Jackson, and P. R. Subramanian. 1999. Processing high-temperature refractory-metal silicide *in situ* composites. *JOM* 51(4):32–36.
42. Balsone, S. J., B. P. Bewlay, M. R. Jackson, P. R. Subramanian, J.-C. Zhao, A. Chatterjee, and T. M. Heffernan. 2001. Materials beyond superalloys: Exploiting high-temperature composites. GE Research and Development Report No. 2001CRD098.
43. Balsone, S. J., B. P. Bewlay, M. R. Jackson, P. R. Subramanian, J.-C. Zhao, A. Chatterjee, and T. M. Heffernan. 2001. Materials beyond, superalloys: Exploiting high-tempertature composites. In *Proceedings of the 2001 Intermetallics Symposium on Structural Intermetallics*, eds. K. J. Hemker, D. M. Dimiduk, H. Clemens, R. Darolia, H. Inui, J. M. Larsen, V. K. Sikka, M. Thomas, and J. D. Whittenberger, 99–108. Warrendale, PA: TMS.
44. Bewlay, B. P., M. R. Jackson, J. C. Zhao, and P. R. Subramanian. 2002. A review of very high-temperature Nb-silicide based composites. GE Research and Development Report No. 2002GRC172.
45. Bewlay, B. P., M. R. Jackson, J. C. Zhao, P. R. Subramanian, M. G. Mendiratta, and J. J. Lewandowski. 2003. Ultrahigh-temperature Nb-silicide-based composites. *MRS Bull.* 28(9):646–653.
46. Bewlay, B. P., M. R. Jackson, J. C. Zhao, and P. R. Subramanian. 2003. A review of very-high-temperature Nb-silicide-based composites. *Metall. Mater. Trans. A* 34A(10):2043–2052.
47. Zhao, J.-C., B. P. Bewlay, M. R. Jackson, and L. A. Peluso. 2001. Alloying and phase stability in niobium silicide *in situ* composites. In *Proceedings of the 2001 Intermetallics Symposium on Structural Intermetallics*, eds. K. J. Hemker, D. M. Dimiduk, H. Clemens, R. Darolia, H. Inui, J. M. Larsen, V. K. Sikka, M. Thomas, and J. D. Whittenberger, 483–491. Warrendale, PA: TMS.
48. Chattopadhyay, K., R. Sinha, R. Mitra, and K. K. Ray. 2007. Effect of Mo and Si on morphology and volume fraction of eutectic in Nb–Si–Mo alloys. *Mater. Sci. Eng. A* 456:358–363.
49. Inui, H., M. Moriwaki, S. Ando, and M. Yamaguchi. 1997. Plastic deformation of single crystals of $CrSi_2$ with the C40 structure. *Mater. Sci. Eng. A* 239–240:63–68.
50. Raj, S. V. 1995. An evaluation of the properties of Cr_3Si alloyed with Mo. *Mater. Sci. Eng. A* 201(1–2):229–241.
51. Villars, P. and L. D. Calvert. 1985. *Pearson's Handbook of Crystallographic Data for Intermetallic Phases*. Metals Park, OH: ASM.
52. Dey, G. K. 2003. Physical metallurgy of nickel aluminides. *Sadhana* 28(1/2):247–262.
53. Stoloff, N. S. and C. T. Liu. 1997. The physical and mechanical metallurgy of Ni_3Al and its alloys. In *Physical Metallurgy and Processing of Intermetallic Compounds*, eds. N. S. Stoloff and V. K. Sikka, 15–211. New Delhi, India: CBS.
54. Hughes T., E. P. Lautenschlager, J. B. Cohen, and J. O. Brittain. 1971. X-ray diffraction investigation of NiAl alloys. *J. Appl. Phys.* 42:3705–3716.
55. Norby, P. and A. N. Christensen. 1986. Preparation and structure of Al_3Ti. *Acta Chem. Scand. A* 40:157–159.
56. Menon, E. S. K., A. G. Fox, and R. Mahapatra. 1996. Accurate determination of the lattice parameters of γ-TiAl alloys. *J. Mater. Sci. Lett.* 15:1231–1233.
57. Gehlen, P. C. 1971. The lattice parameters of ordered Ti_3Al between 100 and 989 K. *Metall. Trans. A* 2(4):1249–1250.

58. Shu, X., W. Hu, H. Xiao, and H. Deng. 2001. Vacancies and antisites in B2 FeAl and DO_3Fe_3Al with a modified analytic EAM model. *J. Mater. Sci. Tech.* 17(6):601–604.

59. Nash, P., M. F. Singleton, and J. L. Murray. 1991. Al-Ni (aluminum-nickel). In *Phase Diagrams of Binary Nickel Alloys*, Vol. 1, ed. P. Nash, 3–11. Materials Park, OH: ASM International.

60. Lui, S. C., J. W. Davenport, E. Plummer, D. M. Zehner, and G. W. Fernando. 1990. Electronic-structure of NiAl. *Phys. Rev. B* 42(3):1582–1597.

61. Dannohl, H. D. and H. L. Lukas. 1974. Calorimetric determination of enthalpies of formation of some intermetallic compounds. *Z. Metallkd.* 65(10):642–649.

62. Henig, E. T. and H. L. Lukas. 1975. Calorimetric determination of enthalpy of formation and description of defect structure of ordered beta-phase (Ni, Cu)1-X Alx. *Z. Metallkd.* 66(2):98–106.

63. Neumann, J. P., Y. A. Chang, and C. M. Lee. 1976. Thermodynamics of intermetallic phases with the triple-defect B2 structure. *Acta Metall.* 24(7):593–604.

64. Kucherenko, L. A. and V. A. Troshkina. 1971. Transformations in NiAl compounds. *Russ. Metall.* 1:115.

65. Guard, R. W. and A. M. Turkalo. 1960. Fractographic studies in NiAl and Ni_3Al. In *Mechanical Properties of Intermetallic Compounds*, ed. J. H. Westbrook, 358. New York: Wiley.

66. Smialek, J. L. and R. F. Hehemann. 1973. Transformation temperatures of martensite in beta-phase nickel aluminide. *Metall. Trans.* 4(6):1571–1575.

67. Chakravorty, S. and C. M. Wayman. 1976. The thermoelastic martensitic transformation in β′ Ni–Al alloys: II. Electron microscopy. *Metall. Trans. A* 7:555–768.

68. Au, Y. K. and C. M. Wayman. 1972. Thermoelastic behavior of the martensitic transformation in β′ NiAl alloys. *Scr. Metall.* 6(12):1209–1214.

69. Enami, K., S. Nenno, and K. Shimizu. 1973. Crystal structure and internal twins of the Ni-36.8 at% Al martensite. *Trans. Jpn. Inst. Met.* 14(2):161–165.

70. Nagasawa, A., K. Enami, Y. Ishino, Y. Abe, and S. Nenno. 1974. Reversible shape memory effect. *Scr. Metall.* 8(9):1055–1060.

71. Gao, F. and D. J. Bacon. 1993. Calculating defect energies in alloys: Point defect and displacement energies in Ni_3Al. *Phil. Mag.* 67:275–288.

72. Yamaguchi, M. and H. Inui. 1994. Al_3Ti and its $L1_2$ variations. In *Intermetallic Compounds, Vol. 2, Practice*, eds. J. H. Westbrook and R. L. Fleischer, 147–173. Chichester, UK: Wiley.

73. McCullough, C., J. J. Valencia, C. G. Levi, and R. Mehrabian. 1989. Phase equilibria and solidification in Ti–Al alloys. *Acta Metall.* 37(5):1321–1336.

74. Huang, S. C. and P. A. Siemers. 1989. Characterization of the high-temperature phase fields near stoichiometric γ-TiAl. *Metall. Trans. A* 20(10):1899–1906.

75. Banerjee, D. 1994. Ti_3Al and its alloys. In *Intermetallic Compounds, Vol. 2, Practice*, eds. J. H. Westbrook and R. L. Fleischer. New York: Wiley.

76. Konitzer, D. G., I. P. Jones, and H. L. Fraser. 1986. Site occupancy in solid solutions of Nb in the intermetallic compounds TiAl and Ti_3Al. *Scr. Metall.* 20(2):265–268.

77. Nandy, T. K., D. Banerjee, and A. K. Gogia. 1988. Site substitution behavior in Ti_3Al (DO19) and Ti_2AlNb (B2) intermetallics. In *Proceedings of 6th World Conference on Titanium*, Vol. II, eds. P. Lacombe, R. Tricot, and G. Beranger. *J. de Physique*, 943–948. Paris: Les Éditions de Physique.

78. Arrell, D. J., H. M. Flower, and S. Kerry. 1992. The role of silicon in a Ti$_3$Al based alloy. In *Titanium '92: Science and Technology*, Vol. II, eds. F. H. Froes and I. L. Caplan, 1259–1266. Warrendale, PA: TMS.
79. Music, D. and J. M. Schneider. 2006. Effect of transition metal additives on electronic structure and elastic properties of TiAl and Ti$_3$Al. *Phys. Rev. B.* 74(17):174110–174114.
80. Liu, Y. L., L. M. Liu, S. Q. Wang, and H. Q. Ye. 2007. First-principles study of shear deformation in TiAl alloys. *J. Alloy. Compd.* 440:287–294.
81. Schwaighofer, E., H. Clemens, S. Mayer, J. Lindemann, J. Klose, W. Smarsly, and V. Güther. 2014. Microstructural design and mechanical properties of a cast and heat treated intermetallic multi-phase γ-TiAl based alloy. *Intermetallics* 44:128–140.
82. Schwaighofer, E., M. Schloffer, T. Schmoelzer, S. Mayer, J. Lindemann, V. Guether, J. Klose, and H. Clemens. 2012. Influence of heat treatments on the microstructure of a multi-phase titanium aluminide alloy. *Prac. Metallog.* 49:124–137.
83. Clemens, H., B. Boeck, W. Wallgram, T. Schmoelzer, L. M. Droessler, G. A. Zickler, H. Leitner, and A. Otto. 2008. Experimental studies and thermodynamic simulations of phase transformations in Ti–(41 45)Al–4Nb–1Mo–0.1B alloys. *Materials Research Society Symposium and Proceedings*, 115–120. Warrendale, PA: MRS.
84. Tetsui, T., K. Shindo, S. Kobyashi, and M. Takeyama. 2002. A newly developed hot worked TiAl alloy for blades and structural components. *Scr. Mater.* 47(6):399–403.
85. Rowe, R. G., D. Banerjee, K. Muraleedharan, M. Larsen, E. L. Hall, D. G. Konitzer, and A. P. Woodfield. 1992. Phase equilibria in Ti-Al-Nb alloys near Ti$_2$NbAl. In *Titanium '92: Science and Technology*, Vol. II, eds. F. H. Froes and I. L. Caplan, 1259–1266. San Diego, CA.
86. Banerjee, D., A. K. Gogia, T. K. Nandy, and V. A. Joshi. 1988. A new ordered orthorhombic phase in a Ti$_3$Al–Nb alloy. *Acta Metall.* 36:871–882.
87. Wua, Y., D. Z. Yang, and G. M. Song. 2000. The formation mechanism of the O phase in a Ti$_3$Al–Nb alloy. *Intermetallics* 8:629–632.
88. Muraleedharan, K., T. K. Nandy, D. Banerjee, and S. Lele. 1995. Phase stability and ordering behavior of the O phase in Ti–Al–Nb alloys. *Intermetallics* 3:187–199.
89. Nandy, T. K. and D. Banerjee. 1997. The mechanical behavior of the intermetallic Ti$_2$AlNb. In *Structural Intermetallics 1997*, eds. M. V. Nathal, R. Darolia, C. T. Liu, P. L. Martin, D. B. Miracle, R. Wagner, and M. Yamaguchi, 777–786. Warrendale, PA: TMS.
90. American Society for Metals. *Metals Handbook*, 8th edn, Vol. 8, ed. ASM Handbook Committee, 260. Materials Park, OH: ASM.
91. Vedula, K. and J. R. Stephens. 1987. B2 aluminides for high temperature applications. In *High Temperature Ordered Intermetallic Alloys II, Proceedings of the Second Symposium*, 2–4 December, Boston, MA, eds. N. S. Stoloff, C. C. Koch, C. T. Liu, and O. Izumi, 381–392. Pittsburgh, PA: MRS.
92. Ho, K. and R. A. Dodd. 1978. Point defects in FeAl. *Scr. Metall.* 12(11):1055–1058.

2

Methods of Processing

2.1 Introduction

The microstructure and properties of any material depend on the processing method along with the conditions used. As some of the intermetallic alloys have complex crystal structures, their properties are strongly influenced by stoichiometry, impurities, and defects. Usually, the intermetallic compounds have an ordered structure with a unique composition. A range of compositions is not permitted, and a distinct atomic ratio of constituents is required in these alloys for obtaining a given crystal structure and mechanical properties. In the alloys with a range of stoichiometry, the microstructures and properties are strongly dependent on the atomic ratios of the elemental constituents.

It is often challenging to obtain a defect-free multicomponent intermetallic alloy having a microstructure containing the desirable combination of phases with the specified composition, morphology, and grain size. Moreover, many of the intermetallic compounds of interest have congruent melting points much higher than those of the individual elements. In such cases, reactive sintering or melting of the elemental constituents in a calculated ratio is preferred to lower the processing temperature, but obtaining the correct composition and the desired amount of densification becomes the key challenge. Particularly if the alloying elements in an intermetallic with a wide solubility range possess significantly different vapor pressures, accurate control of the composition becomes difficult. The other option is sintering of powder having the stoichiometric composition of the intermetallic alloy. Such powders could be either commercially available or prepared by grinding of a porous lump obtained by either arc-melting or reactive sintering. Due to the high melting point, consolidation of the intermetallic powder to achieve complete densification through pressureless sintering, or hot pressing at elevated temperatures becomes extremely challenging.

The processing technique adopted depends on the nature of the product desired. For example, a thin-film processing technique is adopted for a coating, whereas for near-net-shaped bulk products ingot metallurgy processing involving melting and casting or suitable powder metallurgy processing may be preferred. For processing of the intermetallics, usually vacuum or inert gas environments are preferred in order to avoid oxidation of constituent elements or contamination of the products, which may lead to incomplete densification as well as the formation of undesirable or detrimental phases as impurities. Furthermore, minor concentrations of interstitial impurities may stabilize a nonequilibrium or undesirable crystal structure, or transform one structure to another, as well as have an adverse effect on the properties. In some of the alloys, the interstitial solute atoms may segregate at grain boundaries or interphase interfaces. Therefore, the efforts of researchers have been directed toward developing special methods of processing to tailor the level of purity and to control the grain size or sizes of dispersed phases, all of which have an impact on the mechanical properties.

Near-net-shaping thermomechanical processing is possible only if the intermetallic alloy shows sufficient room-temperature ductility or a relatively lower brittle-to-ductile transition temperature. As-cast or powder-processed aluminides and silicides may be subjected to subsequent heat treatments, or deformation processing for microstructural refinement, which improves the mechanical properties. Some intermetallic alloys with a suitable microstructure have been found to be amenable to superplastic forming. As the intermetallics are hard and brittle, machining by conventional tools to obtain products with required exact dimensions is almost impossible. However, as the intermetallics possess reasonable electrical conductivity, electrodischarge machining is universally possible.

2.2 Ingot Metallurgy Processing

The ingot metallurgy processing methods used to fabricate the intermetallics include arc-melting,[1,2] induction skull melting,[3] XD processing,[4–7] directional solidification,[8] single-crystal growth,[9–11] investment casting,[12,13] etc. Conventional methods of single-crystal growth such as the Bridgman or Czochralski techniques cannot be used because of the high melting point of silicides. Hence, alternative processes involving an electron-beam float zone, inductively coupled plasma pedestal growth, and an induction float zone in atmospheric and pressurized chambers have been developed for producing single crystals of the high-melting silicides.[9–11]

The melting of intermetallic compounds may not be trivial for alloy systems having constituents with large differences in melting points. Therefore, for an alloy of a given composition, in order to select the appropriate melting technique and understand the evolution of microstructure during solidification, one needs to examine the equilibrium phase diagram very carefully. For example, it may not be possible to obtain an intermetallic alloy without segregation, particularly if the solidification path includes a peritectic reaction, as in the case of TiAl with equiatomic composition (Figure 1.10). On the Ti-rich side, solidification of the Ti–Al melt leads to the formation of β-Ti through a peritectic reaction, such that formation of Ti_3Al directly from the melt is impossible. On cooling of the alloys with Al concentration in the range of 20%–35%, α-Ti first forms from β-Ti, and on further cooling, it is expected to transform into Ti_3Al.

Vacuum induction melting is one of the most common techniques for melting of intermetallic alloys in ceramic crucibles. It is absolutely neces-sary to choose the crucible and mold materials very carefully in order to avoid undesirable impurities. The crucible or mold material is chosen so as to avoid any kind of chemical reaction with the melt. For example, an alumina crucible is preferred for relatively less reactive Fe- and Ni-based melts. On the other hand, for the intermetallics based on more reactive metals such as Ti, regular ceramic crucibles are not suitable. A study by Degawa has shown that TiAl is contaminated due to oxygen pickup of 700 and 12,000 wppm when CaO and ZrO_2 crucibles are used, respectively, whereas the amount of oxygen contamination is relatively lower when MgO and Al_2O_3 crucibles are used.[14] Therefore, melting techniques using water-cooled metallic crucibles are used to melt the reactive intermetallic alloys. Vacuum arc-remelting has been used to process large ingots (≈150–1000 kg) of TiAl and Ti_3Al.

Recently, an induction skull melting technique has been developed to cast small ingots (≈30 kg) of high-purity intermetallics.[15] In this process, melting is carried out in a water-cooled metallic crucible in a clean atmosphere, and induction stirring aids in obtaining a uniform composition in a single opera-tion. Another technique, known as plasma hearth melting, involves the use of a plasma torch with high power density to melt refractory alloys. The melt, on a thin skull of its own composition, is stirred by the intense energy of the beam.

In all the abovementioned methods, the solidification path for the conven-tional melting technique is controlled by the thermodynamic parameters of the intermetallic alloy as well as kinetic considerations. The presence of peritectic reactions in the equilibrium phase diagram makes it difficult to avoid segrega-tion. Segregation of low-melting alloying elements is also expected to cause inhomogeneous deformation or failure during thermomechanical processing.

2.2.1 Nonequilibrium Solidification

Nonequilibrium processing such as rapid solidification processing and ther-mal spraying can be used to bypass the peritectic reaction under equilibrium conditions and significantly reduce the extent of segregation. Moreover, it is possible to obtain higher than equilibrium solid solubility as well as amor-phous or fine-grained microstructures.

2.2.1.1 Rapid Solidification

Rapid solidification of a melt involves rapid heat extraction, which in turn requires the thickness of the solid to be sufficiently small. In this manner, it is possible to prepare rapidly solidified intermetallics in the form of powders, splat/ribbon, and self-substrate quenching.[16] These products have different end uses, and the methods used to process them are quite different from one another and have been commercialized for practical applications.

The rapidly solidified powders need to be further consolidated for obtaining bulk products. Two major techniques used for rapid solidification processing of powders are: (i) high-pressure gas atomization and (ii) the rotating electrode process.[17–19] In the high-pressure gas-atomization process, convective cooling with the help of an inert gas such as argon is used to split the molten metal stream and to quench the droplets. On the other hand, the rotating electrode process uses a plasma arc to melt the end of a rotating electrode so that small droplets are ejected and subsequently cooled by convective heat transfer to solidify into powders, which are collected in an inert gas chamber. It is well accepted that gas atomization causes the formation of relatively finer powders than those obtained by the rotating electrode. In the gas-atomization process, the application of high-pressure gas jets leads to the formation of powder particles of hollow character or irregular shape, or both. Of course, some studies have shown the production of defect-free powders by gas-atomization techniques.[20,21]

Rapid heating of the surfaces using pulsed lasers is used for the formation of surface films comprising intermetallic alloys.[22,23] In this technique, a pulsed laser is used for deposition of one component of the intermetallic alloy on the surface of the other component. The surface region is melted and then rapidly solidified as the heat is extracted by the underlying substrate.

Rapid solidification processing has the following advantages:

1. It is possible to obtain fine microstructure, which increases the degree of homogeneity.
2. Microstructures containing desirable metastable phases, with some having a higher than equilibrium concentration of solute atoms, can be obtained.
3. The products obtained can be used as raw materials for subsequent processing of bulk materials and coatings.

The following disadvantages are also associated with rapid solidification processing:

1. The gases dissolved or entrapped in the powders during atomization may be liberated during subsequent consolidation at higher temperatures, leading to formation of porosities in the final product.
2. The intermetallic powders formed through this process have oxide scale, which persists during the subsequent consolidation operation.

This oxide scale can be broken up by deformation processing methods such as extrusion.

2.2.1.2 Thermal Spraying

In thermal spray processing, the melt is sprayed by acceleration to high velocity on to a substrate. This process is termed as *splat*. This method of processing, including flame spraying, arc-wire spraying, and plasma spraying, leads to the formation of a stream of molten particles, which are accelerated toward a substrate in a high-velocity gas stream. In the process of flame spraying, a combustion flame is used to melt powder, rod, or wire. On the other hand, arc spraying involves the formation of an electric arc between two wire electrodes used as feedstock. In the plasma-spraying process, a direct current or radiofrequency field discharge creates a plasma in inert gas or inert gas + hydrogen or nitrogen, which is used to melt the powder. The substrate for thermal spraying can be either at ambient temperature or, more likely, preheated to facilitate bonding between the splats.

Some of the commercialized thermal spray processes are the *Osprey* process and *low-pressure plasma spraying (LPPS)*. In the Osprey process, the material is melted in a bottom-pour crucible. As the melt flows from the crucible, it is atomized by a high-velocity gas stream, which also accelerates the molten droplets toward the substrate. Plasma spraying carried out in a low-pressure inert/reducing environment is useful for thermal spraying of intermetallic alloys, which are sensitive to interstitial contamination by pickup of oxygen or nitrogen.

Thick films of near-stoichiometric NiTi with shape memory properties have been processed by LPPS using arc-wire spraying of NiTi, whereas cosputtering of Ni and Ti has led to the formation of other Ni_xTi_y compositions $(x \neq y)$.[24] The LPPS of $MoSi_2$ onto mild steel substrates has led to the formation of a phase mixture containing primarily $MoSi_2$ along with a minor amount of Mo_5Si_3, which indicates that some Si is lost during processing.[25] The $MoSi_2$ deposited by the LPPS process has been found to have higher indentation fracture toughness than the hot-pressed prealloyed powder. An increase in crack-path tortuosity due to preferential cracking along the splat boundaries is responsible for the increase in toughness.

The thermal spraying process has the following advantages:

1. Almost any material can be deposited by this technique.
2. The process can be tailored to deposit a mixture of phases with different compositions, or a material having a gradient in composition, structure, or both.
3. It is possible to produce thick deposits, because the deposition rates are high.

The major limitations of this process are

1. It is a line-of-sight process, and therefore deposition can be done in only one direction.
2. The as-formed deposits are porous.
3. There is a possibility of contamination by both the thermal spray environment and handling of the powder used as feedstock.

2.2.2 Directional Solidification

The popular methods of producing single crystals of intermetallics are the Bridgman method, the zone-refining or floating zone method, and Czochralski pulling.[26] In the Czochralski technique, a seed crystal is slowly raised from a molten bath to grow a single crystal of a given intermetallic alloy in a chosen crystallographic direction. It has been used to grow a few rare-earth/transition-metal intermetallics. The Bridgman and floating zone methods can also be used for directional solidification. The alloys are usually melted in a water-cooled hearth in vacuum or argon atmosphere. In the Bridgman method, the entire specimen is in the liquid state. In this method, the process of solidification is initiated at one end of the melt by moving the specimen and heat source relative to one another. On solidification, columnar grain structure is obtained. By introducing a constriction in the crucible, it is possible to choose a specific orientation for growth. On introduction of a seed crystal with a desired orientation, it is possible to grow single crystals with similar orientation.

In the process of zone refining, the hot zone of the furnace is moved across the specimen in a given direction, resulting in localized melting. The impurities are progressively rejected into the melt as the molten zone resolidifies. By repeated passage of the molten zone in the same direction, it is possible to purify the sample progressively with a directionally solidified structure. As the molten zone is somewhat narrow in a sample held vertically, the partially molten sample is self-supporting, and therefore the method is known as the *floating zone* method. This method has been used by Hirano to directionally solidify stoichiometric Ni_3Al with preferred orientation of $\langle 111 \rangle$, which has shown significant ductility.[27] The floating zone technique has been modified for operation in high vacuum, where heating is carried out by an optical technique for the purpose of melting.

2.3 Powder Metallurgy Processing

2.3.1 Conventional Methods

The powder metallurgy processing techniques involve consolidation of the powders to form densified intermetallic alloys of a specified composition as the end product. The powders could either have the specified composition of the intermetallic alloy or be synthesized reactively from elemental precursors. For some of the intermetallic alloys, reactive synthesis from the blend of elemental powders is deliberately combined with application of pressure using either uniaxial hot pressing or hot isostatic pressing. In such cases, the reactions taking place during synthesis are exothermic and release sufficient heat to drive the reaction forward toward completion.

The conventional powder metallurgy process schedule involving cold compaction followed by pressureless sintering or hot pressing in either vacuum or inert atmosphere can be used for processing intermetallic alloy powders as well. As the diffusion kinetics is sluggish in many of the intermetallics, longer durations and higher temperatures are often required for complete densification. Moreover, as these materials are meant for high-performance applications at elevated temperatures, it is necessary to avoid impurities by using clean raw materials and environments.

Increase in the temperature of consolidation leads to grain growth, as has been observed for $MoSi_2$.[28] Plotting grain size against reciprocal of absolute temperature has exhibited two distinct regimes, with activation energy ≈ 46 kJ mol^{-1} at lower temperatures and ≈ 348 kJ mol^{-1} at relatively higher temperatures. On increasing the temperature beyond 1900°C, the rate of grain growth with increasing temperature is found to be very rapid. The results in this plot show that grain growth at temperatures ≤ 1900°C is accommodated by viscous flow of the intergranular glassy film, whereas grain rotation and creep appear to be operative at higher temperatures.

2.3.2 Reactive Consolidation

Nonconventional processing methods include processes used to prepare intermetallics in the form of either powders or bulk products starting from elemental powders as raw materials. Some examples of these processes are self-propagating high-temperature synthesis (SHS) and reactive sintering.[29–31] The process of reactive sintering can be carried out either under pressureless conditions or with the application of external pressure, as in reactive hot pressing or reactive hot isostatic pressing. Several reviews on reactive sintering are available in the literature.[32–37] In these processes, rapid reaction of the elemental powders occurs exothermically to form intermetallic alloys with the desired compositions. Although heating is required to initiate the reaction between the given elements, X and Y, evolution of significant additional heat equal to the heat of formation of the intermetallic compound occurs and drives the reaction to completion. Here, the reactants are ignited in order to spontaneously transform to products in an exothermic reaction. The process of reactive synthesis can be categorized on the basis of the progress of the reaction front, that is, whether it travels from one end to the other, as in the SHS process, or whether it occurs throughout the volume of the compact.

In the SHS process, the reaction is ignited by applying heat in the form of either a laser pulse or an electric discharge to one of the surfaces of the elemental powder compact.[37] The initial heating rate is fast enough to trigger a rapid reaction between the elemental raw materials, which in turn leads to adiabatic heating with the temperature rising to the ignition temperature. As the reaction becomes self-sustaining due to its own enthalpy, the external heating source is no longer required. The heat generated in this manner not only sustains the reaction but also greatly increases the reaction rate. Once the reaction is ignited, it propagates rapidly in the form of a combustion wave through the compact to its other surface. The maximum temperature, known as the combustion temperature (T_c), reached during the reaction could be either less than or equal to the adiabatic temperature (T_{ad}), which can be theoretically evaluated. For a given intermetallic, XY, with melting point, T_m, formed by the reaction of elements, the heat of formation can be considered as ΔH. This reaction can be written as[37]

$$mX + nY \rightarrow X_mY_n + \Delta H \tag{2.1}$$

If $T_{ad} < T_m$,

$$\Delta H_{f,298}\left(X_mY_n\right) + \int_{298}^{T_R} C_p.\left(mx+ny\right).dT = \int_{T_R}^{T_{ad}} C_p.\left(X_mY_n\right).dT \tag{2.2}$$

If $T_{ad} > T_m$,

$$\Delta H_{f,298}\left(X_mY_n\right) + \int_{298}^{T_R} C_p.\left(mX+nY\right).dT$$

$$= \int_{T_R}^{T_{ad}} C_p.\left(X_mY_n\right).dT + v.H_m + \int_{T_m}^{T_{ad}} zC_p.\left(X_mY_n, melt\right).dT \tag{2.3}$$

where:
 T_R = the reaction temperature
 v = the volume fraction of molten XY
 H_m = the enthalpy of fusion

For $T_{ad} = T_m$, $0 < v < 1$ and $z = 0$; whereas for $T_{ad} > T_m$, $v = z = 1$. $C_p(X_mY_n)$ and $C_p(mX+nY)$ are the specific heat capacities of X_mY_n and $(mX+nY)$ elemental reactant mixtures, respectively (J mol^{-1} K^{-1}). The maximum reaction temperature obtained experimentally is usually found to be $\leq T_{ad}$.

As the reaction rate is very rapid, the process is beyond any type of control, and therefore the product is usually porous. However, a product of uniform density and composition can be processed if the entire compact is uniformly ignited, and this process is known as *bulk* SHS. The rate of cooling from the processing temperature can be reduced significantly to minimize the damage caused by thermal shock. The thermal shock is more damaging in the case of intermetallics with noncubic structure, which possess coefficient of thermal expansion anisotropy.

If the reaction initiation temperature is higher than the melting point of one of the reactants, then the presence of transient liquid phase aids in the process of densification, as observed in the case of nickel aluminides. It has been possible to achieve ≥95% of theoretical density in Ni_3Al by reactive sintering process.[38] Through the results of this study, it has been shown that the density of the sintered product depends on both amount and connectivity of the transient liquid phase.

In the case of $MoSi_2$ or Mo–Si–B alloys, reactive hot pressing at a temperature (≈1500°C) slightly higher than the melting point of Si (≈1410°C) has led to complete reaction of the elemental powders as well as maximum densification.[39,40] The presence of liquid Si enhances the kinetics of mass transport, and therefore increases the reaction rate as well as the densification kinetics. Whereas the free energy of formation remains negative, the negative heat of formation of $MoSi_2$ increases sharply at temperatures slightly higher than the melting point of Si, as is obvious from the results in Figure 2.1. In other words, the reaction between Mo and Si becomes more exothermic at temperatures greater than the melting point of the latter element, and therefore becomes self-sustaining in this manner. Both reaction rate and densification kinetics are further enhanced by the application of external pressure, which

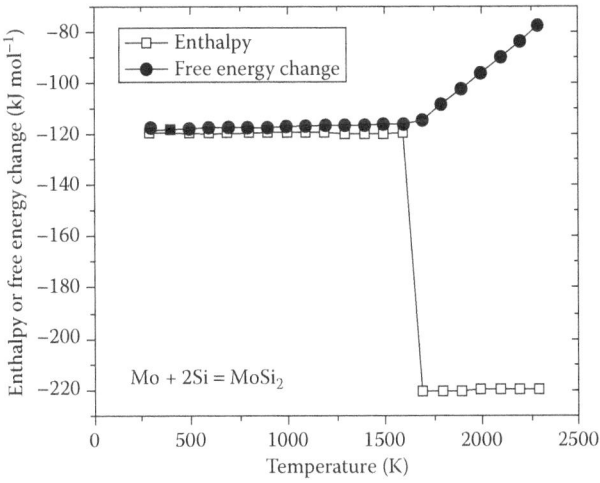

FIGURE 2.1
Plots showing the change in free energy and heat of formation of $MoSi_2$ with temperature.

aids both partial rearrangement of particles and flow of liquid through inter-particle spaces. The gaps between the Mo powder particles are filled by liquid Si, aided by capillary forces.

In an earlier study,[41] it has been shown that the reaction forming $MoSi_2$ in a solid–solid Mo–Si mixture progresses in steps as Si diffuses through the successive silicide layers on the surface of Mo. Formation of $MoSi_2$ is preceded by that of lower Si-containing silicides such as

$$Mo + Si \rightarrow Mo_3Si + Si \rightarrow Mo_5Si_3 + Si \rightarrow MoSi_2 \qquad (2.4)$$

In this case, the progress of reaction in each layer is dependent on the diffusivity of Si and is therefore expected to follow a parabolic rate law. However, Deevi has shown that in the SHS process the Si atoms diffuse through the Mo lattice and $MoSi_2$ forms in a single step reaction, as the rates of heating and diffusion are high.[42] As the reaction for the formation of $MoSi_2$ proceeds from the surface of the Mo particle to its center, it is evident that smaller particle size would reduce the diffusion distance and lead to faster diffusion kinetics. Interestingly, reactive hot pressing of $MoSi_2$ has led to products with much lower oxygen content than that in the raw materials, suggesting that *in situ* processing is self-purifying.[39,43] The oxygen content of *in situ* reaction-processed $MoSi_2$ has been found to be less than that of the elemental powders. The oxygen content of the $MoSi_2$ produced by Deevi through the SHS process has been found to contain 2200 wppm, in contrast to 4700 wppm and 10,000 wppm, respectively, in Mo and Si powders used as raw materials.[43]

Reactive processing is pursued for the following advantages:

1. The process is simple and requires relatively lower energy.
2. It is possible to obtain products with higher purity and lower oxygen content.
3. It involves simultaneous formation and densification of the intermetallic alloy.
4. Near-net shaping is possible.

However, the process of reactive sintering can also have a few problems, as follows:

1. Contaminants in powders which were not adequately removed prior to sintering
2. Undesirable reactions between the powders and die materials
3. Nonuniform reaction within the powder mixture leading to generation of internal stresses and cooling
4. Cracking of the final product due to nonuniform cooling from the maximum temperature

2.3.3 Mechanical Alloying

Mechanical alloying (MA) is a technique based on high-energy milling of elemental powder mixtures to obtain intermetallics with desired compositions and microstructures.[44–46] In the process of MA, a powder charge is placed inside vials with a suitable grinding medium containing spheres made of hard material (which should be harder than the material to be milled). Collision of the powders with balls and container walls results in powder fragmentation and rewelding, which in turn leads to microstructural refinement. The powders are also subjected to severe plastic deformation, which has a profound effect on structural refinement and phase transformations. Of course, much of the imparted energy is converted into heat, which also affects the above phenomena. The process of MA is usually carried out using hexane or toluene as the medium in order to inhibit agglomeration. Moreover, the material for balls needs to be chosen carefully, so as to avoid possibilities of contamination by surface erosion.

The length scale in the microstructures of MA powders varies from micrometer to nanometer scale, while the structure can be either partly or fully amorphous. The powders subjected to the high-energy MA possess high internal energy, which aids in the formation of extended solid solutions (with higher than equilibrium solute concentration). The intermetallic alloys formed by MA result from a balance between the tendency toward equilibrium and the accumulated disorder caused by the severe deformation during milling. The diffusion distances in the MA powders result from the nature of powder particle size and morphology. Whereas the brittle constituents are repeatedly fragmented, the ductile constituents are deformed severely to form ribbon-like morphologies. The process of MA can also be scaled up in order to produce large nanostructured amounts of different intermetallic alloy powders, which can be subsequently consolidated. Titanium aluminides such as Ti_3Al (+Nb)[47] and Al_3X (X = Ti, Zr, and Hf),[48] as well as silicides[49–51] including $MoSi_2$, Ti_5Si_3, $NbSi_2$, Nb_5Si_3, $TaSi_2$, and Ta_5Si_3, have been processed by the MA process.

The process of MA has the following advantages:

1. The formation of nanometric powders can reduce the sintering temperature and enhance the densification kinetics.
2. It is possible to obtain microstructures with nanocrystalline phases or a mixture of amorphous and nanocrystalline phases.

On the other hand, the disadvantages are as follows:

1. There is a strong possibility of contamination of the MA powders by erosion of the grinding media.
2. Oxygen and nitrogen contamination of the powders can occur through exposure to air.

3. Additional precautions are required for highly reactive metallic powders, which can catch fire due to increased surface area in an oxidizing environment.

2.3.4 Shock Consolidation and Synthesis

The intermetallic alloy powders prepared by rapid solidification processing can be consolidated by shock consolidation. As exposures at elevated temperatures for long durations can be avoided, it is possible to retain the fine microstructures obtained through rapid solidification. For example, the product obtained by shock consolidation of the rapidly solidified Ti–48 at.% Al powder has been found to retain the metastable α_2 phase in addition to the equilibrium γ phase formed by melting and resolidification of the interparticle boundaries.[52] A study on shock consolidation of Ni–Al powder mixtures has shown that a self-sustaining exothermic reaction could be initiated on attaining a critical shock-induced temperature.[53] It has been also possible to carry out shock-induced synthesis of silicides such as $NbSi_2$, $MoSi_2$, and Ti_5Si_3.[54]

One of the major shortcomings of the shock consolidation of hard and brittle intermetallics is the formation of both micro- and macro-scale cracks. Cracking is believed to occur due to the reflected components of the initial shock wave, where reflections occur from the interfaces in the regions surrounding the powder compact. The reflected waves generate tensile stresses within the compact, which in turn leads to the formation of cracks. Such cracking has been widely noticed in case of TiAl and Ti_3Al processed by dynamic shock consolidation.[55] However, this problem could be overcome by carrying out hot isostatic pressing after shock consolidation by using an explosive with a relatively low detonation velocity.

2.3.5 Displacement Reaction-Based Synthesis

Direct reduction of oxide precursors or displacement reactions on other compounds can lead to the formation of intermetallics with desirable compositions. In a study by Suzuki et al., it has been possible to process powders of TiAl and Ti_3Al from TiO_2 powders through the following calciothermic reduction reactions[56]:

$$2TiO_2 + Al_2O_3 + 7Ca \rightarrow 2TiAl + 7CaO \tag{2.5}$$

$$6TiO_2 + Al_2O_3 + 15Ca \rightarrow 2Ti_3Al + 15CaO \tag{2.6}$$

The powders of TiAl and Ti_3Al obtained in this manner have been found to be homogeneous with stoichiometric compositions and reasonably low oxygen contents ($\approx$1200 wppm). In a similar way, powders of Al_3Ti have also been prepared.[57]

Excellent mechanical properties have been observed in the $MoSi_2$–SiC composites processed by a displacement reaction of Mo_2C and Si[52]:

$$Mo_2C + Si \rightarrow 2MoSi_2 + SiC \qquad (2.7)$$

Some other examples of displacement reactions resulting in the reduction of SiO_2 present in $MoSi_2$ as an impurity are as follows[58–64]:

$$SiO_2\ (s) + 3C\ (s) \rightarrow SiC\ (s) + 2CO\ (g) \qquad (2.8)$$

$$SiO_2\ (s) + Al\ (l) \rightarrow Al_2O_3\ (s) + Si\ (s\ or\ l) \qquad (2.9)$$

2.4 Homogenization and Thermomechanical Processing

Homogenization of the as-cast intermetallics by heat treatment is a very slow process because of slow diffusion kinetics. For example, the as-cast Nb–Si and Nb–Si–X-based multicomponent alloys (with Nb_{ss}–Nb_5Si_3 composite microstructure) are usually homogenized by exposing to elevated temperatures for several hours.[8,65] On the other hand, homogenization of chemical composition and grain refinement can be carried out by thermomechanical processing for selected intermetallic alloys under optimized conditions of temperature and strain rate. Forging, extrusion, and rolling can be used for thermomechanical processing.[66,67]

Isothermal forging at 1200°C has been used for thermomechanical processing of wrought intermetallic alloys. Testing of compressive or tensile hot workability prior to isothermal forging has been used to determine the appropriate conditions for defect-free forging, as has been shown for TiAl.[68] The criterion for critical work of fracture has been reported to be used for estimating optimum conditions of temperature and strain rate, as well as the maximum strain attainable for the formation of intermetallic alloys.[69] Conventional nonisothermal hot-die forging has been used for shaping TiAl at temperatures <260°C.[70] In near γ-TiAl alloys, chemical inhomogeneity persists even after isothermal forging. For homogenization, the hot-worked alloy may be annealed at a higher temperature, or it can be forged at a suitable temperature to cause dynamic recrystallization.

Extrusion is widely used for thermomechanical processing of intermetallics due to the compressive stress state developed by reaction with the die walls during processing. Extrusion of γ-TiAl ingots has been carried out at temperatures between 1000°C and 1350°C with a reduction ratio of 6:1.[71] Furthermore, extrusion of Ni_3Al, NiAl, and FeAl has also been demonstrated by different research groups.[72–74] Hydrostatic extrusion has also been attempted on intermetallics. Extrusion of the *in situ* Nb_{ss}–Nb_5Si_3-based multicomponent composites has been carried out by encapsulating homogenized cast ingots in Mo cans at 1350°C or 1400°C with a ratio of 6:1.[8,65] This has led to a significant improvement in the mechanical properties of these composites. It has also been reported that extrusion of a β phase-containing TiAl alloy (Ti–42Al–9V–0.3Y alloy) in the range of 1275°C–1325°C leads to excellent hot workability and impressive mechanical properties.[75]

Rolling has been carried out on selected intermetallics with the object of processing foils. Foils of Ti_3Al have been obtained successfully by a combination of hot rolling and cold rolling. Cold rolling of twinned TiAl with special orientations has been demonstrated by Nishitani et al.[76] Moreover, Ni_3Al made ductile through B addition has been successfully cold rolled to 50% reduction.[77] Furthermore, hot rolling of FeAl, Ni_3Al, and TiAl is also reported in the literature.[78] With increasing brittleness of the intermetallics, ultrafine microstructures are required.

Some of the Ti-aluminides and Ni-aluminides with fine grain size are amenable to superplastic forming. The presence of multiphase microstructures ensures the stability of ultrafine grain size in these materials, where grain growth of the matrix phase is restricted by the presence of other phases.[79,80] It has been possible to process complex engineering shapes using a combination of diffusion bonding and superplastic forming of thin sheets. For example, rolled Ti_3Al alloy sheet with a fine grain size can be successfully diffusion bonded to itself and then superplastically formed to process large truss-core sections suitable for use in the hot-end components of gas turbines.

2.5 Summary

The methods used for processing intermetallics include both ingot and powder metallurgy-based routes. Although melting and casting appears to be a simple and inexpensive method of obtaining near-net-shaped composites, difficulties arise due to very high melting points as well as microstructural inhomogeneities caused by solute segregation. Melting of high melting-point intermetallics is usually carried out using either suitable ceramic crucibles or water-cooled metallic crucibles, as in the case of the skull melting technique. Further homogenization heat treatment is usually required to reduce compositional heterogeneities and to refine the dendritic microstructure. Directional solidification leading to growth of single crystals with desirable orientations has been attempted to obtain silicides and intermetallics with superior high-temperature strength-retention ability.

Nonequilibrium processing such as rapid solidification and mechanical alloying has often been used for producing intermetallic powders. Rapid solidification processing involving melt-spinning, thermal spraying, and atomization can significantly reduce the scale of inhomogeneity. On the other hand, mechanically alloyed powders exhibiting solid solubilities for alloying elements are also fine grained. Self-propagating high-temperature synthesis and XD processing uses the heat of reaction to complete the reaction. Besides hot isostatic pressing, advanced consolidation techniques such as shock compaction and spark plasma sintering have been developed for both aluminides and silicides. Reactive sintering of elemental powders under pressure causes the simultaneous formation and densification of intermetallic products, including aluminides and silicides. Further challenges are faced by researchers in forming complex shapes due to their poor formability and machinability, along with susceptibility to environmental embrittlement. Suitable alloying additions, as well as control of microstructure and grain size, have repeatedly been used to partially overcome such problems.

References

1. Schneibel, J. H., C. T. Liu, L. Heatherly, and M. J. Kramer. 1998. Assessment of processing routes and strength of a 3-phase molybdenum boron silicide (Mo_5Si_3–Mo_5SiB_2–Mo_3Si). *Scr. Mater.* 38:1169–1176.
2. Mendiratta, M. G. and D. M. Dimiduk. 1991. Phase relations and transformation kinetics in the high Nb region of the Nb–Si system. *Scr. Metall. Mater.* 25:237–242.
3. Bewlay, B. P., M. R. Jackson, and P. R. Subramanian. 1999. Processing high-temperature refractory-metal silicide *in situ* composites. *JOM* 51:32–36.
4. Brupbacher, J. M., L. Christodoulou, and D. C. Nagale. 1987. U.S. Patent 4,710,348.
5. Christodoulou, L., D. C. Nagale, and J. M. Brupbacher. 1988. U.S. Patent 4,774,052.
6. Nagale, D. C., L. Christodoulou, and J. M. Brupbacher. 1990. U.S. Patent 4,916,029.
7. Suzuki, M., S. R. Nutt, and R. M. Aikin Jr. 1993. Creep behavior of an SiC-reinforced XDTM $MoSi_2$ composite. *Mater. Sci. Eng. A* 162:73–82.
8. Bewlay, B. P., M. R. Jackson, and H. A. Lipsitt. 1996. The balance of mechanical and environmental properties of a multielement niobium-niobium silicide-based *in situ* composite. *Metall. Mater. Trans. A* 27:3801–3808.
9. Hirano, T., M. Nakamura, K. Hirano, and Y. Umakoshi. 1991. *Ceram. Eng. Sci. Proc.* 12:1619.
10. Stoloff, N. S. 1999. An overview of powder processing of silicides and their composites. *Mater. Sci. Eng. A* 261:169–180.
11. Weiser, M. M., S. R. Smelser, and J. J. Petrovic. 1990. Reaction sintering of molybdenum disilicide based composites. In *Intermetallic Matrix Composites*, eds. D. L. Anton, R. McMeeking, D. Miracle, and P. Martin. *Mater. Res. Soc. Symp. Proc.*, 194:53. Pittsburgh, PA: MRS.
12. Bewlay, B. P., M. R. Jackson, J. C. Zhao, P. R. Subramanian, M. G. Mendiratta, and J. J. Lewandowski. 2003. Ultrahigh-temperature Nb-silicide-based composites. *MRS Bull.* 28:646–653.

13. Bewlay, B. P., M. R. Jackson, J. C. Zhao, and P. R. Subramanian. 2003. A review of very-high-temperature Nb-silicide-based composites. *Metall. Mater. Trans. A* 34:2043–2052.

14. Degawa, T., K.-Y. Kamata, and Y. Nagashima. 1991. In *Proceedings of International Symposium on Intermetallic Compounds: Structure and Mechanical Properties (JIMIS-6)*, ed. O. Ijumi, 1003–1008. Sendai, Japan: Japan Institute of Metals.

15. Breig, P. G. and S. W. Scott. 1989. Induction skull melting of titanium aluminides. *Mater. Manuf. Proc.* 4:73–83.

16. Koch, C. C. 1988. Rapid solidification of intermetallic compounds. *Int. Mater. Rev.* 33:201–219.

17. McCullough, C., J. J. Valencia, C. G. Levi, and R. Mehrabian. 1990. Microstructural analysis of rapidly solidified Ti_3Al–X powders. *Mater. Sci. Eng. A* 124:83–101.

18. Valencia, J. J., C. McCullough, C. G. Levi, and R. Mehrabian. 1987. Microstructure evolution during conventional and rapid solidification of a Ti–50 at % Al alloy. *Scr. Metall.* 21:1341–1346.

19. Shih, D. S., G. K. Scarr, and J. C. Chestnut. On microstructural evolution of gas atomized Ti–50 at % Al–2 at % Nb powder. In *High Temperature Ordered Intermetallic Alloys III. Mater. Res. Soc. Symp. Proc.* 133:167–174, eds. C. C. Koch, C. T. Liu, N. S. Stoloff, and A. I. Taub. Pittsburgh, PA: MRS.

20. Huang, S. C. and A. M. Ritter. 1989. Microstructure of atomized Ni_3Al-B powder. *J. Mater. Res.* 4:288–293.

21. Laag, R., W. A. Kaysser, and G. Petzow. 1991. A comparative study on the influence of Nb and Ti additions to different processed atomized NiAl powders. In *High Temperature Ordered Intermetallic Alloys IV*, eds. L. A. Johnson, D. P. Pope, and J. O. Stiegler. *Mater. Res. Soc. Symp. Proc.* 213:821–826. Pittsburgh, PA: MRS.

22. D'Anna, E., A. V. Drigo, G. Leggieri, A. Luches, G. Majni, and P. Mengucci. 1990. Synthesis of chromium silicide with laser pulses. *Appl. Phys. A* 50:411–415.

23. Yang, G. Q. and S. U. Campisano. 1987. Phase formation and redistribution of Xe implanted in the Ni–Si system after fast melting and solidification. *Appl. Phys. A.* 44:279–284.

24. Jardine, A. P., Y. Horan, and H. Herman. 1991. Cavitation-erosion resistance of thick-film thermally sprayed NiTi. In *High Temperature Ordered Intermetallic Alloys IV*, eds. L. A. Johnson, D. P. Pope, and J. O. Stiegler. *Mater. Res. Soc. Symp. Proc.* 213:815–820. Pittsburgh, PA: MRS.

25. Tiwari, R., H. Herman, and S. Sampath. 1991. Spray forming of $MoSi_2$ and $MoSi_2$ based composites. In *High Temperature Ordered Intermetallic Alloys IV*, eds. L. A. Johnson, D. P. Pope, and J. O. Stiegler. *Mater. Res. Soc. Symp. Proc.* 213:807–812. Pittsburgh, PA: MRS.

26. Goodman, C. H. L. 1974. *Crystal Growth Theory and Techniques*. New York: Plenum.

27. Hirano, T. 1990. Improvement of room-temperature ductility of stoichiometric Ni_3Al by unidirectional solidification. *Acta Metall. Mater.* 38:2667–2671.

28. Sadananda, K., C. R. Feng, R. Mitra, and S. C. Deevi. 1999. Creep and fatigue properties of high temperature silicides and their composites. *Mater. Sci. Eng. A* 261:223–238.

29. German, R. M. 1990. Powder injection molding. In *Advances in Powder Metallurgy, Part 2*, 115–132. Princeton, NJ: Metal Powder Industries Federation.

30. Yi, H. C. and J. J. Moore. 1990. Self-propagating high-temperature (combustion) synthesis (Shs) of powder-compacted materials. *J. Mater. Sci.* 25:1159–1168.

31. Munir, Z. A. and U. Anselmi-Tamburini. 1989. Self-propagating exothermic reaction. *Mater. Sci. Rep.* 3:277–358.
32. Subrahmanyam, J. and M. Vijayakumar. 1992. Self-propagating high-temperature synthesis. *J. Mater. Sci.* 27:6249–6273.
33. Munir, Z. A. 1992. Reaction synthesis processes: Mechanisms and characteristics. *Metall. Trans. A* 23:7–13.
34. Dunand, D. C. 1995. Reactive synthesis of aluminide intermetallics. *Mater. Manuf. Process* 10:373–403.
35. Moore, J. J. and H. J. Feng. 1995. Combustion synthesis of advanced materials. 1. Reaction parameters. *Prog. Mater. Sci.* 39:243–273.
36. Moore, J. J. and H. J. Feng. 1995. Combustion synthesis of advanced materials. 2. Classification, applications and modeling. *Prog. Mater. Sci.* 39:275–316.
37. Morsi, K. 2001. Review: Reaction synthesis processing of Ni–Al intermetallic materials. *Mater. Sci. Eng. A* 299:1–15.
38. Bose, A., B. H. Rabin, and R. M. German. 1988. Reactive sintering nickel-aluminide to near full density. *Powder Metall. Int.* 20:25–30.
39. Mitra, R., Y. R. Mahajan, N. E. Prasad, and W. A. Chiou. 1997. Processing–microstructure–property relationships in reaction hot-pressed $MoSi_2$ and $MoSi_2$/SiCp composites. *Mater. Sci. Eng. A* 225:105–117.
40. Mitra, R., A. K. Srivastava, N. E. Prasad, and S. Kumari. 2006. Microstructure and mechanical behaviour of reaction hot pressed multiphase Mo–Si–B and Mo–Si–B–Al intermetallic alloys. *Intermetallics* 14:1461–1471.
41. Ivanov, V. E., E. P. Nechiporenko, and V. I. Zmiy. 1964. Study on the reaction diffusion in the Mo–Si system. *Fiz. Metal. Metalloved.* 17:94–99.
42. Deevi, S. C. 1992. Diffusional reactions in the combustion synthesis of $MoSi_2$. *Mater. Sci. Eng. A* 149:241–251.
43. Deevi, S. C. 1991. Self-propagating high-temperature synthesis of molybdenum disilicide. *J. Mater. Sci.* 26:3343–3353.
44. Maurice, D. R. and T. H. Courtney. 1990. The physics of mechanical alloying: A first report. *Metall. Trans. A* 21:289–303.
45. Patankar, S. N., S. Q. Xiao, J. J. Lewandowski, and A. H. Heuer. 1993. The mechanism of mechanical alloying of $MoSi_2$. *J. Mater. Res.* 8:1311–1316.
46. Srinivasan, S. R. and R. B. Schwarz. 1992. Synthesis of $MoSi_2$-based alloys by mechanical alloying. In *Novel Powder Processing, Advances in Powder and Particulate Materials*, 7:345. Princeton, NJ: Metal Powder Industries Federation.
47. Christman, T. and M. Jain. 1991. Processing and consolidation of bulk nanocrystalline titanium-aluminide. *Scr. Metall. Mater.* 25:767–772.
48. Srinivasan, S., P. B. Desch, and R. B. Schwarz. 1991. Metastable phases in the $Al_3x(X = Ti, Zr, and Hf)$ intermetallic system. *Scr. Metall. Mater.* 25:2513–2516.
49. Kumar, K. S. and S. K. Mannan. 1991. Mechanical alloying behavior in the Nb-Si, Ta-Si, and Nb-Ta-Si systems. In *High Temperature Ordered Intermetallic Alloys III. Mater. Res. Soc. Symp. Proc.* 133:415. Pittsburgh, PA: MRS.
50. Radhakrishnan, R., S. Bhaduri, and C. H. Henager, Jr. 1997. The reactive processing of silicides. *JOM* 49:41–45.
51. Schwarz, R. B., S. R. Srinivasan, J. J. Petrovic, and C. J. Maggiore. 1992. Synthesis of molybdenum disilicide by mechanical alloying. *Mater. Sci. Eng. A* 155:75–83.
52. Vassiliou, M. S., C. G. Rhodes, M. R. Mitchell, and J. Graves. 1989. Metastable microstructure in dynamically consolidated γ titanium aluminide. *Scr. Metall.* 23:1791–1794.

53. Simonsen, I. K., Y. Horie, R. A. Graham, and M. Carr. 1987. Formation of amorphous nickel aluminides under shock-wave loading. *Mater. Lett.* 5:75–78.

54. Yu, L. H. and M. A. Meyers. 1991. Shock synthesis and synthesis-assisted shock consolidation of silicides. *J. Mater. Sci.* 26:601–611.

55. Ferreira, A., M. A. Meyers, N. N. Thadhani, S. N. Chang, and J. R. Kough. 1991. Dynamic compaction of titanium aluminides by explosively generated shock waves: Experimental and materials systems. *Metall. Trans. A* 22:685–695.

56. Suzuki, R. O., M. Ikezawa, T. H. Okabe, T. Oishi, and K. Ono. 1990. Preparation of TiAl and Ti_3Al powders by calciothermic reduction of oxides. *Mater. Trans. Jpn. Inst. Metals* 31:61–68.

57. Suzuki, R. O., T. Ueki, M. Ikezawa, T. H. Okabe, T. Oishi, and K. Ono. 1991. A fundamental study on preparation of Al_3Ti powders by calciothermic reduction of oxides. *Mater. Trans. Jpn. Inst. Metals* 32:272–277.

58. Henager Jr, C. H., J. L. Brimhall, and J. P. Hirth. 1992. Synthesis of a $MoSi_2$/SiC composite *in situ* using a solid state displacement reaction. *Mater. Sci. Eng. A* 155:109–114.

59. Maloy, S., A. H. Heuer, J. Lewandowski, and J. Petrovic. 1991. Carbon additions to molybdenum disilicide: Improved high-temperature mechanical properties. *J. Am. Ceram. Soc.* 74:2704–2706.

60. Jayashankar, S. and M. J. Kaufman. 1992. *In situ* reinforced $MoSi_2$ composites by mechanical alloying. *Scr. Metall. Mater.* 26:1245–1250.

61. Costa e Silva, A. and M. J. Kaufman. 1995. Applications of *in situ* reactions to $MoSi_2$-based materials. *Mater. Sci. Eng. A* 195:75–88.

62. Sadananda, K. and C. R. Feng. 1995. Effect of carbon addition on the creep of molybdenum disilicide composites. *Mater. Sci. Eng. A* 192–193, Part 2:862–867.

63. Costa e Silva, A. and M. J. Kaufman. 1993. Microstructural modification of $MoSi_2$ through aluminum additions. *Scr. Metall. Mater.* 29:1141–1145.

64. Mitra, R., V. V. R. Rao, and A. V. Rao. 1999. Effect of small aluminum additions on microstructure and mechanical properties of molybdenum di-silicide. *Intermetallics* 7:213–232.

65. Mendiratta, M. G., J. J. Lewandowski, and D. M. Dimiduk. 1991. Strength and ductile-phase toughening in the 2-phase Nb/Nb_5Si_3 alloys. *Metall. Trans. A* 22:1573–1583.

66. Summers, E., A. J. Thom, B. Cook, and M. Akinc. 2000. Extrusion and selected engineering properties of Mo–Si–B intermetallics. *Intermetallics* 8:1169–1174.

67. Jehanno, P., H. Kestler, A. Venskutonis, M. Boning, M. Heilmaier, B. Bewlay, and M. Jackson. 2005. Assessment of a powder metallurgical processing route for refractory metal silicide alloys. *Metall. Mater. Trans. A* 36:515–523.

68. Semiatin, S. L., N. Frey, S. M. Elsoudani, and J. D. Bryant. 1992. Flow softening and microstructure evolution during hot-working of wrought near-gamma titanium aluminides. *Metall. Trans. A* 23(6):1719–1735.

69. Seetharaman, V., R. L. Goetz, and S. L. Semiatin. 1991. Tensile behavior of a cast gamma-titanium aluminide. In *High Temperature Ordered Intermetallic Alloys IV*, eds. L. A. Johnson, D. P. Pope, and J. O. Stiegler. *Mater. Res. Soc. Symp. Proc.* 213:895–900. Pittsburgh, PA: MRS.

70. Winkler, P. J., M. A. Daubler, and M. Peters. 1992. Application of Ti alloys in the European aerospace industry. In *Titanium 92, Science and Technology*, eds. F. H. Froes and I. L. Caplan, 3:2877. Warrendale, PA: TMS.

71. Seetharaman, V., J. C. Malas, and C. M. Lomibard. Hot extrusion of Ti–Al–Nb–Mn alloy. In *High Temperature Ordered Intermetallic Alloys IV*, eds. L. A. Johnson, D. P. Pope, and J. O. Stiegler. *Mater. Res. Soc. Symp. Proc.* 213:889–894, Pittsburgh, PA: MRS.

72. Krueger, D. D., B. J. Marquardt, and R. D. Field. Structure and properties of powder metallurgy Ni_3Al alloys. In *High Temperature Ordered Intermetallic Alloys II. Mater. Res. Soc. Symp. Proc.*, 81:309.

73. Baker, I. and E. M. Schulson. 1984. The structure of extruded NiAl. *Metall. Trans. A* 15:1129–1136.

74. Baker, I. and D. J. Gaydosh. The microstructure and tensile properties of extruded melt-spun ribbons of iron-rich B2 FeAl. In *High Temperature Ordered Intermetallic Alloys II*, eds. N. S. Stoloff, C. C. Koch, C. T. Liu, and O. Izumi. *Mater. Res. Soc. Symp. Proc.* 81:315–320. Pittsburgh, PA: MRS.

75. Xu, W. C., D. B. Shan, H. Zhang, X. A. Li, Y. Z. Zhang, and S. Nutt. 2013. Effects of extrusion deformation on microstructure, mechanical properties and hot workability of β containing TiAl alloy. *Mater. Sci. Eng. A.* 571:199–206.

76. Nishitani, S. R., M. H. Oh, A. Nakamura, T. Fujiwara, and M. Yamaguchi. 1990. Cold-rolling of Ti-rich TiAl polysynthetically twinned crystals. *J. Mater. Res.* 5:484–487.

77. Takasugi, T., S. Rikukawa, and S. Hanada. 1991. The boron effect on the superplastic deformation of $Ni_3(Si,Ti)$ alloys. *Scr. Metall. Mater.* 25:889–894.

78. Mirshams, A. R. and V. K. Sikka. 1990. An experimental study on the formability of intermetallic nickel aluminide. *J. Mater. Shaping Technol.* 8:123–132.

79. Yang, H. S., P. Jin, E. Dalder, and A. K. Mukherjee. 1991. Superplasticity in a Ti_3Al-base alloy stabilized by Nb, V and Mo. *Scr. Metall. Mater.* 25:1223–1228.

80. Dutta, A. and D. Banerjee. 1990. Superplastic behavior in a Ti_3Al–Nb alloy. *Scr. Metall. Mater.* 24:1319–1322.

3

Mechanical Behavior

3.1 Introduction

The mechanical behavior of intermetallics is strongly dependent on their crystal structures and grain sizes. As expected, both the strength and the ductility of intermetallics are functions of the complexity of their crystal structures, as well as degree of order and crystallographic anisotropy, which in turn influence the nature of defects formed. Both the stability of dislocations and their mobility depend on the structure and planarity of the dislocation cores in these alloys. Interestingly, certain intermetallics are ductile as single crystals but exhibit brittle behavior in the polycrystalline condition, because fewer than five independent slip systems are operative. However, with increasing temperature additional slip systems become operative or dislocation climb becomes possible, which in turn contributes to brittle-to-ductile transition in these materials. Of course, in multiphase intermetallic alloys, the volume fraction of different phases also has a strong role to play in their deformation and fracture behavior. Superimposed on the intrinsic mechanical properties of intermetallics is the effect of the environment, which has a particularly strong influence on the ductility of Ni- and Fe-aluminides at ambient temperature. Strength retention at elevated temperature is considered as quite challenging and is addressed through either suitable alloying or addition of reinforcement.

3.2 Dislocations in Intermetallics

3.2.1 Antiphase Boundaries and Superdislocations

The ordered nature of intermetallics strongly influences the dislocation formation and mobility.[1,2] A two-dimensional ordered arrangement of atoms is shown in Figure 3.1, in which a superlattice (with unit cell marked as bold lines) is made up of two sublattices (unit cells shown using dotted lines).

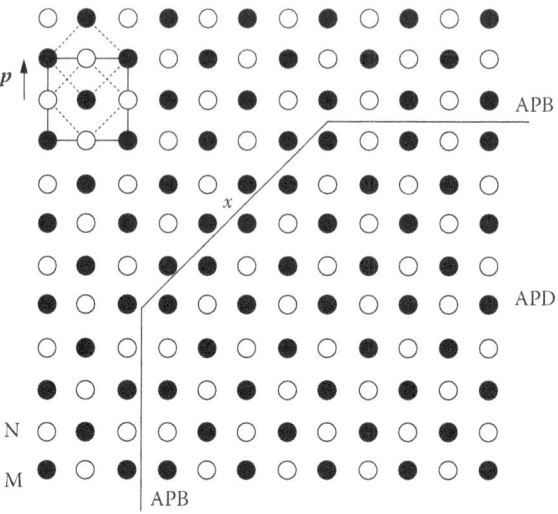

FIGURE 3.1

A square superlattice unit cell in the two-dimensional lattice of an ordered MN alloy. The location of the antiphase boundary (APB) and the antiphase domain (APD) is shown. The part of the boundary marked x is considered to be nonconservative, as N-type atoms are completely absent. The vector p in the unit cell is a typical APB vector.

In Figure 3.1, the sublattices are designated after the types of atoms occupying them, that is, M and N. For example, the $L1_2$ (cP4) superlattice with composition A_3B possesses ordered face-centered cubic (fcc) structure. This crystal structure contains four interpenetrating simple cubic sublattices, one of which comprises A atoms, whereas the other three contain B atoms. A vector (p) within the superlattice unit cell of the ordered alloy, MN, as shown in Figure 3.1, is considered to be an antiphase vector if it translates one type of sublattice into another type. Here, the p vector may be considered as a translation vector of the disordered lattice. Therefore, an interface separating two identically oriented crystals that have been displaced with respect to each other by a vector p is called an antiphase boundary (APB). Furthermore, the region surrounded by a continuous APB within an ordered crystal is known as the antiphase domain. The APBs are formed at the interfaces, where the ordered domains are "out of phase" with respect to one another. It should also be noted that there is no restriction on the plane normal (n) of the APB. The antiphase domain being enveloped by an APB is shown in Figure 3.1.

An ideal APB separates two rigidly displaced crystals with uniform composition and degree of order. The APB has a characteristic energy because the nearest-neighbor coordination of the superlattice is altered. Typical APB energies are similar to those of stacking faults, that is, ≈10–100 mJ m^{-2}.[1] The APBs can form during either disorder–order phase transformations or plastic deformation. In disorder–order transformations during cooling, the APBs usually form at a given transition temperature by nucleation of small ordered regions or nuclei. The antiphase domains (APDs) and APBs are formed as these nucleated ordered regions grow and come into contact with one another. In the dislocation-based mechanism, an APB appears in the form of a narrow ribbon coupling two or more partial dislocations. Here, the APB vector *p* is equal to the Burgers vector of the dislocation producing the APB.

The formation of APBs by a dislocation mechanism can be well understood by comparing the dislocation movement in disordered and ordered intermetallic alloys. In disordered alloys with face-centered or body-centered structure, the unit slip vectors are ½⟨110⟩ and ½⟨111⟩, respectively. In the ordered state, the slip vectors operative in the disordered alloys are not lattice translation vectors. For example, ½⟨111⟩ is not the Burgers vector of a perfect dislocation in an ordered body-centered cubic (bcc) lattice, whereas the ⟨100⟩ dislocations can be considered as perfect in both L1$_2$ and B2 structures. The motion of superpartial dislocations (e.g., ½⟨111⟩) leaves behind an APB. A schematic illustration of APB in the MN$_3$ superlattice, as shown in Figure 3.2, presents the atomic arrangement in two adjacent {111} planes. Displacement of one layer by ½⟨110⟩ with respect to the other shifts the atom in position M to that in N. This destroys the order by placing two M atoms in adjacent sites, as shown in Figure 3.1. In this case, the order can be restored by a second displacement of ½⟨110⟩, which takes the atom originally at X to Z, as shown in Figure 3.2. Thus, the perfect dislocation in the superlattice consists of two ½⟨110⟩ dislocations joined by an APB. Such an arrangement can be considered to be a superdislocation with Burgers vector of ⟨110⟩, which is the lattice translation vector in close-packed

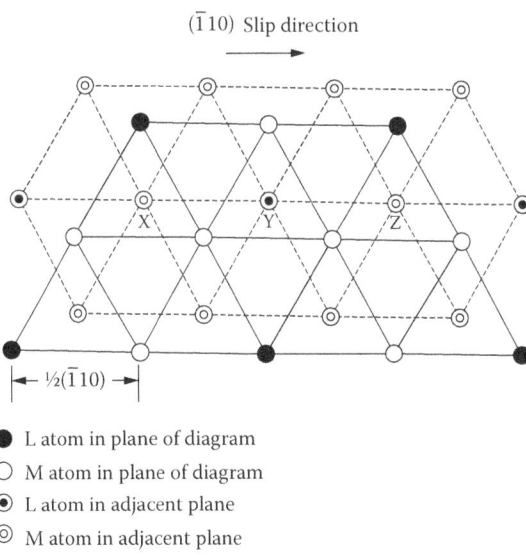

($\bar{1}$10) Slip direction

● L atom in plane of diagram
○ M atom in plane of diagram
⊙ L atom in adjacent plane
◎ M atom in adjacent plane

FIGURE 3.2
Schematic illustration of slip in {111} planes leading to formation of APB in the MN$_3$ superlattice.

direction in the ordered fcc lattice. In a similar manner, the ordered bcc alloy would show a superdislocation with Burgers vector of $\langle 111 \rangle$, which is split into two $\frac{1}{2}\langle 111 \rangle$ components on $\{110\}$ planes. It is important to note that the magnitudes of Burgers vector of the superdislocations, such as $\langle 110 \rangle$ and $\langle 111 \rangle$ in ordered fcc and bcc, respectively, are twice those of dislocations in the disordered lattice. This suggests that the self-energies ($\sim 0.5\ Gb^2$) of the superdislocations would be four times higher than those of the ordinary dislocations. This would reduce the stability of the superdislocations, and would promote their splitting into ordinary dislocations. Similarly to a stacking fault with Shockley partial dislocations, the separation between the component ordinary dislocations is a function of the balance between the elastic repulsive force between the dislocations and the opposing antiphase boundary energy. In some cases, the $\frac{1}{2}\langle 111 \rangle$ dislocations on either side of the APB may be further dissociated into partial dislocations with $b = 1/6\langle 112 \rangle$.

In relatively simpler superlattice structures, $L1_2$ (cP4), $L1_0$ (tP4), and B2 (cP2), there is only one type of antiphase vector. On the other hand, there can be two or more antiphase vectors in intermetallics with more complex crystal structures. For example, there are two antiphase vectors in superlattices with DO_{22} (tI8), DO_3 (cF16), and $L2_1$ (cF16) structures, whereas there are more such vectors in the case of the DO_{19} (hP8) structure. The antiphase vectors for some of the common superlattice structures are shown in Table 3.1. The APBs can be characterized on the basis of $p.n$ value. When $p.n = 0$, the displacement vector lies within the interface, and the APB can be formed by pure shear without change of stoichiometry in the boundary. These APBs are referred to as conservative and are caused by dissociation of dislocations during glide. When $p.n \neq 0$, the APB can be considered as being formed by a combination of two mechanisms: pure shear, and removal or insertion of a layer with thickness equal to $p.n$. In such cases, the APBs are considered as nonconservative. Here, the composition in the APB could be different from that of the bulk, if the removed or inserted layer does not contain atoms in stoichiometric proportion. For example, a layer of N atoms has been removed from the neighborhood of the APB shown in Figure 3.1, and therefore it can be considered to be nonconservative.

3.2.1.1 Observation of Antiphase Boundaries and Measurement of Fault Energies

The APBs and APDs can be observed using transmission electron microscopy (TEM) in dark-field imaging mode. The contrast arises due to the interference of electron waves diffracted from both sides of the APB to form the spot with reciprocal lattice vector g, and the phase difference between the waves is given by $2\pi g.p$. The contrast is nonzero when $g.p \neq$ integer (such as 0, 1, 2, …), and this is possible when g is a superlattice reflection in the diffraction pattern. The image vanishes when $g.p =$ integer, that is, g is a fundamental reflection or is normal to p. Therefore, the contrast due to APB is similar to that of stacking faults. When the plane of APB is inclined to the thin foil

TABLE 3.1

APB Energies (γ) in Selected Intermetallic Compounds

Intermetallic Composition and APB Vector, p	Crystal Structure	APB Energies, γ (mJ m^{-2})		Method	References
Ni$_3$Al	L1$_2$ (cP4)	γ$_{111}$=180±30 (350°C)	γ$_{010}$=140 (350°C)90±5	WB TEM	Veyssière et al.[3]
p=½⟨110⟩				WB TEM	Douin et al.[4]
NiAl	B2 (cP2)	γ$_{110}$=240	γ$_{112}$=380	EAM	Clapp et al.[5]
TiAl	L1$_0$ (tP4)	γ$_{111}$=145±15	γ$_{110}$=100 (600°C)	WB	Hug et al.[6]
Al$_3$Ti	DO$_{22}$ (tI8)	γ$_{111}$=200 (400°C)	γ$_{001}$=25 (400°C)	WB	Hug et al.[6]
FeAl	B2 (cP2)	γ$_{110}$=230		WB	Crawford and Ray[7]
Fe$_3$Al	DO$_3$ (cF16)	γ$_{110}$=70		WB	Crawford and Ray[7]

surface, the image consists of an oscillation of bright and dark bands. Weak-beam TEM imaging has been used extensively to experimentally measure the APB energies. Here, the APB is defined as the space between the dislocations, which appear as sharp bright lines under weak-beam conditions. It is usually impossible to separate the fault contrast from the interference caused by the strain field of the coupled dislocations.

3.2.1.2 Structure and Energies of Antiphase Boundaries

Until the early 1980s, the APB energy was estimated using the Ising model, in which the increase in the internal energy caused by APB formation was derived from the changes in the interaction energies between different atom pairs. Here, it is assumed that the pairwise interaction energies of any two atoms in the system are independent of the presence of other atoms. The methodology of calculation has evolved to include not only the interactions between the nearest-neighbor atoms,[8] but also those between the higher-order neighbors.[9-11] In last two decades, the method of *ab initio* quantum mechanical calculations has been developed to estimate the APB energies.[2] The energy increase per unit area, which is considered as the APB energy, can be expressed as

$$\gamma = \frac{2V}{\sqrt{3a^2}} \tag{3.1}$$

where:
$V = \frac{1}{2}(V_{AA} + V_{BB}) - V_{AB} = $ ordering energy of the nearest neighbor
$a = $ lattice constant

Here, $V_{ij}(i,j=A, B)$ is the interaction energy between the nearest neighbors. Usually, the APB energies are calculated by considering the energies for interactions between second-nearest-neighbor atoms to be only a fraction of that between the nearest neighbors.[12] This approach has shown the APB energy to be anisotropic in the ordered intermetallic alloys with $L1_2$ and B2 structures.[3,4]

Some superlattice structures are related to each other by APBs periodically on specific crystallographic planes. For example, in the case of an intermetallic alloy with the composition A_3B, the $D0_{22}$ and $D0_{23}$ crystal structures are obtainable from the $L1_2$ structure by creating an APB on every second and fourth {001} atomic layer, respectively, as shown in Figure 1.9. It has been shown using pair potential calculations that the [001] APB energy in the $L1_2$ structure would need to be negative for the DO_{22} structure to be stable.[13] If the interatomic interactions are included until the fourth nearest neighbor, the [001] APB energy is found to be the difference between the internal energies of the $L1_2$ and $D0_{23}$ structures.[14,15] For Al_3Ti, it has been shown that enhancement of the structural stability of $D0_{22}$ with respect to that of $L1_2$, along with an increase in the tetragonality ratio, occurs with a decrease in the {001} APB energy.[15] On the other hand, the [001] APB energy of Al_3Sc is significantly high (~450 mJ m^{-2}), and therefore this intermetallic is stable in the $L1_2$ structure.[16]

3.2.2 Dislocation Core and Peierls Stress

The minimum stress required for motion of dislocations in a given crystal is expressed as[1]

$$\tau_p = \left(\frac{2G}{\alpha}\right).\exp\left(\frac{-4\pi\zeta}{b}\right) \tag{3.2}$$

where:
   G = shear modulus
   ζ = width of the core
   b = magnitude of Burgers vector of dislocation

$$\zeta = \frac{d}{2a}$$

where:
   d = spacing between the slip planes
   a = interatomic spacing in the direction of slip

Moreover, $\alpha=1$ and $1-\upsilon$ for screw and edge dislocations, respectively, with υ being Poisson's ratio. Relation 3.1 clearly indicates that the value of τ_p decreases with increasing ζ, and its value is the lowest for dislocation motion on close-packed planes along the close-packed direction. Therefore, dislocations are sessile in the non-close-packed planes due to high Peierls stress, and hence the close-packed planes are the possible slip planes in pure metals. The expression for Peierls stress has been derived assuming the dislocation core to be localized in a close-packed slip plane. In such cases, the Peierls stress is usually low. However, unlike in metals, the dislocation cores in the close-packed intermetallics are nonplanar and therefore exhibit very high Peierls stress. The dislocation core could be spread into several parallel or nonparallel planes by climb or cross-slip, respectively. In the case of cross-slip, the Burgers vectors of the partials or continuously distributed dislocations lie in the planes of core spreading. A typical example of the cross-slip core is found in the bcc metal, in which the core of ½⟨111⟩ screw dislocation is spread into three {110} planes intersecting along the ⟨1$\bar{1}$1] direction.[17]

In the ordered intermetallics, the behavior of core spreading is a strong function of both crystal structure and ordering energy. In other words, the appearance of a nonplanar core in the ordered intermetallic alloys is driven by anisotropy of the elastic constants and planar fault energies, as well as the symmetries of the superlattice structures. For example, the close-packed directions in the {111} planes of intermetallics with L1$_0$ and D0$_{22}$ structures are not equivalent due to their tetragonality. Due to the high Peierls barrier, the dislocations with nonplanar cores have either limited or no mobility on the slip plane of the dislocation, and this is the basis for the intrinsic strengthening mechanism. It has also been shown that the Peierls barrier is overcome by dislocations subjected to thermal activation at elevated temperatures. Therefore, the intermetallics that are brittle at ambient temperature, exhibit brittle-to-ductile transition at elevated temperatures. However, some of the intermetallics show an anomalous increase in yield strength within a given temperature regime, which has been also attributed to spreading of dislocation cores. For example, the core of a ⟨110⟩ dislocation in an L1$_2$-structured intermetallic has been reported to spread simultaneously on the close-packed {111} slip planes and on the {001} planes with the lowest APB energy.[18] It is intuitive that both stoichiometry and alloying would strongly affect the nature of interatomic bonding and therefore the ordering energies, which in turn would influence the dislocation-core structure.

3.2.3 Possible Planar Faults in Intermetallics and Dislocation Dissociation Mechanisms

In the ordered intermetallics, it is often energetically favorable for a moving superdislocation to undergo multiple subdissociation into more than two partials with smaller Burgers vectors according to the b-squared or Frank criterion:

$$b^2 > \sum b_i^2 \tag{3.3}$$

A given superdislocation may dissociate in three different ways, depending on the stability of various faults as well as the length of the Burgers vectors of the superpartials. This may be illustrated by considering {111}-type slip planes of Ni$_3$Al.[19] ⟨110⟩/{111} is the common slip system in these A$_3$B-type alloys with L1$_2$ structure, which reduces to fcc, when A and B represent identical species.

One of the ways of representing APBs is shown in Figure 3.3.[19] The (111) plane, along with different crystallographic directions, is shown in Figure 3.3a. On the other hand, the layers above a given (111) plane are shifted by $b_A = \frac{1}{2}[1\bar{0}1]$, as shown in Figure 3.3b. In this manner, the fault formed in a single layer contains an *fcc* (*abcabc*-type) stacking sequence. However, the *M-M* type nearest neighbors introduced in this manner did not exist in the unfaulted material. The superlattice intrinsic stacking fault (SISF) is produced by shifting the top

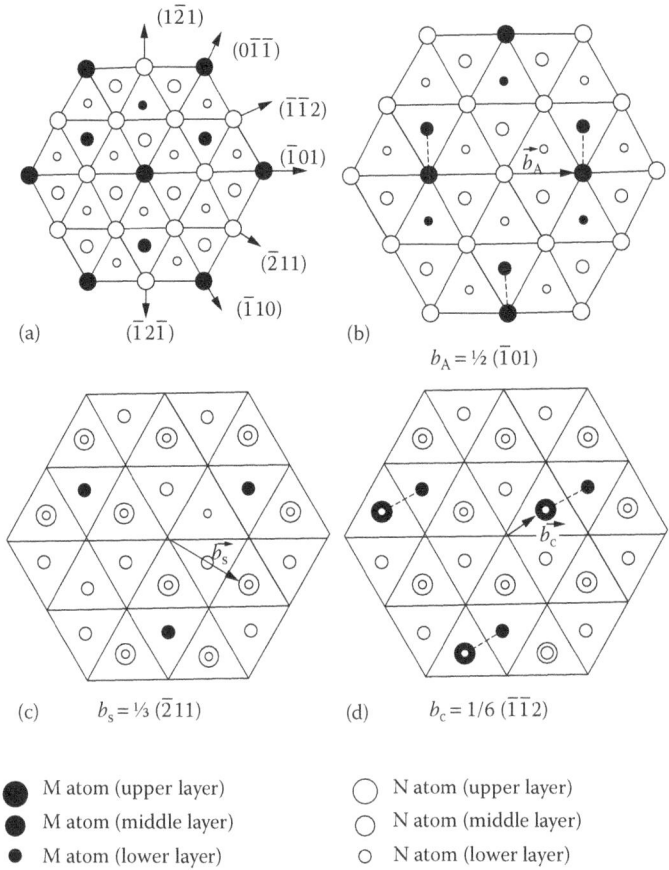

FIGURE 3.3
Schematic illustration of three successive {111} planes in the L1$_2$-structured MN$_3$ alloy. This illustration depicts (a) three {111} planes in a perfect lattice; (b) sliding of the top layer in (a) by $b_A = \frac{1}{2}[1\bar{0}1]$ to produce an APB, causing the formation of incorrect M–M nearest-neighbor bonds (shown using dashed lines); (c) sliding of the top layer in (a) by $b_s = \frac{1}{3}[\bar{2}11]$ to cause the formation of a SISF; and (d) sliding of the top layer in (a) by $b_C = 1/6[\bar{1}\bar{1}2]$ to produce a CSF.

layer, and all those above, by $\frac{1}{3}[2\bar{1}1]$, as shown in Figure 3.3c. Here, the local stacking sequence becomes *hcp* (*abab*-type stacking), but there are no nearest-neighbor violations. This fault is analogous to the ordinary stacking faults observed in fcc materials. However, it should be noted that the SISF has a Burgers vector different in direction and twice the magnitude compared with that found in the disordered fcc material ($1/6[\bar{1}\bar{1}2]$). A complex stacking fault (CSF) is created when the top layer and all layers above it are shifted relative to those below it, such that the M atoms in the top layer lie directly above the N atoms in the bottom layer. Therefore, a typical CSF in L1$_2$ structured Ni$_3$Al contains a combination of localized hcp stacking resembling the SISF and nearest-neighbor violations like that in the APD.

The formation of SISF and APBs by dissociation of the $[10\bar{1}]$ superdislocation involves the following possible dislocation reactions:

$$[10\bar{1}] = \frac{1}{3}[21\bar{1}] + \text{SISF}(1\bar{1}1) + \frac{1}{3}[1\bar{1}\bar{2}] \tag{3.4}$$

$$[10\bar{1}] = \frac{1}{2}[10\bar{1}] + \text{APB}(1\bar{1}1) + \frac{1}{2}[10\bar{1}] \tag{3.5}$$

Furthermore, dissociation of superdislocations with both APB and CSF on {111} planes is also possible in the following manner:[20]

$$[\bar{1}01] = \frac{1}{6}[\bar{1}\bar{1}2] + \frac{1}{6}[\bar{2}11] + \frac{1}{6}[\bar{1}\bar{1}2] + \frac{1}{6}[\bar{2}11] \tag{3.6}$$

Such splitting of the dislocations is promoted by reduction in the net self-energies of the dislocations, as shown in Relation 3.3. However, there is a competition between the dissociation modes shown in Relations 3.4 and 3.5, depending on the energies required for formation of APBs (γ_{APB}) and SISF (γ_{SISF}). For low values of γ_{APB}, as in Ni$_3$Al, the dislocation dissociation according to Relation 3.6 is observed more often.

The dislocations in different types of ordered structures show a variety of dissociation reactions, depending on the symmetry of the parent disordered lattice. Besides the dissociations coupled by APBs, SISFs, or CSFs, superdislocations may also dissociate into dislocations having a Burgers vector equal to perfect lattice translations, which are not held together by a surface defect. Dislocation reactions found in the intermetallics with different crystal structures are shown in Table 3.2.

3.2.4 Relation between Dislocation Core Structure and Crystal Symmetry

3.2.4.1 Dislocations in L1$_2$ Alloys

Although $\langle 100 \rangle$ is the shortest unit translation in the L1$_2$ structure, TEM-based investigations have shown mobile dislocations with $\langle 110 \rangle$-type Burgers

TABLE 3.2

Dislocation Reactions in Different Intermetallic Alloys

Crystal Structure	Possible Dislocation Reactions
$L1_2$ (cP4)	$[110]+[1\bar{1}0] \rightarrow 2[100]$
$D0_{22}$ (tI8)	$[100]+[010] \rightarrow [110]$
	$\frac{1}{2}[112] \rightarrow \frac{1}{3}[112]+\text{SISF}+\frac{1}{6}[112]$
$L1_0$ (tP4)	$[110]+[1\bar{1}\bar{0}] \rightarrow [100]+[100]$
	$\frac{1}{2}[110]+[001] \rightarrow \frac{1}{2}[112]$
	$\frac{1}{2}[110]+\frac{1}{2}[1\bar{1}2] \rightarrow [101]$
	$[100]+[001] \rightarrow [101]$
	$[011] \rightarrow \frac{1}{2}[011]+\text{APB}+\frac{1}{2}[011]$
	$\frac{1}{2}[112] \rightarrow \frac{1}{2}[101]+\text{APB}+\frac{1}{2}[011]$
$D0_{19}$ (hP8)	$\frac{1}{2}[11\bar{2}0]+\frac{1}{2}[1\bar{1}02] \rightarrow [10\bar{1}1]$
	$\frac{1}{6}[11\bar{2}6] \rightarrow \frac{1}{6}[11\bar{2}0]+[0001]$
$B2$ (cP2)	$[100]+[010]+[001] \rightarrow [111]$
	$[100]+[010] \rightarrow [110]$
$L2_1$ (cF16)	$[110]+[001] \rightarrow [111]$
$C11_b$ (tI8)	$[100]+[010] \rightarrow [110]$
	$\frac{1}{2}[111]\ 1\ [\bar{1}00] \rightarrow \frac{1}{2}[\bar{1}11]$

Source: Veyssière, P. and J. Douin, In *Intermetallic Compounds: Principles and Practice,* Vol. 1, eds. J. H. Westbrook and R. L. Fleischer, 519–558, Wiley, Chichester, 1995; Unal, O., J. J. Petrovic, D. H. Carter, and T. E. Mitchell, *J. Am. Ceram. Soc.* 73, 1752–1757, 1990; Mitra, R., N. E. Prasad, S. Kumari, and A. V. Rao, *Metall. Mater. Trans. A* 34A, 1069–1088, 2003.

vectors. The total energy of the ⟨110⟩ dislocation in its dissociated state is found to be greater than that of the undissociated ⟨100⟩ dislocation. The dislocation decomposition reaction (3.5) is most frequently observed in the case of intermetallic alloys with $L1_2$ structure, such as Cu_3Au,[24] Ni_3Fe,[25] and several Ni_3Al-based alloys.[26,27] The plane having the lowest APB energy is not necessarily the most preferred dissociation plane. It is intuitive that it would be energetically favorable for the screw dislocations to be dissociated in the octahedral slip planes ({111} planes), in spite of the fact that the cube planes ({100} planes) have lower APB energy.

Formation of pairs of ⅓⟨112⟩ superpartials coupled by SISF according to Reaction 3.4 has been reported on the basis of TEM studies on some of the $L1_2$ alloys, which include Ni_3Ga,[28] Ni_3Al-based alloys,[29] and Zr_3Al.[30,31] A transformation from the mechanism involving APB formation to that leading to SISF has been reported in the case of Ni_3Ga and Ni_3Al at relatively low and intermediate temperatures. In the Ni_3Al-based alloys, the ½⟨110⟩ dislocations with screw orientations, as formed by Reaction 3.4 or 3.5, have been found to cross-slip into other octahedral slip planes, which are cozonal with (i.e., these planes contain) the screw dislocation line. The ½⟨110⟩ dislocations are further split into 1/6⟨112⟩-type Shockley partial dislocations, thereby forming a CSF. Splitting of ⟨110⟩ dislocations into four 1/6⟨112⟩ partials has been experimentally observed by weak-beam dark-field TEM imaging.[32] As the APB energy on the {100} planes is the lowest, cross-slip of the 1/6⟨112⟩-type Shockley partial dislocations into these planes is energetically preferred. In this manner, the core of the ⟨110⟩ dislocations is spread simultaneously on close-packed {111} planes, being the slip planes, and on the {100} planes having the lowest APB energy, as shown schematically in Figure 3.4. This mechanism results in nonplanar core spreading of dislocations and is referred to as the Kear–Wilsdorf (KW) configuration in the literature. As the lattice friction stresses are high on the non-close-packed {001} planes, these partial dislocations are sessile and therefore restrict the glide of other lattice dislocations.

3.2.4.2 Dislocations in D0₂₂ Alloys

The shortest perfect lattice translations of the $D0_{22}$ (tI8) structure are ⟨100⟩, ⟨110⟩, and ½⟨112⟩, which is obvious on examination of its unit cell (Figure 1.9b). The deformation of the $D0_{22}$ alloys has been studied with reference to the relationship between the stability of derived phases and plasticity. The transformation from $D0_{22}$ to $L1_2$ or vice versa occurs through shear by ½⟨110⟩ over every other (001) plane, as the APB energy on this plane is the lowest.[33]

A significant amount of experimental effort has been directed to investigation of the dislocation structure in Al₃Ti with either $D0_{22}$ or $L1_2$ stabilized structure. In the $D0_{22}$ Al₃Ti, the ⟨100⟩ dislocations are dominant in the microstructure of the samples deformed between room temperature and 800°C.[34] During the glide of [100] dislocations in the basal (001) plane, the [110] superdislocation is formed through the reaction with [010] dislocation, as shown in Table 3.2. The subsequent reaction involving dissociation of the [110] superdislocation into a pair of ⟨100⟩ dislocations separated by an APB

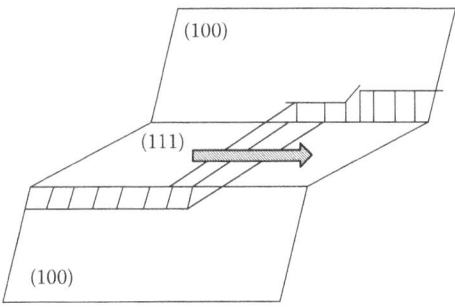

FIGURE 3.4
Schematic illustration of a typical Kear–Wisdorf lock in an L1₂-structured intermetallic alloy: superdislocation in dissociated form in an octahedral (111) plane, which has further cross-slipped into the (100) plane. The *arrow* in the (111) plane shows the direction of glide in the (111) plane.

is also reported.[27] Due to the anisotropy of APB energy, the dislocation line prefers to lie in the {100} planes with the lowest γ_{APB}. Therefore, the ⟨110⟩ dislocations are kinked with a staircase-type appearance, such that the length fraction in the {111} planes remains shorter than that in the {100} planes. The presence of dislocations with $b = \frac{1}{2}\langle 112 \rangle$ has been noticed in Al_3X-type alloys deformed at all possible temperatures. Moreover, in Al_3Ti deformed at high temperatures (400°C) the following reaction has been observed:

$$\frac{1}{2}\langle 112 \rangle = \frac{1}{3}\langle 112 \rangle + SISF + \frac{1}{6}\langle 112 \rangle \tag{3.7}$$

3.2.4.3 Dislocations in $L1_0$ Alloys

Due to tetragonality ($c/a > 1$) of the $L1_0$ (face-centered tetragonal unit cell, tP4 as in TiAl) structure, the magnitudes of all ⟨100⟩, ⟨110⟩, or ⟨112⟩-type lattice translation vectors are not equal. The shortest translation vectors in the $L1_0$ structure can be arranged in ascending order as follows: ½⟨110⟩, ⟨100⟩, ½⟨112⟩, and ⟨011⟩, provided $c > (1 - 1/\sqrt{2})a$. The dislocation structure has been investigated experimentally or theoretically and modeled experimentally by various researchers.[35–39] The stability of dislocations with different Burgers vectors in these intermetallic alloys is dependent on temperature, stoichiometry, ternary alloying, and oxygen content. The presence of directional bonding has also been used to rationalize the nature of the dislocation structure observed in these alloys. The ½⟨110⟩ dislocations are found at all temperatures, whereas ⟨001] and ½⟨112] dislocations are observed between low and ambient temperatures. The ½⟨112] dislocations are sessile at intermediate temperatures, but these become mobile above 700°C. The following dislocation reactions have been observed:

$$\frac{1}{2}[112] + \frac{1}{2}[110] \rightarrow [011] \tag{3.8}$$

$$\frac{1}{2}[110] + [001] \rightarrow \frac{1}{2}[112] \tag{3.9}$$

Both ⟨011] and ½⟨112] dislocations may split as

$$[011] \rightarrow \frac{1}{2}[011] + APB + \frac{1}{2}[011] \tag{3.10}$$

$$\frac{1}{2}[112] \rightarrow \frac{1}{2}[101] + APB + \frac{1}{2}[011] \tag{3.11}$$

Furthermore, there could be subdissociation of ½[011] superpartials into 1/6⟨112⟩ Shockley partial dislocations, which may lead to a configuration such as

$$[011] \rightarrow \frac{1}{6}[\bar{1}21] + CSF + \frac{1}{6}[112] + APB + \frac{1}{6}[\bar{1}21] + SISF + \frac{1}{6}[112] \tag{3.12}$$

Dissociation of dislocations with a Burgers vector of ½⟨011⟩ dislocation into 1/6⟨112⟩ superpartials separated by SISF is observed more commonly as $\gamma_{SISF} < \gamma_{CSF}$. Moreover, twinning on {111} planes is an important mode of deformation at ambient and intermediate temperatures, where twin activity increases with increasing temperature.

Deformation of TiAl is also dependent on ternary alloying additions, as expected. The presence of oxygen in interstitial solid solution adversely affects the ductility by reducing the activity of ½⟨110⟩ dislocations.[40,41] This observation has been attributed to enhancement of bond directionality by the presence of oxygen atoms in solid solution.

3.2.4.4 Dislocations in B2 Alloys

The shortest Burgers vectors in the B2-structured ordered lattice are ⟨100⟩, ⟨110⟩, and ⟨111⟩. Whereas some of the B2 alloys deform by motion of ⟨100⟩ dislocations, others, such as CuZn and FeAl, deform by ⟨111⟩ dislocations.[27] In the B2 alloys with high ordering energy, the slip occurs along the ⟨001⟩ direction, but it is preferred along ⟨111⟩ directions in weakly ordered alloys.[42,43] It has been observed that deformation is preferred along the ⟨111⟩ direction in the B2 alloys that exhibit order–disorder transformation below the melting point. However, in materials like FeAl and AgMg, a transition from ⟨111⟩ to ⟨100⟩ slip has been observed with increase in temperature.[12] The slip transition temperature depends on composition. Slip by movement of ⟨111⟩ dislocations is preferred in the alloys with relatively lower APB energies, when the following dissociation reaction is energetically favorable:

$$[111] \rightarrow \frac{1}{2}[111] + APB + \frac{1}{2}[111] \tag{3.13}$$

As the self-energy of ⟨100⟩ dislocation is always less than that of ½[111], the former type of dislocation is energetically more favorable. Moreover, the preference for slip direction in the B2 alloys is dependent on mobility of these dislocations. No stacking fault is reported to be stable in the B2 structure.[44] However, the existence of a stacking fault with a Burgers vector of 1/6⟨111⟩ has been observed by Song et al.[45]

It has been observed that B2-structured NiAl usually deforms primarily by ⟨100⟩{011} slip, and to a smaller extent by ⟨110⟩{011}.[46] In the NiAl alloys, the APB energies are large, and therefore dissociation of ⟨111⟩ dislocations is not reported to be resolved using weak-beam TEM. Moreover, TEM studies have shown evidence for slip of ⟨111⟩ dislocations on {112} planes.[47] Due to core distortions, ⟨111⟩ dislocations are less mobile than ⟨100⟩ dislocations. No evidence of nonplanarity is found for the ⟨100⟩ dislocations through high-resolution transmission electron microscopy (HRTEM) study. In the single crystals of NiAl compressed along ⟨100⟩ at temperatures >400°C, deformation is found to involve the motion of ⟨110⟩ dislocations.[48,49] It has been observed that $\langle 110 \rangle$ dislocations are split by climb into two $\frac{1}{2}\langle 111 \rangle$ partial dislocations according to the following reaction:

$$[110] \rightarrow \frac{1}{2}[111] + APB + \frac{1}{2}[11\bar{1}] \tag{3.14}$$

3.2.4.5 Dislocations in L2₁ and D0₃ Structures

The crystal structures of L2₁ (cF16) and D0₃ (cF16) can be considered as being built from eight unit cells of B2 (cP2), which implies that $\frac{1}{2}\langle 111 \rangle_{D03} = \frac{1}{2}\langle 111 \rangle_{L21} = \langle 111 \rangle_{B2}$. The major lattice translation vectors, in increasing order of magnitude, are as follows: $\frac{1}{2}\langle 110 \rangle$, $\langle 001 \rangle$, and $\langle 111 \rangle$. No stacking fault is reported to be stable in the L2₁ and D0₃-structured alloys, similarly to B2 alloys.[44] The APBs can be formed in these alloys through the following dislocation dissociation reactions:

$$\langle 111 \rangle \rightarrow \frac{1}{2}\langle 111 \rangle + APB_1 + \frac{1}{2}\langle 111 \rangle \tag{3.15}$$

$$\langle 111 \rangle \rightarrow \frac{1}{4}\langle 111 \rangle + APB_2 + \frac{1}{4}\langle 111 \rangle + APB_1 + \frac{1}{4}\langle 111 \rangle + APB_2 + \frac{1}{4}\langle 111 \rangle \tag{3.16}$$

$$\langle 001 \rangle \rightarrow \frac{1}{2}\langle 001 \rangle + APB_1 + \frac{1}{2}\langle 001 \rangle \tag{3.17}$$

The habit planes of APB_1 have been determined as {110} and {112}. When APB_2 is stable, fourfold dissociation of $\langle 111 \rangle$ dislocation, as in Reaction 3.16, has been reported in the case of D0₃-structured Fe₃Al.[50] However, twofold dissociation of the $\langle 111 \rangle$ dislocation through Reaction 3.15 has been reported when APB_2 is unstable. The L2₁ alloys have shown slip in the $\langle 001 \rangle$ and $\langle 110 \rangle$ directions.[51]

3.2.5 Effect of Alloying, Elastic Anisotropy, and Bonding on Dislocation-Core Structure

Besides crystal structure, the dislocation-core structure is affected by alloying, elastic anisotropy, and directionality of the chemical bond. Addition of different alloying elements has been found to strongly influence the mechanical behavior of single-phase intermetallic alloys, and such observations have been reported for Ni₃Al,[52,53] TiAl,[54] Ti₃Al,[55] and NiAl or FeAl-based alloys.[56] As alloying additions influence both APB and CSF energies, the separation between dissociated dislocations and the stability of dislocation segments along different directions are altered. Intuitively, the type of influence that temperature has on APB and CSF energies of various intermetallic alloys is also dependent on composition.[57]

The intrinsic elastic anisotropy in a given crystal structure is known to affect the dislocation core structure quite strongly. In an anisotropic crystal, the shear modulus varies with crystallographic direction. Therefore, the dislocation self-energies and interaction energies change depending on crystallographic plane and direction, which in turn strongly affects the core spreading in different planes. In other words, the dislocation configuration is unstable in certain slip planes, which in turn may lead to segmentation of the dislocation line.

Directionality of bonding also affects the dislocation-core structure. With an increase in bond directionality, the distortion imposed by the formation of a given dislocation is enhanced, which in turn increases the stresses required for dislocation motion. In the $L1_0$-structured TiAl, anisotropy of charge distribution leads to directional d–d bonds in planes containing Ti atoms.[58] Therefore, the dislocation segments lying perpendicular to these planes are reported to have reduced widths, leading to the requirement of high values of Peierls stress for their movement.[36] The presence of a strong Al (p)–Ti (d) bond has also been shown by a few researchers.[59,60]

3.2.6 Slip Systems

For the deformability of polycrystalline materials, it is absolutely necessary that five independent slip systems are operative. This condition ensures strain compatibility at the grain boundaries with formation of geometrically necessary dislocations during deformation, and thereby premature fracture is avoided. The absence of five independent slip systems is responsible for the brittleness of several intermetallic alloys in their polycrystalline form. Interestingly, the single crystals of these alloys have been found to be ductile at selected orientations. In such cases, slip systems with an applied shear stress component exceeding their critical resolved shear stress (CRSS) are found to operate. Therefore, single crystals having a stress axis oriented for deformation through the operation of slip systems with a lower CRSS are likely to deform more easily at lower stress and are known to have soft orientations. On the other hand, the crystal orientations for operation of slip systems with higher values of CRSS are considered to be hard. Although single crystals may be deformable even at subzero temperatures, polycrystalline samples may be not deformable even at higher temperatures because five independent slip systems are not operative. Such intermetallic materials are known to have a brittle-to-ductile transition temperature (BDTT) when an adequate number of slip systems are activated or intergranular strain is accommodated by diffusive mass transfer.

There have been several attempts to design intermetallic alloys with additional slip systems to reduce brittleness at ambient temperature. Some of the examples are $L1_2$-stabilized $D0_{22}$ Al_3TiX or NiAl-based alloys with their superdislocations activated. The most favored slip system contains one of the shortest lattice translation vectors and operates in one of the most densely packed planes. Having the Burgers vector equal to the shortest lattice translation vector ensures that the dislocation self-energy is lowest. This criterion is violated if the mobility of the dislocation in a given slip plane is restricted. For example, if the core of the dislocation is narrow, its mobility is significantly reduced. Furthermore, the activity of a given dislocation segment is reduced if its core is spread out of the slip plane.

The most preferred dislocations have the capability to dissociate into components with smaller Burgers vectors, which experience a smaller Peierls-type barrier. For example, $\langle 110 \rangle$ slip is favored in the $L1_2$-structured alloys, where $\langle 110 \rangle$ superdislocations are dissociated into two $\frac{1}{2}\langle 110 \rangle$ superpartials. In the intermetallics with B2 structure, $\langle 111 \rangle$ is found to be the preferred slip direction, if the APB energy is either low or moderate as in FeAl. However, if APB energy or line energy is high due to elastic anisotropy, slip along the $\langle 001 \rangle$ direction may be favored, as in NiAl.

3.3 Twinning and Displacive Transformation

In the intermetallic alloys with fewer than five independent slip systems, as well as higher self-energy and lower mobility of the existing dislocations, twinning is reported to play an important role in deformation.[61,62] Here, deformation twinning is observed not only at high strain rates/low temperatures, as for bcc/hexagonal close-packed (hcp) metals, but also at lower strain rates/higher temperatures. For example, contribution of deformation twinning to plastic straining has been observed in intermetallic alloys with various types of crystal structures, such as $L1_2$ (cP4, Cu_3Au type), B2 (cP2, CsCl type), DO_3 (cF16, BiF_3 type), tetragonal $L1_0$ (tP4, CuAu type), DO_{22} (tI8, Al_3Ti type), and hexagonal DO_{19} (hP8,Ni_3Sn type).[56] Microtwinning has been abundantly observed in TiAl, Al_3Ti, Al_3Nb, and NiTi. It has been shown that a dispersion of fine microtwins leads to toughening of a material, whereas massive (large-size) macrotwins could cause brittle fracture. The effects of slip–twin and twin–twin interactions on plastic flow or crack initiation in intermetallics need to be further investigated. It has been observed that the ductility of TiAl increases with an increase in the propensity of twinning at elevated temperatures, induced by alloying additions of Mn and Ga.[63] The compressive stress–strain curves obtained for single crystals of Al_3Ti show serrations caused by twinning. To improve the ductility of Al_3Ti, it is necessary to refine the twins, as the presence of massive macrotwins is known to cause brittle fracture.[64] For Al_3Ti, it has been shown that the addition of alloying elements (X) forming an orthorhombic DO_a phase (oP8) with a composition of Al_3X or X_3Ti lower the SISF energy and therefore increase the probability of twinning.[65]

3.4 Tensile and Compressive Deformation Behavior

The tensile and compressive tests on the intermetallic alloys have shown wide variations in strength and ductility, depending on their composition, test temperature, strain rates, and environment. For single crystals, orientation of the stress axis with respect to crystallographic directions has been found to have a strong effect on yield strength and flow behavior.[66–69] This observation is attributed to the dependence of operative slip systems on orientation of the single-crystal samples. For deformation occurring with operation of slip systems having higher values of CRSS, the flow stress is found to be high. The composition affects the amount of directional bonding and elastic properties of the intermetallics, which in turn influence the dislocation-core structure and Peierls stress. As the mobile dislocation density is relatively lower in most of the intermetallics, both temperature and strain rate have a significant effect on their flow behavior. In some of the silicide-based intermetallics, such as $MoSi_2$ and $NbSi_2$, yield-point phenomenon and serrated plastic flow have been observed.[66–68,70] The serrated plastic flow in these materials has been attributed to dynamic strain aging involving interaction of dislocations with point defects and interstitial solute atoms.[70–72] This mechanism has been confirmed by *in situ* dynamic TEM studies involving straining of samples, which has shown jerky movement of dislocations.[71,72]

Temperature has a strong effect on the deformation behavior of polycrystalline intermetallics, because of the insufficient number of independent slip systems being operative at temperatures less than the BDTT. At or above the BDTT, the fracture toughness increases significantly, and the mode of fracture changes from cleavage to ductile, involving microvoid coalescence. In some cases, the fracture mode remains cleavage in spite of increase in the fracture toughness. For example, γ-TiAl remains ordered until its melting point of 1460°C and shows <1% strain until 700°C, with strain to failure increasing up to 40% at temperatures >800°C.[73]

The influence of environment on ambient-temperature ductility is well established. This dependence is due to the embrittlement caused by the interaction of moisture or oxygen with the grain boundaries of the intermetallics, particularly aluminides.[74–77] $L1_2$-structured polycrystalline Ni_3Al shows poor ductility with typical intergranular mode of fracture when tested at ambient temperature in moist air, but shows reasonably high strain to failure if tested in dry air, vacuum, or dry oxygen. On exposure of aluminides to a moist environment, the Al atoms react with H_2O. The following reaction is expected:

$$2Al + 3H_2O \rightarrow Al_2O_3 + 6H \tag{3.18}$$

The atomic hydrogen (H) penetrates the crack tips as well as the grain boundaries and thereby leads to intergranular fracture. Segregation of H at the crack tips of FeAl promotes propagation of cracks along the cleavage planes.

Polycrystalline samples with a composition of Ni–22.65Al–0.26Zr have shown tensile ductility of 6%–9% in water, 11%–13% in air, and 48%–51% in oxygen.[78] All these specimens have shown intergranular fracture. Alloying of polycrystalline Ni_3Al-based alloys containing <25 at.% Al with B has been found to increase the strain to failure significantly and lead to a transgranular mode of fracture with suppression of the intergranular failure process.[79–81] It has been possible to achieve >50% tensile ductility in Ni_3Al by microalloying with 0.1 wt.% B. Studies using a scanning auger microprobe and an atom probe have shown strong evidence for segregation of B at the Ni_3Al grain boundaries. The ductilizing effect of B segregation has been attributed to enhancement in grain-boundary cohesive strength[82] and ease of slip transfer across grain boundaries.[83] Considering that both B and H occupy similar grain-boundary sites, it is appropriate to suggest that prior segregation of B would inhibit H diffusion and thereby prevent hydrogen embrittlement.

3.5 Creep and Superplasticity

The creep properties of the single-crystal intermetallic alloys have shown dependence on the crystallographic direction of the stress axis.[66,84–86] Of course, the high-temperature creep strength of single-crystal samples has been found to be higher than that of polycrystalline samples.[61] Dislocation creep (stress exponent ≈ 4–5) has usually been observed in these alloys. Therefore, intermetallic single crystals with hard orientations have been found to be significantly more creep resistant compared with those oriented differently. In order to understand the deformation mechanism during creep, it is necessary to examine the slip systems responsible for gliding of dislocations at the test temperature. However, the creep rate is controlled by dislocation climb, which involves lattice diffusion of constituent atoms. The lattice diffusion is restricted by the necessity to maintain the ordered structure of the intermetallic alloy.

In polycrystalline intermetallic alloys, the grain size is reported to have a significant effect on creep behavior.[67,87] The role of grain-boundary diffusion becomes more significant primarily because of the absence of an adequate number of slip systems. The purity of grain boundaries and the presence of glassy film are known to have a strong influence on the creep performance of polycrystalline intermetallic alloys such as $MoSi_2$.[87–89] As expected, the effect of grain size on creep is more predominant in the diffusion-creep regime (stress exponent = 1), in which Newtonian viscous flow involving grain-boundary sliding is the operating mechanism.[87] In such cases, damage is initiated at grain boundaries. In the multiphase intermetallic alloys, the presence of hard phases enhances the creep resistance by acting as an obstacle to dislocation motion. If the aspect ratios of the constituent phases are large, their orientation with respect to the stress axis also plays an important role. Both the composition and the heat-treatment procedure need to be optimized for such multiphase alloys in order to obtain the most desirable microstructure for achieving the specified level of creep resistance.

The creep resistance of the intermetallics is decreased sharply with an increase in temperature above the BDTT, because of thermally activated motion of dislocations through enhancement in diffusion kinetics, as well as grain-boundary sliding. The creep resistance of the intermetallics is usually enhanced by solid-solution strengthening as well as through addition of hard ceramic reinforcements. For example, the creep resistance of $MoSi_2$ can be significantly enhanced by alloying with W, which substitutes in the Mo sites.[90] A significant increase in creep resistance has been observed following the addition of a suitable amount of ceramic reinforcement.[89,90]

Superplasticity has been demonstrated in single-phase Ni_3Si and Ni_3Al alloys with fine grains (<6 μm).[91–94] The Ti_3Al-based alloys containing a mixture of α_2 (hcp D019, hP8) + β (bcc, cP2) phases exhibit fine-structure superplasticity over a wide temperature range.[95,96] In recent years, superplastic deformation behavior has been observed in coarse-grained intermetallics such as Fe_3Al and FeAl alloys,[97–101] Ni_3Al-based intermetallic alloy,[102–104] and γ-TiAl-based alloy,[105] with their initial grain sizes being 100–300 μm, 10–30 μm, and 95 μm, respectively. Furthermore, Ni–40Al,[106] Ni–42Al,[107] Ni–45Al,[108] and Ni–48Al[109–111] with coarse grain sizes have also exhibited superplasticity in the temperature range of 1000°C–1100°C. This behavior has been attributed to continuous dynamic recovery and recrystallization during deformation.

3.6 Fracture Behavior

Most of the ordered intermetallics with low-symmetry crystal structures usually exhibit poor fracture toughness along with transgranular cleavage fracture due to poor cleavage strength, inadequate slip systems, low dislocation density, weak grain boundaries, and environmental effects.[112] Whereas brittle intergranular fracture is observed in polycrystalline Ni- and Fe-aluminides in moist air,[74–77] alloying of Ni_3Al with B leads to ductile failure with fracture surfaces containing dimples.[112] Brittle intergranular fracture is also observed due to the segregation of impurities (e.g., sulfur, phosphorus, tin, and antimony) at grain boundaries. These impurity elements are more electronegative than the host metals and weaken the metal–metal bonds through electronic charge transfer. Different approaches to enhancing ductility and toughness include alloying to stabilize a relatively simpler high-symmetry crystal structure, microstructural/grain refinement, and tailoring of multi-phase microstructures. Transformation of $D0_{22}$ or $D0_{23}$ structures to $L1_2$ has not led to much increase in ductility, and the mechanism of failure is found to be transgranular cleavage. Even polycrystalline intermetallics with five independent slip systems, such as γ-TiAl ($L1_0$), Ti-48 at.% Al ($\gamma + \alpha_2$), Ni_3Al ($L1_2$), and $(Al,Cu)_3Ti$, have shown hardly any ductility.[112]

In TiAl and NiAl, the development of duplex microstructures containing mixed granular + lamellar morphology is known to be desirable for the enhancement of ductility and toughness.[112] Interestingly, the occurrence of cleavage or intergranular fracture is not necessarily accompanied by completely brittle failure. Fracture toughness of the intermetallic samples has been measured through four-point bend tests on the single-edge notch-bend (SENB) specimens.[113] The fracture-toughness values for a few intermetallic alloys at ambient temperature are shown in Table 3.3. A sharp increase in fracture toughness is usually observed with increase in temperature above the BDTT. For example, polycrystalline NiAl has exhibited fracture toughness of ≈ 5 MPa $m^{1/2}$ at ambient temperature, which rises to 10 MPa $m^{1/2}$ at 250°C and is found to be in the range of 40–50 MPa $m^{1/2}$ at 400°C.[113] The behavior observed depends on grain size as well as method of processing.

For NiAl (45 at.% Al), the fracture toughness has been found to be ≈ 5.4 MPa $m^{1/2}$, and only a modest increase has been reported on addition of ceramic reinforcement.[114] However, in the NiAl–32Cr–6Mo (at.%) alloy with eutectic microstructure comprising lamellae with alternating plates of NiAl and Cr(Mo), the fracture toughness has been found to be 23.7 MPa $m^{1/2}$.[122] The fracture toughness of Ni_3Al is found to rise from 20 MPa $m^{1/2}$ with the formation of a large plastic zone and intergranular failure, in spite of poor

TABLE 3.3

Fracture Toughness of Selected Intermetallics

Intermetallic	Fracture Toughness (MPa $m^{1/2}$)	References
Ni_3Al	18.7–20.9	Rigney and Lewandowski[114]
$Ni_3Al + B$	28.1–33.1	Rigney and Lewandowski[114]
NiAl	4.1–6.6	Rigney and Lewandowski[114]
γ-TiAl	10–22	Dogan et al.[115] Huang and Chestnut[116]
TiAl–Cr	14.5–16.2	Dogan et al.[115,117]
Ti_3Al-based alloys	15–20	Chan[118,119]
FeAl	10 (slow in air)	Inoue et al.[120]
	40 (rapid/in oil)	
$MoSi_2$	2.5–4	Mitra[121]

tensile ductility.[109] Significant R-curve behavior has been observed. The fracture toughness of Ni_3Al rises significantly to 30 MPa $m^{1/2}$ on alloying with B. Addition of B has been reported to change the mode of fracture in Ni_3Al from brittle intergranular to ductile tearing, as the grain-boundary cohesive strength is significantly enhanced.

Tests carried out on compact tension (CT) and SENB specimens of two-phase alloys based on γ-TiAl and TiAlCr with duplex microstructure have shown the fracture toughness to be ≈10 MPa $m^{1/2}$.[115] However, the results obtained for these specimens at 700°C have shown the K_{IC} test to be invalid. Under cyclic loading conditions, the stress intensity factor for crack initiation was found to be 9 and 20 MPa $m^{1/2}$ for SENB and CT specimens, respectively. The initiation and growth of cracks in multiphase TiAl (γ + $α_2$) alloys are strongly dependent on the orientation of the lamellae at the notch tip. The presence of large lamellar grains leads to higher toughness, caused by crack deflection by the lamellae, which in turn increases the crack-path tortuosity. R-curve behavior has been reported in the case of β-stabilized TiAl alloys having >80 vol.% $α_2$ + γ colonies, with the mean fracture toughness for crack propagation being in the range of ≈10–15 MPa $m^{1/2}$.[123] Crack deflection, ligament bridging, large grain size, the presence of globular γ-grains, and the formation of shielding dislocations or twins have been found to contribute positively to toughening in these alloys. Fully lamellar (L) structure with a coarse grain size (>500 μm) gives almost the highest fracture toughness (≈30 MPa $m^{1/2}$) but the lowest tensile ductility (<1%).[124] The resistance to fatigue-crack growth rate has been found to be higher in alloys with fully lamellar microstructure than in alloys with duplex microstructure.[116]

A strong effect of the environment on the fatigue and fracture behavior of aluminides has been noticed. The fracture toughness of Fe-aluminides has been found to be considerably lower in air than in oxygen or vacuum.[125,126] The susceptibility to environmental embrittlement is found to be greater in $D0_3$-structured Fe_3Al than in B2-FeAl, and therefore the fatigue-crack growth curve of the former material exhibits a diminished Paris law regime. Fatigue cracks initiated at surfaces or grain boundaries propagate along a transgranular path. Studies on the fatigue-crack-growth behavior in γ-TiAl with duplex, as well as refined and coarse, lamellar microstructures at temperatures ranging from ambient to 800°C have shown the crack-growth rates in air to be higher than that in ultrahigh vacuum (UHV) by an order of magnitude.[127] The highest crack-growth rate has been observed in the temperature range of 600°C–700°C. Examination of the closure-corrected fatigue-crack-growth behavior of γ-TiAl with coarse lamellar microstructure has shown the fatigue threshold at the ambient temperature to be ≈10 MPa $m^{1/2}$ in UHV and ≈6 MPa $m^{1/2}$ in air. The influence of the environment has also been observed in high-temperature fatigue-crack-growth behavior of $MoSi_2$ and $MoSi_2$ matrix composites.[128]

3.7 Summary

The deformation and fracture behavior of various intermetallics depend on their composition, crystal structure, nature and directionality of bonding, and grain size as well as microstructural variables such as the nature of constituent phases and their volume fractions and morphologies. The preference for operating slip systems depends on the packing density of the planes, the magnitudes of lattice translation vectors, and elastic anisotropy, as well as energies for formation of antiphase domains and stacking faults. As the nearest-neighbor atoms are altered on formation of APBs, the APB energy is found to be a strong function of the ordering energy. The superdislocations in ordered intermetallics are split, giving rise to the formation of APD, SISF, or CSF. An anomalous increase in yield strength is observed in Ni_3Al due to thermally activated cross-slip of screw dislocations into the less close-packed planes with lower APB energies. For some intermetallics, such as $MoSi_2$, single crystals are found to be ductile, whereas the lack of five independent slip systems causes brittleness in polycrystalline samples.

Although $L1_2$-structured Ni_3Al has been found to be ductile at room temperature, only a limited amount of compressive plasticity has been reported for Al_3Ti–Cr in its $L1_2$-stabilized form. The aluminides of Ni and Fe are also susceptible to environmental embrittlement in the presence of moisture due to the formation of atomic H by reaction of H_2O with Al. The segregation of H atoms at the grain boundaries and its ability to diffuse to crack tips is considered to cause embrittlement, and this can be reduced by suitable alloying, such as B in Ni-aluminides and C in Fe-aluminides. The fracture toughness can be significantly enhanced by tailoring microstructures with the formation of phases for accommodation of strain.

References

1. Hull, D. and D. J. Bacon. 1984. *Introduction to Dislocations*, 3rd edn, 112–140. Oxford, England: Pergamon.
2. Sun, Y.-Q. 1995. Structure of antiphase boundaries and domains. In *Intermetallic Compounds, Principles and Practice*, Vol. 1, eds. J. H. Westbrook and R. L. Fleischer, 495–517. Chichester, UK: Wiley.
3. Veyssière, P., J. Douin, and P. Beauchamp. 1985. On the presence of super lattice intrinsic stacking faults in plastically deformed Ni_3Al. *Phil. Mag. A* 51:469–483.
4. Douin, J., P. Veyssière, and P. Beauchamp. 1986. Dislocation line stability in Ni_3Al. *Phil. Mag. A* 54:375–393.
5. Clapp, R. C., M. J. Rubins, S. Charpency, J. A. Rifkin, Z. Z. Yu, and A. F. Voter. 1989. Some thermodynamic properties of NiAl calculated by molecular dynamics simulations. In *High Temperature Ordered Intermetallic Alloys III*, eds. C. T. Liu, A. I. Taub, N. S. Stoloff, and C. C. Koch, Materials Research Society Symposium Proceedings, Vol. 133, 29. Pittsburgh, PA: MRS.

6. Hug, G., J. Douin, and P. Veyssière. 1989. Surface energies of planar defects and dislocation processes in Al₃Ti and TiAl. In *High Temperature Ordered Intermetallic Alloys III,* eds. C. T. Liu, A. I. Taub, N. S. Stoloff, and C. C. Koch, Materials Research Society Symposium Proceedings, Vol. 133, 125. Pittsburgh, PA: MRS.

7. Crawford, R. C. and I. L. F. Ray. 1977. Antiphase boundary energies in iron-aluminium alloys. *Phil. Mag.* 35:549–565.

8. Flinn, P. A. 1960. Theory of deformation of superlattices. *Trans. AIME* 218:145–154.

9. Paidar, V. 1985. The structure and energy of antiphase boundaries in Ll₂ alloys. *Acta Metall.* 33:1803–1811.

10. Beauchamp, P., J. Douin, and P. Veyssiere. 1987. Dependence of the antiphase boundary energy upon orientation in the L1₂ structure. *Phil. Mag. A* 55:565–581.

11. Lasalmonie, A., B. Chenal, G. Hug, and P. Beauchamp. 1988. Structure and anisotropy of antiphase boundaries in rapidly solidified Ni₃Al. *Phil. Mag. A* 58:543–554.

12. Yamaguchi, M. and Y. Umakoshi. 1990. The deformation-behavior of intermetallic superlattice compounds. *Prog. Mater. Sci.* 34:1–148.

13. Kikuchi, R. and J. W. Cahn. 1979. Theory of interphase and antiphase boundaries in f.c.c. alloys. *Acta Metall.* 27:1337–1353.

14. Yodogawa, Y., D. M. Wee, Y. Oya, and T. Suzuki. 1980. The morphology of antiphase domains of Cu₃Pt and Cu₃Au-Ni alloys. *Scr. Metall.* 14:849–854.

15. Nicholson, D. M., G. M. Stocks, W. M. Temmerman, P. Sterne, and D. G. Pettifor. 1989. Structural energy differences in Al₃Ti: The role of tetragonal distortion in APB and twin energies. In *High Temperature Ordered Intermetallic Alloys III,* eds. C. T. Liu, A. I. Taub, N. S. Stoloff, and C. C. Koch, Materials Research Society Symposium Proceedings, Vol. 133, 17. Pittsburgh, PA: MRS.

16. Fu, C. L. 1990. Electronic, elastic, and fracture properties of trialuminide alloys—Al₃Sc and Al₃Ti. *J. Mater. Res.* 5:971–979.

17. Vitek, V. 1985. Effect of dislocation core structure on the plastic properties of metallic materials. In *Dislocations and Properties of Real Materials*, 30–50. London: The Institute of Metals.

18. Kear, B. H. and H. G. F. Wilsdorf. 1962. Dislocation configurations in plastically deformed Cu₃Au alloys. *Trans. Metall. Soc. AIME* 224:382–386.

19. Liu, C. T. and D. P. Pope. 1995. Ni3Al and its alloys. In *Intermetallic Compounds: Principles and Practice*, Vol. 2, eds. J. H. Westbrook and R. L. Fleischer, 17–51. Chichester, UK: Wiley.

20. Marcinkowski, M. J., N. Brown, and R. M. Fisher. 1961. Dislocation configurations in AuCu₃ and AuCu type superlattices. *Acta Metall.* 9:129–137.

21. Veyssière, P. and J. Douin. 1995. Dislocations. In *Intermetallic Compounds: Principles and Practice*, Vol. 1, eds. J. H. Westbrook and R. L. Fleischer, 519–558. Chichester, UK: Wiley.

22. Unal, O., J. J. Petrovic, D. H. Carter, and T. E. Mitchell. 1990. Dislocations and plastic-deformation in molybdenum disilicide. *J. Am. Ceram. Soc.* 73:1752–1757.

23. Mitra, R., N. E. Prasad, S. Kumari, and A. V. Rao. 2003. High-temperature deformation behavior of coarse- and fine-grained MoSi₂ with different silica contents. *Metall. Mater. Trans. A* 34A:1069–1088.

24. Sastry, S. M. L. and B. Ramaswami. 1976. Fault energies in ordered and disordered Cu₃Au. *Phil. Mag.* 33:375–380.

25. Korner, A. and H. P. Karnthaler. 1985. Fault energies in ordered and disordered Ni₃Fe. *Phil. Mag. A* 52:29–38.

26. Pope, D. P. and S. S. Ezz. 1984. Mechanical properties of Ni_3Al and nickel-base alloys with high volume fraction of γ'. *Int. Met. Rev.* 29:29–136.

27. Veyssière, P., M. H. Yoo, J. A. Horton, and C. T. Liu. 1989. Temperature effect on superdislocation dissociation on a cube plane in Ni_3Al. *Phil. Mag. Lett.* 59:61–68.

28. Takeuchi, S. and E. Kuramoto. 1973. Temperature and orientation dependence of the yield stress in Ni{in3}Ga single crystals. *Acta Metall.* 21:415–425.

29. Kear, B. H., J. M. Oblak, and A. F. Giamei. 1970. Stacking faults in gamma prime $Ni_3(Al,Ti)$ precipitation hardened nickel-base alloys. *Metall. Trans.* 1:2477–2486.

30. Howe, L. M., M. Rainville, and E. M. Schulson. 1974. Transmission electron-microscopy investigations of ordered Zr_3Al. *J. Nucl. Mater.* 50:139–154.

31. Douin, J. 1991. A dissociation transition in Zr_3Al deformed at room-temperature. *Phil. Mag. Lett.* 63:109–116.

32. Hemker, K. J. and M. J. Mills. 1993. Measurements of antiphase boundary and complex stacking fault energies in binary and B-doped Ni_3Al using TEM. *Phil. Mag. A* 68:305–324.

33. Carlsson, A. E. and P. J. Meschter. 1989. Relative stability of L12, Do22, and Do23 structures in Ma13 compounds. *J. Mater. Res.* 4:1060–1063.

34. Morris, D. G. and R. Lerf. 1991. Plastic deformation of the intermetallic Al_3Ti. *Phil. Mag. A* 63:1195–1206.

35. Hug, G., A. Loiseau, and P. Veyssière. 1988. Weak-beam observation of a dissociation transition in TiAl. *Phil. Mag. A* 57:499–523.

36. Greenberg, B. F., V. I. Anisimov, Y. N. Gornostirev, and G. G. Taluts. 1988. Possible factors affecting the brittleness of the intermetallic compound TiAl. II. Peierls manyvalley relief. *Scr. Metall.* 22:859–864.

37. Vasudevan, V. K., S. A. Court, P. Kurath, and H. L. Fraser. 1989. Effect of grain size and temperature on the yield stress of the intermetallic compound TiAl. *Scr. Metall.* 23:467–469.

38. Court, S. A., J. P. A. Löfvander, M. H. Loretto, and H. L. Fraser. 1989. The nature of c-component dislocations in samples of a polycrystalline Ti_3Al-based alloy deformed at room temperature and at 300°C. *Phil. Mag. A* 59:379–399.

39. Court, S. A., V. K. Vasudevan, and H. L. Fraser. 1990. Deformation mechanisms in the intermetallic compound TiAl. *Phil. Mag. A* 61:141–158.

40. Vasudevan, V. K., M. A. Stucke, S. A. Court, and H. L. Fraser. 1989. The influence of second phase Ti_3Al on the deformation mechanisms in TiAl. *Phil. Mag. Lett.* 59:299–307.

41. Sriram, S., V. K. Vasudevan, and D. M. Dimiduk. 1991. Deformation mechanisms in TiAl-based alloys containing low oxygen. In *High Temperature Ordered Intermetallic Alloys IV, Materials Research Society Symposium Proceedings*, eds. L. A. Johnson, D. P. Pope, and J. O. Stiegler, Vol, 213:375. Pittsburgh, PA: MRS.

42. Baker, I. and P. R. Munroe. 1990. Properties of B2 compounds. In *High Temperature Aluminides and Intermetallics*, eds. S. H. Wang, C. T. Liu, D. P. Pope, and J. O. Stiegler, 425. New York: TMS.

43. Yoo, M. H., T. Takasugi, S. Hanada, and O. Izumi. 1990. Slip modes in B2-type intermetallic alloys. *Mater. Trans. Jap. Inst. Met.* 31:435–442.

44. Yamaguchi, M., D. P. Pope, V. Vitek, and Y. Umakoshi. 1981. Planar faults and dislocation dissociations in body-centred-cubic-derivative ordered structures. *Phil. Mag. A* 43:1265–1275.

45. Song, Z. Y., H. Hashimoto, C. T. Chou, and H. Endoh. 1991. Stacking faults in plastically deformed FeAl alloy with the B2 structure. *Phil. Mag. A* 64:333–339.

46. Field, R. D., D. F. Lahraman, and R. Darolia. 1991. Comparison of TEM observations with dislocation core structure calcuiations in B2 ordered compounds. In *High Temperature Ordered Intermetallic Alloys IV*, eds. L. A. Johnson, D. P. Pope, and J. O. Stiegler, Materials Research Society Symposium Proceedings, Vol. 213, 223. Pittsburgh, PA: MRS.

47. Veyssière, P. and R. Noebe. 1992. Weak-beam study of <111> superlattice dislocations in NiAl. *Phil. Mag. A* 65:1–13.

48. Field, R. D., D. F. Lahrman, and R. Darolia. 1991. Slip systems in (001) oriented NiA1 single-crystals. *Acta Metall. Mater.* 39:2951–2959.

49. Mills, M. J. and D. B. Miracle. 1993. The structure of a(100) and a(110) dislocation cores in NiAl. *Acta Metall. Mater.* 41:85–95.

50. Ray, I. L. F., R. C. Crawford, and D. J H. Cockayne. 1970. The weak-beam technique applied to superlattice dislocations in an iron-aluminium alloy. *Phil. Mag.* 21:1027–1032.

51. Umakoshi, Y., M. Yamaguchi, and T. Yamane. 1985. Effects of non-stoichiometry on the high-temperature deformation of Ni_2AlTi. *Phil. Mag. A* 52:357–367.

52. Dimiduk, D. M. 1991. Dislocation structures and anomalous flow in $L1_2$ compounds. *J. de Physique III.* 1:1025–1053.

53. Suzuki, T., Y. Mishima, and S. Miura. 1989. Plastic behaviour in Ni3 (Al, X) single crystal: Temperature, strain-rate, orientation and composition. *ISIJ Intl.* 29:1–23.

54. Kawabata, T. and O. Izumi. 1990. *High Temperature Aluminides and Intermetallics*, eds. S. H. Whang, C. T. Liu, D. P. Pope, and J. O. Stiegler, 403. New York: TMS.

55. Court, S. A., J. P. A. Löfvander, M. H. Loretto, and H. L. Fraser. 1990. The influence of temperature and alloying additions on the mechanisms of plastic deformation of Ti_3Al. *Phil. Mag. A* 61:109–139.

56. Fliescher, R. L. 1991. Effects of composition on the mechanical properties of tough, high-temperature intermetallic compounds. *ISIJ Intl.* 31:1186–1191.

57. Douin, J. and P. Veyssière. 1991. On the effect of deviation from stoichiometry on the deformation microstructure of binary Ni_3Al. *Phil. Mag. A.* 64:807–817.

58. Anisimov, V. I., G. Ganin, V. R. Galachov, and E. Z. Kurmaev. 1987. *Fiz. Metall. Metalloved.* 63:414.

59. Woodward, C., J. M. Maclaren, and S. Rao. 1992. Electronic-structure of planar faults in TiAl. *J. Mater. Res.* 7:1735–1750.

60. Fu, C. L. and M. H. Yoo. 1990. Elastic constants, fault energies, and dislocation reactions in TiAl: A first-principles total-energy investigation. *Phil. Mag. Lett.* 62:159–165.

61. Yoo, M. H. 1989. Deformation twinning in superlattice structures. *J. Mater. Res.* 4:50–54.

62. Yoo, M. H., S. L. Sass, C. L. Fu, M. J. Mills, D. M. Dimiduk, and E. P. George. 1993. Deformation and fracture of intermetallics. *Acta Metall. Mater.* 41:987–1002.

63. Chan, K. S. and Y. W. Kim. 1992. Influence of microstructure on crack-tip micromechanics and fracture behaviors of a 2-phase TiAl alloy. *Metall. Trans. A* 23:1663–1677.

64. Yoo, M. H. 1998. Twinning and mechanical behavior of titanium aluminides and other intermetallics. *Intermetallics* 6:597–602.

65. Yamaguchi, M. and H. Inui. 1994. Al_3Ti and its $L1_2$ variations. In *Intermetallic Compounds, Vol. 2, Practice*, eds. J. H. Westbrook and R. L. Fleischer, 147–173. Chichester, UK: Wiley.

66. Mitra, R. 2006. Mechanical behaviour and oxidation resistance of structural silicides. *Int. Mater. Rev.* 51:13–64.
67. Maloy, S. A., T. E. Mitchell, and A. H. Heuer. 1995. High-temperature plastic anisotropy in MoSi₂ single-crystals. *Acta Metall. Mater.* 43:657–668.
68. Moriwaki, M., K. Ito, H. Inui, and M. Yamaguchi. 1997. Plastic deformation of single crystals of NbSi₂ with the C40 structure. *Mater. Sci. Eng. A* 239–240:69–74.
69. Ito, K., T. Yano, T. Nakamoto, H. Inui, and M. Yamaguchi. 1999. Plastic deformation of single crystals of WSi₂ with the Cl1(b) structure. *Acta Mater.* 47:937–949.
70. Mitra, R., N. E. Prasad, S. Kumari, and A. V. Rao. 2003. High-temperature deformation behavior of coarse- and fine-grained MoSi₂ with different silica contents. *Metall. Mater. Trans. A* 34:1069–1088.
71. Messerschmidt, U., M. Bartsch, S. Guder, D. Haussler, R. Haushalter, and M. Yamaguchi. 1998. Dynamic dislocation behaviour in the intermetallic compounds NiAl, TiAl and MoSi₂. *Intermetallics* 6:729–733.
72. Guder, S., M. Bartsch, M. Yamaguchi, and U. Messerschmidt. 1999. Dislocation processes during the deformation of MoSi₂ single crystals in a soft orientation. *Mater. Sci. Eng. A* 261:139–146.
73. Lipsitt, H. A., D. Shechtman, and R. E. Schafrik. 1975. Deformation and fracture of TiAl at elevated-temperatures. *Metall. Mater. Trans. A* 6:1991–1996.
74. Liu, C. T. and B. F. Oliver. 1989. Effect of grain shape on environmental embrittlement in Ni₃Al tested at elevated-temperatures. *J. Mater. Res.* 4:294–299.
75. Liu, C. T., E. H. Lee, and C. G. McKamey. 1989. An environmental effect as the major cause for room-temperature embrittlement in FeAl. *Scr. Metall.* 23:875–880.
76. Liu, C. T., C. G. McKamey, and E. H. Lee. 1990. Environmental effects on room-temperature ductility and fracture in Fe₃Al. *Scr. Metall. Mater.* 24:385–389.
77. Fu, C. L. and G. S. Painter. 1991. 1st principles investigation of hydrogen embrittlement in FeAl. *J. Mater. Res.* 6:719–723.
78. George, E. P., C. T. Liu, and D. P. Pope. 1992. Environmental embrittlement: The major cause of room-temperature brittleness in polycrystalline Ni₃Al. *Scr. Metall. Mater.* 27:365–370.
79. Aloki, K. and O. Ijumi. 1979. Improvement in room temperature ductility of the L12 type intermetallic compound Ni3Al by boron addition. *Nippon Kinzoku Gakkaishi* 43:1190–1196.
80. Taub, A. I., S. C. Huang, and K. M. Chang. 1984. Improved strength and ductility of Ni₃Al by boron modification and rapid solidification. *Metall. Trans. A* 15:399–402.
81. Liu, C. T. 1991. Intergranular fracture and boron effects in Ni₃Al and other intermetallics: Introductory paper. *Scr. Metall. Mater.* 25:1231–1236.
82. Liu, C. T., C. L. White, and J. A. Horton. 1985. Effect of boron on grain-boundaries in Ni₃Al. *Acta Metall.* 33:213–229.
83. Schulson, E. M., T. P. Weihs, D. V. Viens, and I. Baker. 1985. The effect of grain size on the yield strength of Ni₃Al. *Acta Metall.* 33:1587–1591.
84. Maloy, S. A., T. E. Mitchell, J. J. Petrovic, A. H. Heuer, and J. J. Lewandowski. 1994. The temperature and strain rate dependence of the flow stress in MoSi₂ single crystals. In *High-Temperature Silicides and Refractory Alloys,* Materials Research Society Symposium Proceedings, Vol. 322, 21. Pittsburgh, PA: MRS.
85. Umakoshi, Y., T. Nakashima, T. Nakano, and E. Yanagisawa. 1994. Plastic behavior and deformation structure of silicide single crystals with transition metals at high temperatures. In *High-Temperature Silicides and Refractory Alloys,* Materials Research Society Symposium Proceedings, Vol. 322, 9. Pittsburgh, PA: MRS.

86. Inui, H., K. Ishikawa, and M. Yamaguchi. 2000. Creep deformation of single crystals of binary and some ternary $MoSi_2$ with the $C11_b$ structure. *Intermetallics* 8:1159–1168.
87. Sadananda, K., C. R. Feng, R. Mitra, and S. C. Deevi. 1999. Creep and fatigue properties of high temperature silicides and their composites. *Mater. Sci. Eng. A* 261:223–238.
88. Sadananda, K. and C. R. Feng. 1995. Effect of carbon addition on the creep of molybdenum disilicide composites. *Mater. Sci. Eng. A* 192–193, Part 2:862–867.
89. Mitra, R., K. Sadananda, and C. R. Feng. 2004. Effect of microstructural parameters and Al alloying on creep behavior, threshold stress and activation volumes of molybdenum disilicides. *Intermetallics* 12:827–836.
90. Sadananda, K., C. R. Feng, H. Jones, and J. Petrovic. 1992. Creep of molybdenum disilicide composites. *Mater. Sci. Eng. A* 155:227–239.
91. Valiev, R. Z., R. M. Gayanov, H. S. Yang, and A. K. Mukherjee. 1991. TEM study of a superplastically deformed Ni_3Al alloy doped with boron. *Scr. Metall. Mater.* 25:1945–1950.
92. Mukhopadhyay, J., G. Kaschner, and A. K. Mukherjee. 1990. Superplasticity in boron doped Ni_3Al alloy. *Scr. Metall. Mater.* 24:857–862.
93. Nieh, T. G., C. M. McNally, and J. Wadsworth. 1989. Superplasticity in intermetallic alloys and ceramics. *JOM* 41:31–35.
94. Nieh, T. G. and W. C. Oliver. 1989. Superplasticity of a nickel silicide, Ni_3Si. *Scr. Metall.* 23:851–854.
95. Yang, H. S., P. Jin, E. Dalder, and A. K. Mukherjee. 1991. Superplasticity in a Ti_3Al-base alloy stabilized by Nb, V and Mo. *Scr. Metall. Mater.* 25:1223–1228.
96. Dutta, A. and D. Banerjee. 1990. Superplastic behavior in a Ti_3Al-Nb alloy. *Scr. Metall. Mater.* 24:1319–1322.
97. Lin, D. L., A. D. Shan, and D. Q. Li. 1994. Superplasticity in Fe_3Al-Ti alloy with large grains. *Scr. Metall. Mater.* 31:1455–1460.
98. Li, D., A. D. Shan, Y. Liu, and D. Lin. 1995. Study of superplastic deformation in an FeAl based alloy with large grains. *Scr. Metall. Mater.* 33:681–685.
99. Lin, D., T. L. Lin, A. Shan, and M. Chen. 1996. Superplasticity in large-grained Fe_3Al alloys. *Intermetallics* 4:489–496.
100. Lin, D., D. Q. Li, and Y. Liu. 1998. Superplasticity in large-grained FeAl-based intermetallic alloys. *Intermetallics* 6:243–256.
101. Lin, D. and Y. Liu. 1999. Microstructural evolution and mechanisms of superplasticity in large-grained iron aluminides. *Mater. Sci. Eng. A* 268:83–89.
102. Gu, Y., D. Lin, A. Shan, Y. Liu, and W. Ye. 1996. *Acta Metall. Sinica* 32:1144.
103. Gu, Y., D. Lin, A. Shan, and Z. Liu. 1997. *Acta Metall. Sinica* 33:325.
104. Gu, Y., D. Lin, A. Shan, and Y. Liu. 1998. *Acta Metall. Sinica* 34:351.
105. Sun, F. and D. Lin. 2001. Superplastic phenomenon in a large-grained TiAl alloy. *Scr. Mater.* 44:665–670.
106. Jiang, D. and D. Lin. 2002. Superplasticity of single-phase Ni-40Al intermetallics with large grains. *J. Mater. Sci. Lett.* 21:505–508.
107. Hu, J. and D. Lin. 2006. Superplasticity of single phase Ni–42Al intermetallics with large grains. *Mater. Sci. Eng. A* 441:142–148.
108. Jiang, D. and D. Lin. 2002. Superplasticity of single-phase Ni–45Al intermetallics with large grains. *Mater. Lett.* 57:747–752.
109. Hu, J. and D. Lin. 2004. Microstructural evolution during super-plastic deformation in large-grained single phase Ni-48Al intermetallics. *Acta Metall. Sinica* 40(5):489–495.
110. Hu, J. and D. L. Lin. 2004. Superplasticity of single-phase Ni–48Al intermetallics with large grains. *Mater. Lett.* 58:1297–1301.

111. Hu, J. and D. Lin. 2004. Microstructural evolution of superplasticity in large-grained Ni–48Al intermetallics. *Mater. Sci. Eng. A* 371:113–118.
112. Varin, R. A. and M. B. Winnicka. 1991. Plasticity of structural intermetallic compounds. *Mater. Sci. Eng. A* 137:93–103.
113. Reuss, S. and H. Vehoff. 1990. Temperature dependence of the fracture toughness of single phase and two phase intermetallics. *Scr. Metall. Mater.* 24:1021–1026.
114. Rigney, J. D. and J. J. Lewandowski. 1992. Fracture toughness of monolithic nickel aluminide intermetallics. *Mater. Sci. Eng. A* 149:143–151.
115. Dogan, B., D. Schoneich, K. H. Schwalbe, and R. Wagner. 1996. Fatigue precracking and fracture toughness testing of TiAl intermetallics. *Intermetallics* 4:61–69.
116. Huang, S. C. and J. C. Chestnut. 1994. Gamma TiAl and its alloys. In *Intermetallic Compounds, Vol. 2. Practice*, eds. J. H. Westbrook and R. L. Fleischer, 73–90. Chichester, UK: Wiley.
117. Dogan, B. D. and K.-H. Schwalbe. 1997. Fracture toughness testing of TiAl base intermetallic alloys. *Eng. Fract. Mech.* 56(2):155–165.
118. Chan, K. S. 1990. Fracture and toughening mechanisms in an α_2 titanium aluminide alloy. *Metall. Trans. A* 21:2687–2699.
119. Chan, K. S. 1992. Influence of microstructure on intrinsic and extrinsic toughening in an alpha-two titanium aluminide alloy. *Metall. Trans. A* 23:183–189.
120. Inoue, M., K. Suganuma, and K. Niihara. 1998. Fracture properties at ambient temperatures of iron aluminides fabricated by reactive hot pressing. *Scr. Mater.* 39(10):1477–1482.
121. Mitra, R. 2006. Mechanical behavior and oxidation resistance of structural silicides. *Inter. Mater. Rev.* 51:13–64.
122. Shang, Z., J. Shen, J. Zhang, L. Wang, L. Wang, and H. Fu. 2014. Effect of microstructures on the room temperature fracture toughness of NiAl–32Cr–6Mo hypereutectic alloy directionally solidified at different withdrawal rates. *Mater. Sci. Eng. A* 611:306–312.
123. Leitner, T., M. Schloffer, S. Mayer, J. Esslinger, H. Clemens, and R. Pippan. 2014. Fracture and R-curve behavior of an intermetallic beta-stabilized TiAl alloy with different nearly lamellar microstructures. *Intermetallics* 53:1–9.
124. Liu, C. T. 1995. Recent advances in ordered intermetallics. *Mater. Chem. Phys.* 42:77–86.
125. McKamey, C. G. 1996. Iron aluminides. In *Physical Metallurgy and Processing of Intermetallic Compounds*, eds. N. S. Stoloff and V. K. Sikka, 351–391. London: Chapman & Hall.
126. McCamey, C. G., J. A. Horton, and C. T. Liu. 1987. Effect of aluminum addition on ductility and yield strength of Fe_3Al alloys with 0.5 wt.% TiB_2. In *High Temperature Ordered Intermetallic Alloys II*, eds. N. S. Stoloff, C. C. Koch, C. T. Liu, and O. Izumi, Materials Research Society Symposium Proceedings, Vol. 81, 321–328. Pittsburgh, PA: MRS.
127. Rosenberger, A. H., B. D. Worth, and J. M. Larsen. 1997. Effects of microstructure, temperature and environment on fatigue crack growth in Ti–46.5Al–3Nb–2Cr–0.2W γ titanium aluminide. In *Structural Intermetallics 1997*, eds. M. V. Nathal, R. Darolia, C. T. Liu, P. L. Martin, D. B. Miracle, R. Wagner, and M. Yamaguchi, 555–561. Warrendale, PA: TMS.
128. Ramamurty, U., A. S. Kim, S. Suresh, and J. J. Petrovic. 1993. Micromechanisms of creep-fatigue crack-growth in a silicide-matrix composite with SiC particles. *J. Am. Ceram. Soc.* 76:1953–1964.

4

Oxidation Behavior

4.1 Introduction

Resistance against environmental degradation by oxidation is one of the key properties of interest for materials to be used at elevated temperatures. Basic information on the oxidation behavior of structural intermetallics can be found in some of the earlier reviews.[1-4] Some of the materials, particularly the aluminides, are also subjected to various types of corrosive environments, where grain-boundary embrittlement, decarburization, and sulfidation are investigated. The intermetallics used at elevated temperatures should be of optimum composition so as to form a protective passive scale to restrict environmental damage. Even if a protective scale is able to form at elevated temperatures, its growth rate at intermediate temperatures is too sluggish to ensure protection against oxidation. The absence of a stable and protective scale, along with defects in the intermetallic alloy, leads to accelerated or catastrophic oxidation. The possible mechanisms of oxidation at different temperatures need to be well understood for devising ways of restricting the amount of environmental degradation. Suitable oxidation-resistance coatings have also been developed for the intermetallic alloys, where the formation of a passive scale is not possible. In this chapter, the focus is on the oxidation behavior of intermetallics in general along with the mechanisms involved in the formation of a stable and protective scale, which should also be strongly adherent to the alloy substrate underneath.

4.2 Mechanism of Oxidation and Oxidation Protection

The intermetallics being considered for application at elevated temperatures are protected against environmental degradation by the formation of a passivating oxide scale, which restricts further oxidation of the underlying

substrate. For the oxide scale to be fully protective, the following requirements should be met:[1-3]

1. The oxide in the outer scale should have higher thermodynamic stability compared with that of the oxides of other alloying elements.
2. The rate of growth should be slow to ensure that it is continuous,
   dense, and pore free.
3. The oxide scale should have impressive adherence to the alloy
   underneath and resist spallation.
4. The oxide scale should have the ability to heal itself on spallation or
   damage.

4.2.1 Thermodynamic Stability of Oxidation Products

High thermodynamic stability is ensured by high negative free energy of formation as well as low vapor pressure. Figure 4.1 shows the variation of free energy
of formation with absolute temperature for different oxides.[5] It is observed that
the free energy of formation is lowest in the case of Al_2O_3, whereas those of TiO_2
and SiO_2 are slightly higher and similar. At temperatures higher than 1000°C,
the oxidation resistance of Al- or Si-bearing intermetallic alloys or compounds
is known to be superior due to the formation of passivating scales of Al_2O_3 or
SiO_2, respectively.[2,3,6] Of course, the free energy of formation of Cr_2O_3 is higher
than that of Al_2O_3 or SiO_2. At temperatures ≤1000°C, the scale of Cr_2O_3 is also
found to be protective because of its impervious and adherent character, as
commonly observed in the case of stainless steels. However, at still higher temperatures, Cr_2O_3 is further oxidized to form CrO_3 with high vapor pressure at

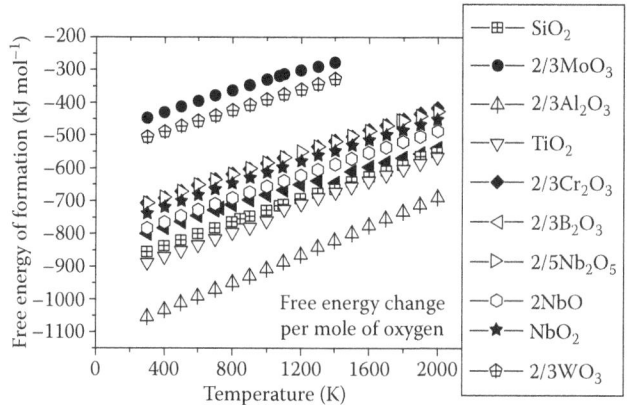

FIGURE 4.1
Plots depicting the variation of free energy per mole of oxygen with absolute temperature for
the formation of selected oxides. (From Barin, I., *Thermochemical Data of Pure Substances*, Vols. 1
and 2, VCH, Weinheim, Germany, 1989.)

the ambient oxygen partial pressure and temperature. Hence, Cr_2O_3 scales are considered to be unprotective at temperatures $\geq 1000°C$.

4.2.2 Kinetics of Oxide Scale Growth

Oxidation reactions leading to the formation of protective oxide scales generally follow a parabolic rate law, as follows:

$$(\Delta w)^2 = k_p.t \tag{4.1}$$

where:
  Δw = mass gain due to oxidation
  k_p = parabolic rate constant
  t = duration of exposure

This reaction can also be written as

$$x^2 = k_p.t \tag{4.2}$$

where x is the thickness of the oxide scale. Experimental results often show minor deviations from the parabolic rate law. A more general form of the abovementioned rate equation is

$$(\Delta w)^n = k.t \tag{4.3}$$

where:
  n = oxidation exponent
  k = rate constant

  The kinetics can be considered as linear, parabolic, or slower than parabolic if $n \approx 1, 2,$ or >2, respectively. Typical examples of kinetics of oxidation that are slower than parabolic and faster than linear, as observed for Ti_5Si_3 at 1200°C and 1300°C, respectively, are shown in Figure 4.2.[7] The rate of oxide scale growth is controlled by the diffusion of either oxygen anions or oxidizing species comprising the metallic atoms in the intermetallic alloy. The process of diffusion is driven by an electrochemical potential gradient across the oxide scale. Therefore, the diffusion of the slowest species through the oxide scale can be considered as the rate-controlling step in the formation of the oxide scales. The diffusion kinetics of a given species through any oxide scale is dependent on the latter's crystallographic and defect structure as well as its electronic characteristics (n-type or p-type). Plots depicting the variation of the parabolic rate constants of some of the oxide scales of interest with temperature are shown in Figure 4.3. The low diffusivities of anions and cations manifest in low parabolic

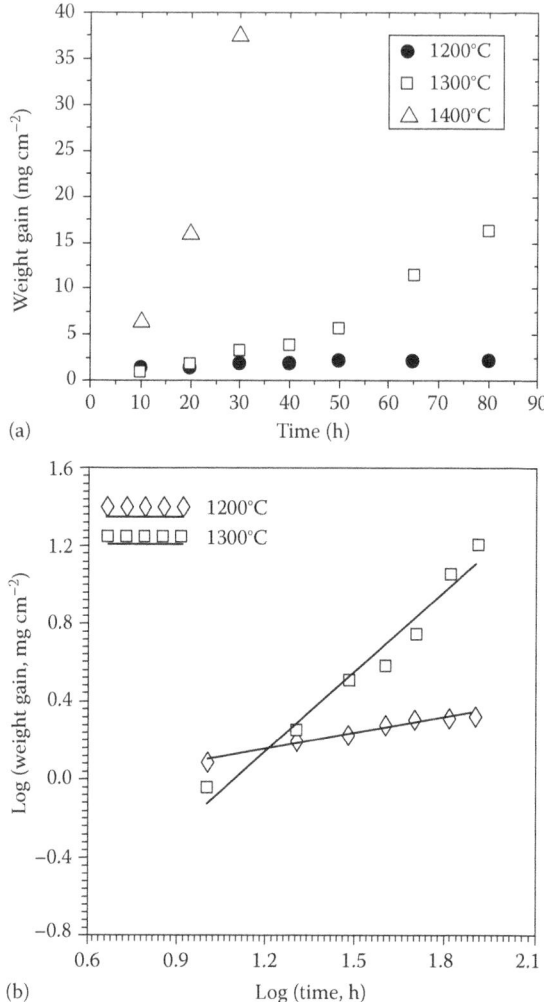

FIGURE 4.2
Plots depicting the variation of (a) mass gain against temperature for Ti₅Si₃ and (b) variation of ln (mass gain) with ln (temperature) to find values of n and k. (With kind permission from Springer Science+Business Media: *Metall. Mater. Trans. A*, Elevated-temperature oxidation behavior of titanium silicide and titanium silicide based alloy and composite, 29, 1998, 1665–1675, Mitra, R., et al., figure 6.9, and any original (first) copyright notice displayed with material.)

rate constants for the growth of SiO_2 and Al_2O_3 when compared with other oxides (Figure 4.3).[8,9] Therefore, it is appropriate to infer that the scales of Al_2O_3 and SiO_2 are the most protective, because of their high thermodynamic stability and low diffusivities for both cations and anions. At temperatures ≥900°C, the growth rate of SiO_2 is less than that of Al_2O_3, whereas the opposite is observed at lower temperatures. Therefore, at temperatures ≥900°C, the SiO_2 scale

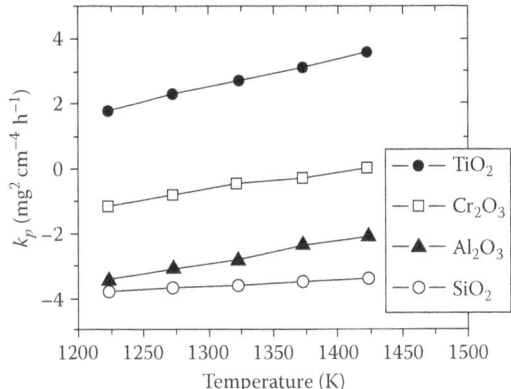

FIGURE 4.3
Plots depicting the variation of parabolic rate constants for the growth of selected oxide scales with temperature. (From Raj, S. V., *Mater. Sci. Eng. A*, 201, 229–241, 1995; Nesbitt, J. M. and C. E. Lowell, Materials Research Society Symposium Proceedings, Vol. 288, 107–118, 1993.)

formed on the silicides offers much better oxidation resistance compared with the Al_2O_3 films formed on aluminide-based intermetallics.

In the case of the oxidation-resistant silicides, the amorphous SiO_2 scale is chiefly responsible for their protection, as shown in Figure 4.4.[10] An oxide scale with an amorphous structure is preferred to a scale with a crystalline character, because grain boundaries provide short-circuit paths for diffusion in the latter type of oxide scale.[6] Besides the low diffusivity for molecular oxygen and its anions, the protectiveness of amorphous SiO_2 films is also due to their high flexibility in being able to form modified glasses or silicates as a result of the uptake of elements from the substrate or environment, and their self-healing properties at high temperatures.

4.2.3 Adhesion of Oxide Scale

For a given oxide scale to be protective, it is necessary for it to be strongly adherent to the underlying intermetallic alloy substrate. Oxide scales are prone to cracking or spallation due to residual stresses caused by either volume expansion or mismatch between the coefficients of thermal expansion of the alloy substrate and the oxide scale. Compressive residual stresses are induced in the oxide scale through volume expansion during formation, or if the alloy substrate shows greater thermal contraction than the oxide scale during cooling. Buckling of the oxide scale caused by the compressive residual stresses is usually followed by delamination or cracking of the oxide scale. The volume change due to the formation of a metal–oxide scale on a pure metallic substrate is usually expressed in terms of the Pilling–Bedworth ratio (molar volume of oxide/molar volume of pure metal).[6,11] Usually, cyclic exposure of a given intermetallic alloy at a given temperature subjects the oxide scale to

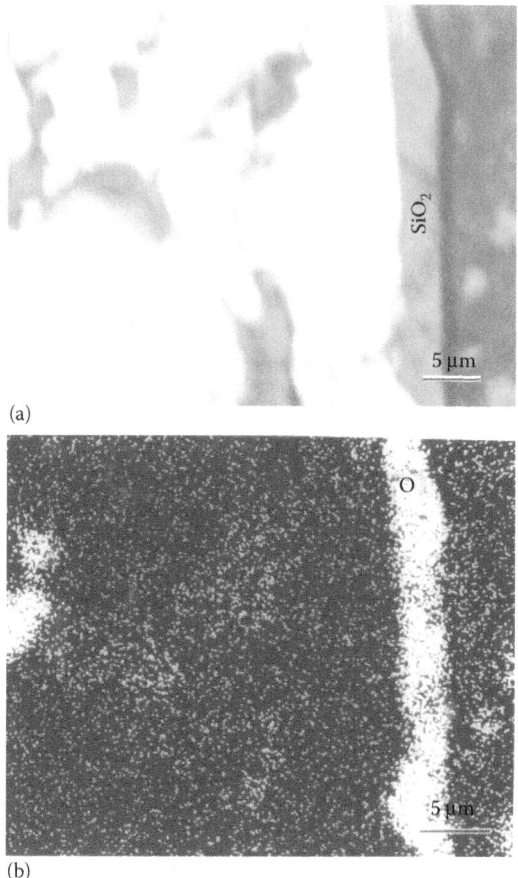

(a)

(b)

FIGURE 4.4
Oxide scale of MoSi$_2$ after exposure in air at 1200°C: (a) scanning electron microscopy (SEM) (backscattered electron) image and (b) wavelength-dispersive x-ray map of O. (From Mitra, R. and V. V. Rama Rao, *Mater. Sci. Eng. A*, 260, 146–160, 1999. Reprinted with permission from Elsevier.)

repeated thermal shock. In other words, the oxide scale is subjected to thermal fatigue, which in turn causes its delamination from the base alloy followed by spallation. It is intuitive that nonadherent oxide scales are more susceptible to spallation, which is manifested as mass loss. Under these circumstances, it is well expected that, during each thermal cycle, there will be initial parabolic oxidation followed by mass loss through scale spallation. This mechanism of oxidation eventually leads to rapid degradation of the intermetallic alloy.

4.2.4 Self-Healing Ability of Oxide Scale

The protective character of the oxide scale is strongly dependent on its ability to heal itself promptly after damage. The ability of a given intermetallic alloy to reform its protective scale after damage depends on the relative thermo-dynamic stabilities of the oxide phases, as well as the rates of their formation, which in turn depend on both the concentration and diffusivities of the oxi-dizing species. This issue can be examined by considering an intermetallic alloy, XY, with element X being less reactive than element Y. On exposure of such an alloy to an oxidizing environment, oxygen is first chemisorbed on the surface. If oxygen partial pressure and the activities of both the alloying elements are high at the oxide–air interface, the following reaction is likely to happen:

$$XY + O_2 \text{ (g)} \rightarrow XO + YO \tag{4.4}$$

With a decrease in oxygen partial pressure, YO would become more sta-ble compared with XO. In such a situation, an oxide scale comprising only YO would form. However, if XO is unstable and the activity of Y is low at the surface, internal oxidation with the formation of YO precipitates is expected. In this context, it has been proposed[2,3] that a critical concentration of the more reactive solute component is required to ensure the formation of a continuous and protective oxide scale. The critical mole fraction N^* is given as

$$N^* = \left(\frac{\pi g^* N_0 D_0 V_m}{2 \upsilon D_Y V_{ox}} \right)^{1/2} \tag{4.5}$$

where:
$N_0 D_0$ = oxygen permeability in the alloy
D_Y = diffusivity of Y in the alloy
g^* = a constant found experimentally as ≈ 0.3
V_m and V_{ox} = molar volume of alloy and oxide, respectively
υ = stoichiometry factor in the oxide YO_n

Even if the critical mole fraction of the element responsible for the formation of a protective oxide scale is present, a transient stage involving the formation of multiple oxide phases on the exposed surface is often observed. As discussed in Section 4.2.2, the less protective oxide, XO, grows faster than YO, if the parabolic rate constant of the former oxide formation is higher, thereby inhibiting the for-mation of a self-healing oxide scale. Therefore, significant degradation by the formation of the less reactive constituent of the intermetallic alloy precedes the steady-state regime accompanied by the formation of a stable and protective scale of YO. This type of behavior has been noticed in the case of Mo–Si–B alloys containing α-Mo, Mo_3Si, and Mo_5SiB_2 in the microstructure. Here, the transient stage involves mass loss due to the rapid formation and vaporization of volatile MoO_3 (g). Subsequently, with the depletion of Mo from near-surface locations, a stable and protective scale of borosilicate (B_2O_3–SiO_2) is able to form.

Occasionally, an SiO_2- or Al_2O_3-based scale may form, but it may not be protective, particularly if there is competitive growth of the oxides of different constituent elements[12] or if spallation is predominant, as in cyclic oxidation.[13] In the case of $MoSi_2$, the high volatility of MoO_3 allows the development of a protective film of SiO_2 at temperatures $\geq 600°C$. On the other hand, protective scales are able to form on $NbSi_2$ and $TaSi_2$ only at temperatures $\geq 1000°C$.[2,3]

4.3 Conditions of Exposure at High Temperatures

It is often found that, in a given intermetallic alloy, the formation of a protective oxide scale of SiO_2 or Al_2O_3 is possible only in a particular temperature range.[14–16] At temperatures below or above the range, the alloys could be susceptible to oxidation to varying degrees at different temperatures. The kinetics of oxidation is not fast enough at lower temperatures for the formation of a stable oxide scale. However, at higher temperatures, the viscosity of the amorphous or glassy oxide scale can be reduced significantly, thereby leading to an increase in the diffusion rate of oxygen anions. For certain compositions of the alloy, the glassy oxide scale devitrifies with the evolution of a crystalline microstructure, in which the grain boundaries provide short-circuit paths for the diffusion of species. At higher temperatures or in reducing atmospheres, the protective scale of SiO_2 either cannot form or is unstable. Under such circumstances, the Si-containing intermetallic phase undergoes active oxidation, whereby SiO (g) is formed instead of SiO_2. Therefore, it is necessary to study the oxidation behavior of specific multiphase intermetallic alloys at different temperatures.

The oxidation behavior of a given intermetallic alloy also depends strongly on the nature of the exposure at elevated temperature, which could be non-isothermal, isothermal, or cyclic.[13,17–20] The formation of an oxide scale is initiated during heating of the sample in air or an oxidizing environment. By heating the sample in air inside a thermogravimetric analyzer, it is possible to find the temperatures corresponding to the initiation of mass gain or changes in the rates of mass gain.[12,21] It is also observed that the mass-gain initiation temperature and the net mass gain at a given temperature vary with the heating rate, which suggests that both the reactions as well as nucleation of the reaction products have a strong activation barrier. The causes for such an activation barrier could be the following requirements for the formation of reaction products: (i) diffusion of oxygen anions through the outer scale as well as through the interphase boundaries; (ii) nucleation and growth of various reaction products, accompanied by a rise in the free energy of the system. On nucleation and growth of oxidation products, there is an increase in both interfacial or surface free energies and net elastic strain energy due to the increase in volume. At a slower heating rate, more time is available for the diffusive transport of oxygen anions through the oxide scale, increasing the extent of their penetration and promoting the oxidation of the phases present in the intermetallic alloy. Hence, both the lowest temperature for the initiation of oxidation and the highest mass change have been observed in the samples heated at the slowest rate. The nature of the mass change and the composition of the oxide scale formed during heating at any rate are found to differ from those observed during isothermal exposure. During heating of the 76Mo–14Si–10B alloy inside a thermogravimetric analyzer, transient mass gain has been observed prior to mass loss for heating rates between 5 and 40°C min^{-1}.[12] However, such mass gain has not been observed during isothermal oxidation on this alloy, probably because recording of mass change was started after the initial stage of mass gain. The operative mechanism in the early stage of oxidation, as well as the adhesiveness of the oxide scale, has a significant effect on oxidation behavior during isothermal exposure at a given temperature for an extended period.

The extent of environmental degradation observed on cyclic oxidation is found to be usually significantly higher than that recorded on isothermal exposure of identical duration.[13] Although a stable and protective oxide scale may form during isothermal exposure, leading to oxidation kinetics defined by the parabolic rate law, an abruptly different behavior showing mass loss may be observed on cyclic exposure of similar duration. On cooling, the oxide scale is subjected to residual stresses generated on account of the coefficient of thermal expansion (CTE) mismatch between the oxide scale and the substrate underneath. The thermal shock experienced by the oxide scale on being subjected to multiple cycles of heating and cooling leads to the formation of cracks, which provide an easy path for continuous ingress of oxygen. With the increase in the thickness of the oxide scale containing cracks, spallation occurs, and in this manner a fresh surface of the substrate is exposed to air. Therefore, the formation of a protective oxide scale is either not possible or it is delayed on cyclic exposure, which in turn leads to unabated and accelerated oxidation.

4.4 Stresses in the Oxide Scale

During the initial stages of isothermal oxidation, regimes of no detectable mass change are found to interrupt the mass changes, as shown in Figure 4.5.[22] The occurrence of regimes of little mass change in this figure can be attributed to the formation of an oxide scale, which is protective for a limited duration. However, this oxide scale ceases to be protective as it ruptures, allowing further ingress of oxygen to the substrate underneath. This observation suggests that the growing oxide scale contains internal stresses, which often cause its rupture during its growth.

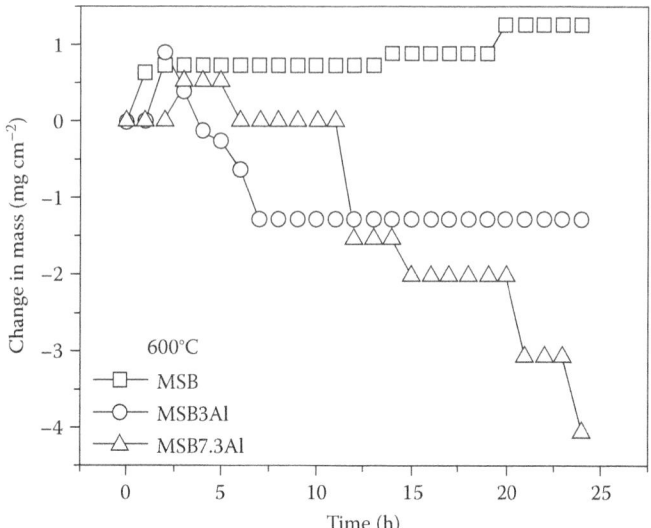

FIGURE 4.5

Plots depicting stepwise mass change with increasing time of exposure of MoSi$_2$ at 500°C. (From Paswan, S., et al., *Mater. Sci. Eng. A*, 424, 251–265, 2006. Reprinted with permission from Elsevier.)

Internal stresses are generated in the oxide scale for several reasons: (i) change in volume with the formation of an oxide scale[6,10,13]; (ii) differences between the CTEs of the oxide scale constituents and the substrate[13]; (iii) diffusion of oxygen anions through the phases in the oxide scale[23]; (iv) selective oxidation and nonreciprocal diffusion of the alloying elements[24]; and (v) formation and escape of volatile oxidation products. The volume expansion with the formation of the oxide scale leads to the generation of compressive residual stress. Therefore, the higher the Pilling–Bedworth ratio,[6,11] the higher would be the magnitude of the compressive residual stress. As the oxide scale is brittle, an increase in the magnitude of the compressive residual stress beyond a certain limit causes buckling of the scale. The CTE of the oxide scale is usually less than that of the intermetallic substrate. Therefore, during heating or cooling, the oxide scale is expected to undergo lower expansion or contraction, respectively, as compared with that of the substrate underneath. This would lead to tensile residual stress in the oxide scale. Selective oxidation and one-directional diffusion of the alloying elements lead to the formation of Kirkendall voids inside the substrate and the alloy–oxide interface. Coalescence of these voids could cause the formation of microporosities in the oxide scale or the oxide–alloy interface, which in turn promote spallation. On the other hand, selective oxidation of the alloying elements causes volume expansion, thereby giving rise to localized compressive residual stresses. The formation of volatile oxides and their escape usually lead to the formation of blister-like features and

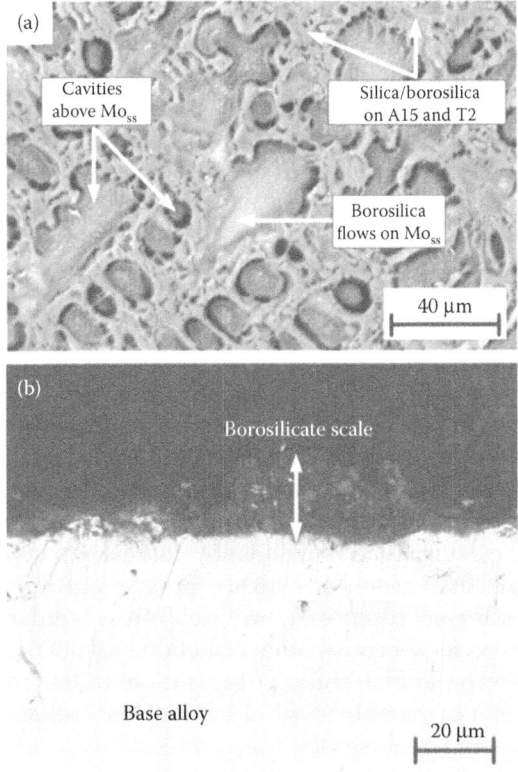

FIGURE 4.6
SEM images of the oxide scale formed after exposure at 1150°C for 20 s: (a) top surface showing pores formed due to the vaporization of MoO_3 (g) and (b) cross section of alloy–oxide interface. (From Roy, B., et al., *Corros. Sci.*, 68, 231–237, 2013. Reprinted with permission from Elsevier.)

pores on the surface of the oxide scales, as shown in Figure 4.6.[25] If the oxide scale constituents form a liquid phase, as in the case of Mo–Si–B alloys, the volatile oxidation products lead to the formation of bubbles on the surface during their escape.[26] Observation of the oxide scale cross sections has indicated the presence of deep channels with one of their ends at the oxide-scale surface. These channels indicate the location for the formation and subsequent vaporization of volatile reaction products.

4.5 Pest Disintegration

The pest-type reaction is known to comprise simultaneous oxidation and cracking of materials, resulting in the complete disintegration of the material into a powdery mass without the application of any external stress.[14] Accelerated oxidation accompanied by complete structural disintegration has been observed in some intermetallic alloys at intermediate or relatively lower temperatures in an oxidizing environment. This phenomenon is primarily promoted by the failure of the passive scale to form under these conditions. Pesting has been reported in the case of several intermetallic alloys, including WSi_2, $NbSi_2$, $NbAl$, $ZrBe_{13}$, and $NbAl_3$.[27-30] The disintegration of these intermetallics has been attributed to grain-boundary embrittlement caused by the short-circuit diffusion of oxygen anions followed by their preferential dissolution near the grain boundaries. Pest disintegration has been found to be rapid in the case of alloy samples containing cracks and porosities.[14,31] It is intuitive that the presence of cracks and porosities introduced during processing would enhance the ingress of oxygen. On the other hand, nearly defect-free samples undergo disintegration after exposure in the critical temperature regime for a much longer duration. The internal oxidation of grain boundaries in intermetallic or reinforcement-intermetallic matrix interfaces is promoted by the short-circuit diffusion of oxygen anions. The formation of these oxides is accompanied by volume expansion, which leads to compressive residual stresses. With the progress of oxidation along the grain boundaries, the net compressive residual stress is enhanced, leading to the formation of cracks. Of course, the formation of flaky oxides at the grain boundaries also leads to intergranular separation. In this manner, the penetration of oxygen into the bulk of the alloy sample is facilitated, followed by internal oxidation and disintegration.

In $NbAl_3$, pesting has been noticed in the temperature range of 600°C–1000°C. Studies by x-ray diffraction and Auger electron spectroscopy have shown evidence of oxygen penetration through the grain boundaries and the formation of Al_2O_3 precipitates at these sites as the mechanisms accompanying the failure of this material.[29,31,32] Interestingly, the rapid disintegration of $NbAl_3$ has been observed with selective oxidation of Al, leading to a depletion in its concentration within the parent phase. As a result, phase transformation leading to the formation of Nb_2Al is observed. The accompanying chemical reaction can be shown as

$$2NbAl_3 \rightarrow Nb_2Al + 5Al \qquad (4.6)$$

This step leads to a considerable reduction in volume, which causes discontinuities at the grain boundaries, thereby adversely affecting the integrity of both its structure and protective scale.[30] As expected from the aforementioned mechanism, the rate of pest disintegration has been found to increase with a decrease in grain size, that is, with an increase in the grain-boundary area.

Pest disintegration has been observed in some of the beryllides, such as $ZrBe_4$ at 700°C, $ZrBe_{13}$ at 700°C, and Nb_2Be_{17} within the range of 500°C–600°C.[29] The powdery mass resulting from accelerated oxidation and the disintegration of ZrBe$_4$ consists of ZrBe$_4$ crystals as well as BeO whiskers and flakes. The preference for the formation of BeO may be attributed to the higher chemical reactivity of Be in an oxidizing environment, as well as its higher atomic mobility. Based on observation of the oxidation products, the following oxidation reactions appear to be possible:

$$ZrBe_{13} + 9/4\ O_2 \rightarrow \tfrac{1}{2}\ Zr_2Be_{17} + 9/2\ BeO \tag{4.7}$$

$$ZrBe_{13} + 15/2\ O_2 \rightarrow ZrO_2 + 13\ BeO \tag{4.8}$$

The amount of volume expansion on oxidation of Be to form BeO is ≈91%. The formation of BeO at the grain boundaries causes substantial volume expansion of 27% and 88% for Reactions 4.6 and 4.7, respectively. Such large volume expansion gives rise to high internal stresses. These stresses are also aggravated by the highly anisotropic character of the BeO whiskers that are formed at the grain boundaries. However, pesting has not been found in the case of NiBe, which is reported to be protected by a continuous scale of BeO between 500°C and 800°C.

The process of pest disintegration is promoted in MoSi$_2$ and WSi$_2$ by the formation of volatile oxidation products, MoO$_3$ and WO$_3$, respectively, through the following reactions:

$$2MoSi_2\ (s) + 7O_2\ (g) \rightarrow 2MoO_3\ (s) + 4SiO_2\ (s) \tag{4.9}$$

$$2WSi_2\ (s) + 7O_2\ (g) \rightarrow 2WO_3\ (s) + 4SiO_2\ (s) \tag{4.10}$$

In polycrystalline samples, the whiskers of MoO$_3$ or WO$_3$ along with clusters of SiO$_2$ are formed at the intergranular locations. With the formation of these oxidation products, there is a significant amount of localized volume expansion, leading to high compressive stress. As the reaction propagates, microcracks form throughout the sample, leading to its disintegration. The time taken for complete disintegration has been found to depend on density, method of processing, concentration of volume defects like microcracks and porosities, as well as composition. Pest disintegration has been found to be rapid in samples with a large quantity of volume defects, whereas it has been found to be very sluggish in the case of single crystals. For example, in the MoSi$_2$ samples processed by powder or ingot metallurgy routes, pest disintegration has taken between 21 and 41 h, whereas in the case of single crystals this duration has been found to be >1000 h.[29] On the surfaces of single crystals, a layer of metastable Si–Mo–O with an amorphous structure is formed first, which is followed by the formation of MoO$_3$ whiskers and SiO$_2$. The whiskers of MoO$_3$ found in the oxide scale of an MoSi$_2$–SiC composite exposed at 500°C for 96 h are depicted in Figure 4.7.[33]

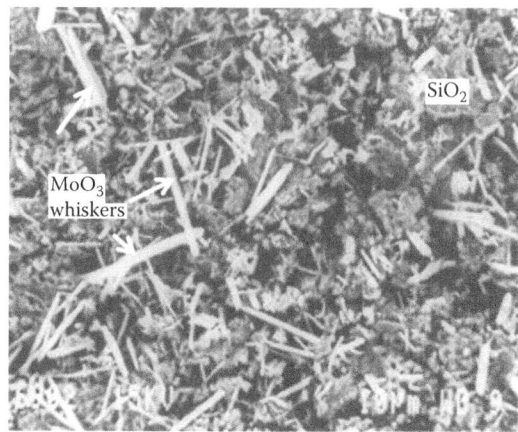

FIGURE 4.7
Top view of the $MoSi_2$ oxide scale formed on exposure in air at 500°C for 250 h showing the presence of MoO_3 whiskers. (From Mitra, R., et al., *Mater. Sci. Tech.*, 13(5), 415–419, 1997.)

Accelerated oxidation similar to pest disintegration has also been observed in Mo–Si–B alloys in the temperature range of 700°C–800°C. The formation of MoO_3 occurs by the oxidation of the constituent phases through the following reactions:[15]

$$2Mo + 3O_2 \rightarrow 2MoO_3 \tag{4.11}$$

$$3Mo_3Si + 11O_2 \rightarrow 6MoO_3 + 2SiO_2 \tag{4.12}$$

$$Mo_5SiB_2 + 10O_2 \rightarrow 5MoO_3 + SiO_2 + B_2O_3 \tag{4.13}$$

The oxidation of Mo_3Si and Mo_5SiB_2 can also take place through the following reactions, which lead to the formation of Mo_5Si_3 first and then MoO_3:

$$Mo_3Si + 6O_2 \rightarrow Mo_5Si_3 + 4MoO_3 \tag{4.14}$$

$$6Mo_5SiB_2 + 39O_2 \rightarrow 2Mo_5Si_3 + 20MoO_3 + 6B_2O_3 \tag{4.15}$$

$$2Mo_5Si_3 + 21O_2 \rightarrow 10MoO_3 + 6SiO_2 \tag{4.16}$$

The oxidation behavior of Mo–Si–B alloys in the temperature range of 700°C–800°C is dominated by the vaporization of MoO_3. It is known that MoO_3 melts at 795°C and vaporizes at 704°C. The vapor pressure of MoO_3 is high in the temperature range of 700°C–800°C, and therefore its vaporization

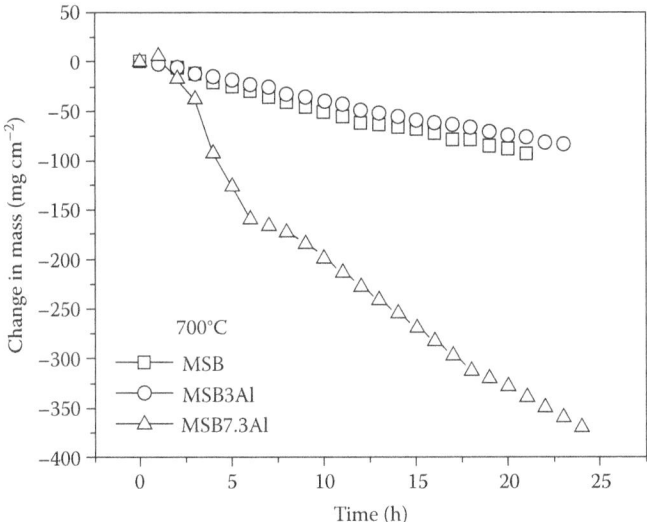

FIGURE 4.8
Plots depicting severe mass loss in the case of Mo–Si–B alloys during exposure at 700°C for 24 h. (From Paswan, S., et al., *Mater. Sci. Eng. A*, 424, 251–265, 2006. Reprinted with permission from Elsevier.)

leads to significant mass loss from Mo–Si–B alloys, as shown in Figure 4.8.[23] Interestingly, prior exposure of the 76Mo–14Si–10B alloy at 1150°C for durations between 1 and 24 h has been found to significantly enhance its resistance to environmental degradation on subsequent exposure at 700°C. Such resistance to environmental degradation at 700°C can be ensured by the formation of a continuous and stable oxide scale comprising B_2O_3–SiO_2 on prior exposure at 1150°C. The formation of this protective scale is not possible at 700°C, as the formation and vaporization of MoO_3 are much faster than the growth rate of SiO_2. The formation of this scale has been dealt with in detail in Chapter 7.

4.6 Summary

The oxidation behavior of intermetallics is controlled by the thermodynamic stability of oxidation products as well as kinetic factors affecting growth. The oxide scales that are known to be protective at elevated temperatures are Cr_2O_3 (<1000°C), SiO_2, and Al_2O_3. For the formation of a protective oxide scale on an alloy, it is necessary that the oxide in the outer scale should be more stable than the oxides of other elements, and its rate of growth should be slow to ensure that it is dense and pore free. Moreover, this oxide scale should adhere strongly to the alloy underneath and have the ability to heal itself on cracking. The ability to heal itself is supported by the viscoplastic flow of the oxide scale at elevated temperatures, as is observed in the presence of borosilicate glass.

One of the deterrents to the formation of a protective oxide scale on both aluminides and silicides is the simultaneous oxidation of different reactive alloying elements in the intermetallic alloy system. Accelerated oxidation, such as pesting, leading to the complete disintegration of samples is observed at intermediate temperatures in selected intermetallics, as the protective scale is not able to form due to relatively slower kinetics. Whereas the ingress of oxygen through pores in the oxide scale leads to unabated oxidation, preferential oxidation at grain boundaries and microcracks leads to the generation of internal compressive stresses, causing extensive damage. Damage is aggravated by the loss of mass due to the vaporization of volatile oxide-scale constituents. Research has focused on finding suitable alloying additions for the sintering of the oxide scale during its growth, so as to restrict the contact of atmospheric oxygen with the alloy underneath, as well as on minimization of internal stress by altering the products of oxidation.

References

1. Doychak, J. 1995. Oxidation behavior of high temperature intermetallics. In *Intermetallic Compounds and Practice, Vol. 1. Principles*, eds. J. H. Westbrook and R. L. Fleischer, 977–1015. Chichester, UK: Wiley.
2. Meier, G. H. and F. S. Pettit. 1992. The oxidation behavior of intermetallic compounds. *Mater. Sci. Eng. A* 153:548–560.
3. Meier, G. H. and F. S. Pettit. 1992. High-temperature oxidation and corrosion of intermetallic compounds. *Mater. Sci. Technol.* 8:331–338.
4. Aitken, E. A. 1967. Corrosion behavior. In *Intermetallic Compounds*, ed. J. H. Westbrook, 491–516. New York: Wiley.
5. Barin, I. 1989. *Thermochemical Data of Pure Substances*, Vols. 1 and 2. Weinheim, Germany: VCH.
6. Chatterjee, U. K., S. K. Bose, and S. K. Roy. 2001. *Environmental Degradation of Metals*, 283. New York: Marcel Dekker.
7. Mitra, R. and V. V. R. Rao. 1998. Elevated-temperature oxidation behavior of titanium silicide and titanium silicide based alloy and composite. *Metall. Mater. Trans. A* 29:1665–1675.
8. Raj, S. V. 1995. An evaluation of the properties of Cr_3Si alloyed with Mo. *Mater. Sci. Eng. A* 201:229–241.
9. Nesbitt, J. M. and C. E. Lowell. 1993. High temperature oxidation of intermetallics. In *High Temperature Ordered Intermetallic Alloys V*, Materials Research Society Symposium Proceedings, Vol. 288, 107–118.
10. Mitra, R. and V. V. Rama Rao. 1999. Effect of minor alloying with Al on oxidation behaviour of $MoSi_2$ at 1200°C. *Mater. Sci. Eng. A* 260:146–160.

11. Pilling, N. B. and R. E. Bedworth. 1923. The oxidation of metals at high temperatures. *J. Inst. Met.* 29:529–591.

12. Mitra, R., R. Khanna, and V. V. Rama Rao. 2004. Microstructure, mechanical properties and oxidation behavior of a multiphase (Mo,Cr)(Si,Al)$_2$ intermetallic alloy–SiC composite processed by reaction hot pressing. *Mater. Sci. Eng. A* 382:150–161.

13. Paswan, S., R. Mitra, and S. K. Roy. 2009. Nonisothermal and cyclic oxidation behavior of Mo–Si–B and Mo–Si–B–Al alloys. *Metall. Mater. Trans. A* 40:2644–2658.

14. Berztiss, D. A., R. R. Cerchiara, E. A. Gulbransen, F. S. Pettit, and G. H. Meier. 1992. Oxidation of MoSi$_2$ and comparison with other silicide materials. *Mater. Sci. Eng. A* 155:165–181.

15. Chou, T. C. and T. G. Nieh. 1993. Pesting of the high-temperature intermetallic MoSi$_2$. *JOM* 45:15–21.

16. Paswan, S., R. Mitra, and S. K. Roy. 2007. Oxidation behaviour of the Mo–Si–B and Mo–Si–B–Al alloys in the temperature range of 700–1300°C. *Intermetallics* 15:1217–1227.

17. Yoshimi, K., S. Nakatani, S. Hanada, S.-H. Ko, and Y.-H. Park. 2002. Synthesis and high temperature oxidation of Mo–Si–B–O pseudo *in situ* composites. *Sci. Technol. Adv. Mater.* 3:181–192.

18. Mendiratta, M. G., T. A. Parthasarathy, and D. M. Dimiduk. 2002. Oxidation behavior of alpha Mo–Mo$_3$Si–Mo$_5$SiB$_2$ (T2) three phase system. *Intermetallics* 10:225–232.

19. Wang, F., A. D. Shan, X. P. Dong, and J. S. Wu. 2007. Microstructure and oxidation resistance of laser-remelted Mo–Si–B alloy. *Scr. Mater.* 56:737–740.

20. Schneibel, J. H. and J. A. Sekhar. 2003. Microstructure and properties of MoSi$_2$–MoB and MoSi$_2$–Mo$_5$Si$_3$ molybdenum silicides. *Mater. Sci. Eng. A* 340:204–211.

21. Chattopadhyay, K., R. Mitra, and K. K. Ray. 2008. Nonisothermal and isothermal oxidation behavior of Nb–Si–Mo alloys. *Metall. Mater. Trans. A* 39:577–592.

22. Paswan, S., R. Mitra, and S. K. Roy. 2006. Isothermal oxidation behaviour of Mo–Si–B and Mo–Si–B–Al alloys in the temperature range of 400–800°C. *Mater. Sci. Eng. A* 424:251–265.

23. Krishnamurthy, R. and D. J. Srolovitz. 2003. Stress distributions in growing oxide films. *Acta Mater.* 51:2171–2190.

24. Suo, Z., D. V. Kubair, A. G. Evans, D. R. Clarke, and V. K. Tolpygo. 2003. Stresses induced in alloys by selective oxidation. *Acta Mater.* 51:959–974.

25. Roy, B., J. Das, and R. Mitra. 2013. Transient stage oxidation behavior of Mo$_{76}$Si$_{14}$B$_{10}$ alloy at 1150°C. *Corros. Sci.* 68:231–237.

26. Parthasarathy, T. A., M. G. Mendiratta, and D. M. Dimiduk. 2002. Oxidation mechanisms in Mo-reinforced Mo$_5$SiB$_2$(T2)–Mo$_3$Si alloys. *Acta Mater.* 50:1857–1868.

27. Westbrook, J. H. and D. L. Wood. 1964. "PEST" degradation in beryllides, silicides, aluminides, and related compounds. *J. Nucl. Mater.* 12:208–215.

28. Lewis, J. R. 1961. Further evaluation of beryllides. *J. Metals* 13:829–882.

29. Chou, T. C., T. G. Nieh, and J. Wadsworth. 1992. Stability of ZrBe$_{13}$, Nb$_2$Be$_{17}$, and NiBe intermetallics during intermediate temperature oxidation. *Scr. Metall. Mater.* 27:897–902.

30. Tolpygo, V. K. and H. J. Grabke. 1993. Mechanism of the intergranular disintegration (pest) of the intermetallic compound NbAl$_3$. *Scr. Metall. Mater.* 28:747–752.

31. McKamey, C. G., P. F. Tortorelli, J. H. Devan, and C. A. Carmichael. 1992. A study of pest oxidation in polycrystalline $MoSi_2$. *J. Mater. Res.* 7:2747–2755.
32. Grabke, H. J., M. Steinhorst, M. Brumm, and D. Wiemer. 1991. Oxidation and intergranular disintegration of the aluminides NiAl and $NbAl_3$ and phases in the system Nb–Ni–Al. *Oxid. Met.* 35:199–222.
33. Grabke, H. J. and G. H. Meier. 1995. Accelerated oxidation, internal oxidation, intergranular oxidation, and pesting of intermetallic compounds. *Oxid. Met.* 44:147–176.
34. Mitra, R., V. V. Rama Rao, and Y. R. Mahajan. 1997. Oxidation behaviour of reaction hot pressed $MoSi_2$/SiC composites at 500°C. *Mater. Sci. Tech.* 13(5):415–419.

5

Alloy Development for Structural Applications

5.1 Introduction

Research on intermetallic alloys and composites has been primarily driven by the demand of the aerospace industry for advanced materials with high specific strength and stiffness, as well as the ability to retain their strength and resist environmental degradation at elevated temperatures. Ordered intermetallics based on silicides and aluminides have exhibited significant promise in the form of attractive properties for high-temperature structural applications, including excellent oxidation and corrosion resistance, relatively light material weight, and superior strength at elevated temperatures.[1] The relevant properties of selected silicides and aluminides are presented in Table 5.1.[2–10] If Young's modulus and densities are considered, it is obvious that the lightweight aluminides exhibit higher specific stiffness than the Ni-based superalloys. The results of studies on the evaluation of the physical and mechanical properties of silicide- and aluminide-based intermetallic alloys and composites have motivated researchers to attempt further development for several potential applications. The primary motivation for such efforts is the distinct superiority of the properties of selected intermetallic materials compared with those of other conventional alloys. The potential applications include hot-end aero-engine and automotive components, tools and dies, corrosion-resistant piping materials, claddings and coatings for chemical industries, heat-treatment fixtures, magnetic and electronic devices, and hydrogen-storage materials.[11] Stoloff has discussed the potential applications of various types of intermetallics, which include primarily aluminides and silicides. An account of potential or ongoing structural applications of MoSi$_2$-based composites may be found in the reviews by Petrovic[12,13] and Yao et al.[14] Furthermore, important design issues, strategies for processing and development, and the advantages and limitations of candidate intermetallic materials including γ-Ti-aluminides, as well as Mo- and Nb-silicides,

TABLE 5.1

Structure and Properties of Selected Intermetallics

Intermetallic	Crystal Structure	Melting Point (°C)	Critical Ordering Temperature (°C)	Density (g ml^{-1})	Young's Modulus (GPa)	BDTT of Polycrystalline Intermetallic (°C)
Ni$_3$Al	fcc (L1$_2$, cP4)	1390	1390	7.5	179	Ductile at room temperature (in absence of moisture)
NiAl	bcc (B2, cP2)	1640	1640	5.86	294	400
Fe$_3$Al	Tetragonal (D0$_3$, cF16)	1540	540	6.72	141	Ductile at room
	bcc (B2, cP2)	1540	760			temperature (in absence of moisture)
FeAl	bcc (B2, cP2)	1250	1250	5.56	261	300–400
Ti$_3$Al	hcp (D0$_{19}$, hP8)	1600	1180	4.2	145	600
TiAl	fcc (L1$_0$, tP4)	1460	1460	3.91	176	700–800
Al$_3$Ti	Tetragonal (D0$_{22}$, tI8)	1350	1350	3.36	215.7	Brittle at all temperatures
	fcc (L1$_2$-stabilized cP4)					Compressive ductility above 750°C
MoSi$_2$	Tetragonal (C11$_b$, tI8)	2020	2020	6.24	439.7	1100–1300

Source: Cinca, N., et al., *J. Mater Res. Technol.*, 2, 75–86, 2013; Liu, C. T., *Scr. Metall. Mater.*, 27, 25–28, 1992; Miracle, D. B., *Acta Metall. Mater.*, 41, 649–684, 1993; Risanti, D., et al., *Intermetallics*, 13, 1337–1342, 2005; Yoshimi, K., et al., *Mater. Trans. JIM*, 35, 51–57, 1994; Lipsitt, H. A., et al., *Metall. Trans. A*, 11, 1369–1375, 1980; Imayev, V. M., et al., *Intermetallics*, 8, 1–6, 2000; Mitra, R., *Inter. Mater. Rev.*, 5, 13–64, 2006; Aikin Jr, R. M., *Scr. Metall. Mater.*, 26, 1025–1030, 1992.

Note: bcc, body-centered cubic; fcc, face-centered cubic; hcp, hexagonal close packed.

have been discussed in a review by Lipsitt et al.[15] The basic objectives of research on intermetallics for structural applications at elevated temperatures include the development of materials with the following properties: (i) high melting points, (ii) density preferably less than those of currently used alloys, (iii) brittle-to-ductile transition temperature (BDTT) as low as possible, (iv) room-temperature fracture toughness of $\geq$15.0 MPa m^{-2}, and (v) impressive oxidation resistance at possible temperatures of exposure. It may be noted that poor fracture toughness adversely affects the resistance to fatigue damage, which in turn is one of the key handicaps in the dynamic load-bearing applications of silicides.

5.2 Development Strategy

The intermetallics considered for load-bearing applications and coatings at extreme environments must possess excellent resistance to oxidation and corrosion, which is ensured by the formation of a protective scale on the exposed surface. These materials include aluminides of titanium, nickel, iron, and niobium, as well as silicides of nickel, molybdenum, and niobium and Laves phases such as Cr_2Nb. The physical properties, such as melting point, density, and elastic modulus, and the mechanical properties, including strength, ductility, and fracture toughness as well as creep and fatigue resistance at both ambient and elevated temperatures, are taken seriously for structural applications. It is also desirable to have lightweight materials in automotive and aeroengine components to increase the thrust-to-weight ratio and thereby improve the energy efficiency. The high thrust-to-weight ratio of various automotive and aero-engine components can be achieved by increasing the specific stiffness of these materials. The results shown in Figure 5.1 demonstrate that the decrease in Young's modulus of NiAl, TiAl, and Ti_3Al with temperature is more gradual than that observed for the Ni-based superalloy.[16,17] Of course, further enhancement in stiffness can be achieved in the aluminides by the addition of ceramic reinforcements with higher Young's moduli. It is necessary that the concentrations of Al and Si in aluminides and silicides, respectively, are sufficient to ensure the formation of the protective oxide scale. The new generation of gas

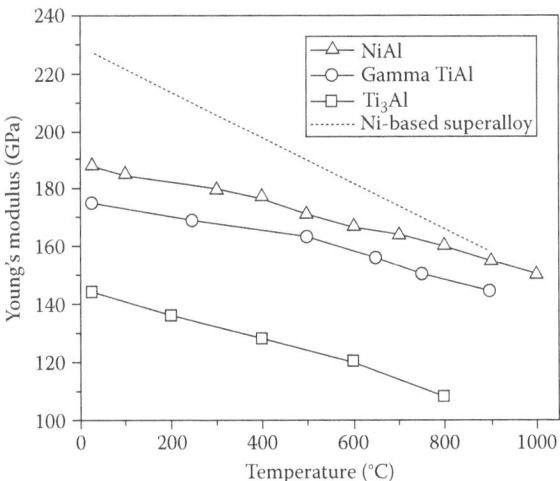

FIGURE 5.1
Plots depicting the variation of Young's modulus of NiAl, TiAl, and Ti_3Al with temperature along with data representing typical nickel-based superalloys. (From Miracle, D. B. and R. Darolia., *Intermetallic Compounds, Principles and Practice, Vol. 2,* 53–72, Wiley, Chichester, UK, 1995; Huang, S. C. and Chesnutt, J. C., *Intermetallic Compounds, Principles and Practice, Vol. 2,* 73–90, Wiley, Chichester, UK, 1995.)

turbine engines requires higher turbine inlet temperatures to ensure a greater extent of fuel combustion. As a result, the gas turbine materials are exposed to an increased service temperature, at which enhanced high-temperature oxidation and hot-corrosion attack are expected. For sea-based aircraft and marine gas turbines, the problem of degradation is more serious due to the aggressive operational environment comprising sulfur, vanadium, and sodium from the fuel and various halides from the seawater. Moreover, conventional MCrAlY (M = Ni, Co, or Ni + Co) types of coating are known to be protective only at temperatures ≤1150°C.[18] Therefore, new types of intermetallic alloy-based protective coatings are being developed for applications at even higher temperatures.

For power-generation equipment such as steam turbines operating for years without any shutdown, the life cycle and reliability are important issues. Resistance to fatigue in a corrosive environment is of interest for such applications. In order to test the performance and reliability for use in aerospace, automotive, and power-generation components, the candidate materials are subjected to stringent tests to avoid any risk or discomfort to human life. These tests are carried out with the objective of evaluating the properties of the candidate materials under extreme environments, which resemble the operational conditions of the prospective components, such as gas turbines.

Considering that most of the intermetallic alloys are brittle, those having even a small amount of ductility at the ambient temperature appear to be attractive for further development. The polycrystalline intermetallics exhibiting ductility at the ambient temperature include Zr_3Al ($L1_2$, cP4), Ni_3Al, Fe_3Al, and Ti_2AlNb ($E1_a$, oC16). On the other hand, the BDTTs of intermetallics such as FeAl and NiAl are reasonably low (Table 5.1). Moreover, single crystals of some of the intermetallics, such as $MoSi_2$ ($C11_b$, tI8) and Ti_3Al ($D0_{19}$, hP8), at certain selected orientations are known to exhibit ductile behavior,[19–21] whereas hardly any plasticity is recorded for the samples having hard orientations. In such cases, highly textured polycrystalline intermetallics with their preferred orientation being the same as the soft orientation of the single-crystal sample have shown significant ductility. For example, 30% elongation to failure has been observed in suitably textured Ti_3Al, which shows hardly any ductility when the grains are randomly oriented.[15] Usually, the single-phase intermetallic alloys exhibit relatively poor strength retention at elevated temperatures, and therefore multiphase alloys or composites are of interest to the scientific community. Suitable alloying additions within the solid solubility range, such as Nb in Ti_3Al[22,23] and transition metals in the case of TiAl,[24] are known to cause an increase in both strength and ductility. However, a further increase in strength is observed if alloying additions reach the limit for the formation of a second phase. Therefore, the formation of β phase in Ti_3Al alloys ($\alpha_2 + \beta$)[21] and the Ti_3Al/TiAl-based two-phase ($\alpha_2 + \gamma$) mixtures[17,25,26] are of interest for obtaining desirable mechanical properties.

In general, the observed ductility of all the intermetallic alloys is found to be the function of strain rate.[15] Under a high strain rate, that is, impact-type conditions, the ductility is reported to be worse even at elevated temperatures. Even under compressive deformation conditions, the amount of uniform strain has been reported to decrease with an increase in the strain rate. Environmental effects can also be damaging at elevated temperatures. For example, B-doped Ni_3Al shows a sharp reduction in fracture strength in humid air at 550°C. To summarize, it has not been possible so far to achieve metal-like fracture behavior in any of the intermetallics. Despite such shortcomings, it is possible to utilize the positive attributes of selected intermetallics for suitable engineering applications in view of their superior properties with respect to those of the conventional alloys.

For use in a variety of engineering components, the key hurdle lies in the processing of near-net shapes with desirable microstructures and mechanical properties. Although the basic methods of processing are the same for both conventional and intermetallic alloys, remarkable developments have taken place in this area, as summarized in Chapter 2. Whereas certain types of brittle intermetallic ingots used to crack due to thermal stresses during cooling, now it is possible to melt and cast ingots weighing 500 kg or more using advanced techniques. Casting is the preferred technique for fabricating engine-related products. The mold needs to be suitably designed for reducing the stresses by eliminating sharp thickness gradients as well as corners and notches. One of the problems during solidification is the segregation of solute elements, which is minimized by homogenization heat treatments and hot working. The as-cast ingots are either hot extruded or isothermally forged. For extrusion, canning is carried out to avoid surface cracking. Extrusion has been carried out on Nb_{ss}–Nb_5Si_3 composites,[27] whereas isothermal forging has been used to shape Ni_3Al-based alloys.[28] Furthermore, nonconventional processing methods including directional solidification, sheet-casting, foil rolling, and superplastic forming have been developed for a reasonably large variety of intermetallic alloys.[15]

In order to decide on the viability of a given material for use in engineering components, sufficient test data are required to establish the minimum achievable properties.[15] Furthermore, the engineering and cost benefits need to be worked out for the replacement of conventional alloys by the selected intermetallic material to meet acceptable economic criteria. Also, scale-up and product certification are required prior to acceptance for mass production. This procedure depends on the nature of the industry. In the aerospace industries, material specification limits are decided by varying processing conditions and composition. Statistically obtained property minima (i.e., minimum of the experimentally obtained properties) are considered in order to arrive at the final design. Further calibration of material properties is carried out by rig testing of selected components taken from production. During this stage, the methods of quality control by inspection of the microstructure and nondestructive testing are finalized. Such tests carried out on TiAl include compressor blades in a military aircraft engine as well as General Electric CF6 low-pressure turbine blades. These tests have shown promising results except for impact toughness. Turbine engine vanes made of NiAl single crystals have been successfully tested for application in jet engines, but further effort has not been pursued due to poor impact properties.[27] Despite the existing knowledge of fatigue mechanisms, it is necessary to understand the nature, probability, and amount of damage of the components in the operating environment in order to predict life. Moreover, the presence of small-scale defects can significantly reduce the fatigue life.

5.3 Applications: Current and Futuristic

5.3.1 Silicides

For several decades, $MoSi_2$-based heating elements, popularly known as superkanthal, have been in use at temperatures up to 1700°C. Superkanthal contains fine grains of $MoSi_2$, bonded together with a network of aluminosilicate ($xAl_2O_3 \cdot ySiO_2$) glass phase.[29] More recently, Kanthal Super 1900 heating elements containing a solid-solution alloy of $MoSi_2$ and WSi_2 have come into use.[14] The life of the heating elements would probably increase with the use of composites and alloys with better creep resistance and fracture toughness. Trials are in progress for several applications of different silicides and silicide–matrix composites at elevated temperatures in the range of 1000°C–1500°C.

5.3.1.1 Gas Turbine Engines of Aerospace Vehicles

In a recent review, Dimiduk and Perepezko have shown, on the basis of data compiled from engines over a period of 70 years, that the overall efficiency actually decreases with increasing turbine rotor inlet temperatures.[30] The decrease in efficiency is attributed to the requirement for advanced cooling schemes. The efficiency of the engine can only be improved by keeping the power of the engine the same or pushing it higher by switching to materials suitable for temperatures beyond those of third-generation Ni-based superalloys, so that cooling is not required.

Molybdenum silicide-based hybrid composites ($MoSi_2$–$Si_3N_{4(p)}$–SiC_f) are candidate materials for the blade outer air seal (BOAS), which is designed to maintain a small gap between itself and the blade and is exposed to high turbine gas temperatures and thermal stresses. The turbine efficiency is adversely affected if the gap between the blade and the seal is increased. The high fracture toughness, impact energy, resistance to pesting, and high-temperature isothermal as well as cyclic oxidation of the $MoSi_2$-based hybrid composites,[31] which are comparable to those of superalloys, contribute to the promise for applications in aircraft engine components. Gas burner testing at Pratt and Whitney has proved that the $MoSi_2$–Si_3N_4 and $MoSi_2$–SiC

composites have significant thermal-shock resistance in the simulated jet fuel combustion environment.[12]

Niobium silicide-based composites are also being developed for applications in airfoils. A prototype investment-cast Nb-silicide-based airfoil has been fabricated.[32] Although investment casting offers a method of producing near-net-shaped aerospace components, including blades and vanes, further research is required for the process development.

5.3.1.2 Molten-Metal Lances

Laminated or functionally graded tubes of $MoSi_2$ and Al_2O_3 undergo graceful failure demonstrating a high energy of fracture, particularly at elevated temperatures.[33] Microlaminate $MoSi_2$–Al_2O_3 composite tubes fabricated by plasma spray forming were tested as lances for the injection of inert gases into molten aluminum alloys at 725°C and molten copper at 1200°C, and were found to perform very well, and much better than tubes of graphite or SiC. The resistance to chemical attack by the molten copper is attributed to the presence of the inert Al_2O_3 phase, while thermal-shock resistance and graceful failure are due to the $MoSi_2$ being above its BDTT.

5.3.1.3 Protective Sheaths of High-Temperature Sensors

Functionally graded tubes of $MoSi_2$ and Al_2O_3 have the potential for use in protective sheaths of high-temperature sensors.[34] At present, expensive platinum coatings are used on alumina sheaths for thermocouples in the glass industry. The failure rate of platinum-coated alumina sheaths is high, which has driven the industry to look for alternative materials. Alumina–zirconia–silica multiphase refractory ceramics are being used, but have the limitation of poor mechanical properties. Studies have shown that $MoSi_2$ possesses excellent resistance to degradation in a molten-glass environment below the glass-line due to the formation of Mo_5Si_3 and above the glass-line due to the formation of SiO_2, while the corrosion rates are high at the glass-line.[12,35-38] It has been reported, however, that anodic protection of $MoSi_2$ lowers the rate of corrosion at the glass-line. The excellent resistance to corrosion, accompanied by impressive mechanical properties at elevated temperatures, has driven Kanthal Corporation to market $MoSi_2$–Al_2O_3 composite immersion tubes for injection of gases into molten glass.[38]

5.3.1.4 Industrial Gas Burners

Industrial gas burners designed for use at higher operating temperatures involving the combustion of oxygen–natural gas mixtures are needed to replace those used in the combustion of air–natural gas mixtures at relatively lower temperatures, so that the NO_x emissions into the environment are reduced. Investigations by Lin and Speyer have shown that $MoSi_2$ possesses excellent resistance to oxidation and environmental damage when subjected to oxygen–natural gas combustion occurring at higher temperatures.[39] $MoSi_2$ has been found to be stable when exposed to stoichiometric as well as fuel-rich mixtures. Hence, $MoSi_2$-based materials are suitable for application in industrial gas burners designed for the combustion of oxygen–natural gas. Prototype gas burners of $MoSi_2$ and its composites prepared by plasma spray processing have performed satisfactorily.[40]

5.3.1.5 Diesel Engine Glow Plugs

Diesel engine glow plugs having a heater element of $MoSi_2$–Si_3N_4 composite with 30–40 wt.% $MoSi_2$ (almost comparable volume percentages) and tungsten electrodes have been developed recently by Denso Corporation in Japan.[41,42] In trial experiments, the heater elements survived in repeated starting tests for 5×10^6 cycles with a thermal shock involving a temperature difference of 1200°C, which suggests a long lifetime. $MoSi_2$–Si_3N_4 composite glow plugs are superior to those made of metal because of resistance to environmental degradation involving diesel fuel combustion and the possibility of heating at higher rates. The inner composite, with a continuous network of $MoSi_2$, has the electrical conductivity required in the glow plug application. The glow plugs can be operated at temperatures as high as 1400°C.

5.3.2 Nickel Aluminides

Ni_3Al has been alloyed with B to obtain adequate room-temperature ductility and with a variety of other alloying elements, such as Mo and Zr, for improving the high-temperature creep resistance.[43,44] For high-temperature load-bearing applications, creep-resistant Ni_3Al-based alloys have been developed. These include IC-221M (Ni8 wt.% Al7.7Cr1.43Mo1.7Zr0.008B) and IC-438 (Ni8.1wt.%Al5.23Cr7.02Mo0.13Zr0.005B) for use up to 1100°C and 1250°C, respectively. Due to its excellent castability, a turbocharger rotor made of the IC-221M alloy has been prepared by investment casting for trials in diesel trucks.[45] This alloy has exhibited superior fatigue resistance and lower cost compared with the IN-713C alloy. The industrial interest in Ni_3Al-based alloys has been reviewed by Sikka et al.[46] and Stoloff et al.[11]

The Ni_3Al-based intermetallic alloys have been found to exhibit excellent resistance to environmental degradation by oxidation and carburization, as well as cavitation and erosion.[11] These properties, along with the ability to retain strength at elevated temperatures, make alloys such as IC-218 and IC-221M suitable for use in high-temperature dies for isothermal forging and molds for the processing of glass.[44] Further applications of these alloys have been proposed in rolls for steel slabs in heat-treatment furnaces, continuous casting guide rolls, radiant burner tubes used in steel-producing industries, and forging dies.[11] As these alloys exhibit impressive strength and oxidation resistance at elevated temperatures, water cooling is not required for their use as rolls for heated steel slabs. Avoiding water-cooling results in significant energy savings and in extending the material life. Due to its resistance to carburization, the IC-221M alloy has been evaluated for use as fixture materials for the heat treatment of auto parts. The Ni_3Al-based alloys are also being considered for application in center posts for pit-carburizing furnaces and link-belts in air heat-treatment furnaces, as well as furnace muffles. A method based on exothermic melt-processing (Exo-Melt™) has been developed by Sikka and coworkers[47] to manufacture transfer rolls of Ni_3Al-based alloy (IC-221M) for application in steel-austenitizing furnaces.

Due to excellent corrosion and erosion resistance, potential applications in chemical industries include reaction vessels operated at high temperatures, pump impellers for slurries, and air deflectors for burning high-sulfur fuel.[11] As the Ni_3Al-based alloys offer excellent vibration-cavitation resistance in water, these materials are also suitable for use in hydroturbines. Furthermore, a directionally solidified alloy having the composition Ni–16.3Al–8.2Mo–0.26B (at.%) with excellent strength and creep resistance at temperatures ≥1000°C in oxidizing environments has been developed as a candidate material for turbine blades and vanes.[44,48] Certain other potential applications are in pistons and valves of automotive diesel engines.

The NiAl-based alloys are better suited than Ni_3Al to high-temperature applications, because their higher Al content contributes to significant oxidation resistance.[2] However, widespread application of NiAl is primarily restricted by its inherent brittleness at the ambient temperature, as well as high BDTT. One of the major applications of NiAl is in the bond-coats applied on the Ni-based superalloys. The presence of such bond-coats lowers the internal stress caused in the outer thermal barrier coating due to its coefficient of thermal expansion mismatch with the Ni-based superalloy substrate.

5.3.3 Titanium Aluminides

Among the intermetallics being examined for turbine applications for aerospace vehicles, TiAl can be considered as the most promising structural material.[15] This intermetallic has a higher melting point (1480°C), a low density (3.8 g ml^{-1}), and creep resistance, as well as greater resistance to both oxidation and fire compared with those of conventional Ti alloys and polycrystalline superalloys used for turbine applications.[11] Moreover, the specific stiffness of the γ-TiAl alloys is reported to be 50% higher than that of the aforementioned conventional materials in use. During operation, the engine airfoils are subjected to vibrations, which are in turn dependent on the material stiffness.[15] For the aero-engine applications, casings and support structures require high stiffness for shape retention, and several smaller ring-shaped structures that require high stiffness exist within an engine. The Young's modulus of γ-TiAl is found to be close to those of some of the Ni-based superalloys at elevated temperatures.[49] Moreover, the specific strength of the cast Ti48Al2Cr2Nb alloy exceeds that of Ti alloys at 500°C and γ'-strengthened Ni-based superalloy at

≥750°C. The intrinsic fatigue resistance of the γ-TiAl alloys subjected to high-cycle fatigue-loading conditions is known to be outstanding.

The potential aircraft applications of TiAl are low-pressure turbine blades in CF6-80C2 and GE90 engines, carbon seal supports, transition duct beams, blade dampers, compressor blades, and high-speed transport components.[11,49] An alloy with the composition Ti47(at.%)Al2.1W0.5Si has been developed for applications in marine turbines used in high-speed ferries, where the maximum inlet temperature is 610°C.[50] The γ-TiAl-based alloys have also been considered for use in turbocharger wheels and turbine blades. Cast TiAl-based alloy ingots can be heated to the α + β phase region, where they can be hot rolled, forged, or both to obtain complex shapes.[51] The alloys containing Nb and Mo solidify into β phase stabilized by the presence of Nb and Mo on solidification, which are then subjected to multistep heat treatment to obtain a microstructure with desirable volume fractions of α_2, β_0 (secondary precipitates), and γ, so as to achieve the optimum combination of room-temperature ductility and high-temperature creep resistance.[52] In this manner, the expensive hot-working steps can be avoided. In 2011, cast TiAl blades were introduced in the low-pressure turbine in General Electric's GEnX jet engine, which powers, for example, the Boeing 787.[53] However, two major issues have to be considered for the industrial exploitation of these alloys on a larger scale: the machinability and the resistance against environmental damage.

The automotive applications of the γ-TiAl-based intermetallics have been reviewed by Isobe and Noda.[54] Such applications arise from the superiority of the specific strength and stiffness of γTiAl compared with those of conventional automobile materials, austenitic heat-resistant steel, and wrought Ni-based superalloy at elevated temperatures. Moreover, the thermal fatigue resistance has been found to be superior to that of 304 grade austenitic stainless steel. The potential automotive components made of γ-TiAl include exhaust engine valves and turbine rotors for turbochargers.[54]

A major shortcoming of γ-TiAl is its poor ductility and fracture toughness. Therefore, the fabrication of the γ-TiAl-based alloys into near-net-shaped components is considered to be challenging, and significant research has been devoted to achieve such goals. It has been possible to manufacture a compressor casing of γ-TiAl by casting. High-pressure compressor stators and civil low-pressure turbine blades are some other important potential applications of this alloy.[55] However, the ability to form TiAl-based alloys superplastically at lower temperatures allows the fabrication of large components.[56,57] It is also possible to spray form TiAl deposits on the substrate.[58]

5.3.4 Iron Aluminides

The iron aluminides, FeAl and Fe_3Al, have attracted attention for their low material cost, ease of fabrication, and corrosion and oxidation resistance.[2,59] In general, the iron-based intermetallic alloys are known to exhibit better resistance to sulfidation than any other iron- or nickel-based alloys. Alloying of Fe-28 at.% Al with Cr and Mo increases the corrosion resistance in acidic, basic, and chloride solutions. Furthermore, FeAl exhibits excellent resistance to catalytic coking, carburization, and wear.[11] The Fe-aluminide-based alloys have been investigated to obtain substitute material for stainless steels for specific applications. As the Fe-aluminides are prone to environmental embrittlement in the presence of moisture, FeAl and Fe_3Al need to be alloyed with B and Cr, respectively, for engineering applications. Alloys based on FeAl (40 at.% Al) have been found to be suitable for use in transfer rolls for hot-rolled steel strip, ethylene crackers, and an air deflector for burning high-sulfur coal. The Fe-aluminides also exhibit high electrical resistivity, which increases with temperature, and good corrosion resistance. Based on the aforementioned properties of Fe-aluminides, other potential industrial applications include heat-treating trays, immersion heaters, porous filters, automotive piston valves and automotive exhaust systems, centrifugally cast tubes, radiant tubes for heat exchangers, and catalytic conversion vessels.[2]

Steels coated with $(Fe,Cr)_3Al$ by a two-step pack-cementation process are found to exhibit enhanced corrosion resistance.[60] Fe-aluminide-based alloys (e.g., Fe–40Al–0.05Zr–50 ppm B with 1 wt.% Y_2O_3) atomized under an argon atmosphere and then subsequently ball-milled to obtain nanocrystalline powders have been used to process coatings on steel substrates by high-velocity oxygen fuel spraying or thermal spraying.[2] Such coatings have been reported to possess 35% higher hardness compared with those with micrometer-size grains, and exhibit excellent adhesion to substrate as well as much higher hardness and wear resistance.

5.4 Summary

The intermetallics based on aluminides and silicides with optimized compositions and microstructures exhibit attractive mechanical properties along with acceptable oxidation resistance, and therefore are being considered for application in aerospace and automotive components. The most promising intermetallics for use in jet engine turbines are multicomponent alloys based on γ-TiAl, whereas Fe-aluminides are considered for applications demanding resistance to hot corrosion. The Ni_3Al-based alloys are also considered in a variety of industrial applications, where strength retention at elevated temperatures, along with oxidation resistance, is expected. The molybdenum silicides with outstanding oxidation resistance are routinely used in the heating elements of high-temperature furnaces, and have been considered for use in BOASs, diesel engine glow plugs, and other industrial applications. Although some of the intermetallic alloys have exhibited appreciable ductility, the fracture toughness and impact energies are not satisfactory, which is a hurdle for widespread application of these materials. Another challenge is in finding inexpensive methods of processing near-net-shaped defect-free components.

References

1. Liu, C. T. 1995. Recent advances in ordered intermetallics. *Mater. Chem. Phys.* 42:77–86.
2. Cinca, N., C. R. C. Lima, and J. M. Guilemany. 2013. An overview of intermetallics research and application: Status of thermal spray coatings. *J. Mater. Res. Technol.* 2:75–86.
3. Liu, C. T. 1992. Environmental embrittlement and grain-boundary fracture in Ni_3Al. *Scr. Metall. Mater.* 27:25–28.
4. Miracle, D. B. 1993. Overview No. 104: The physical and mechanical properties of NiAl. *Acta Metall. Mater.* 41:649–684.
5. Risanti, D., J. Deges, L. Falat, S. Kobayashi, J. Konrad, M. Palm, B. Poter, A. Schneider, C. Stallybrass, and F. Stein. 2005. Dependence of the brittle-to-ductile transition temperature (BDTT) on the Al content of Fe–Al alloys. *Intermetallics* 13:1337–1342.
6. Yoshimi, K., S. Hanada, and H. Tokuno. 1994. Effect of frozen-in vacancies on hardness and tensile properties of polycrystalline B2 FeAl. *Mater. Trans. JIM* 35:51–57.
7. Lipsitt, H. A., D. Shechtman, and R. E. Schafrik. 1980. The deformation and fracture of Ti_3Al at elevated temperatures. *Metall. Trans. A* 11:1369–1375.
8. Imayev, V. M., R. M. Imayev, and G. A. Salishchev. 2000. On two stages of brittle-to-ductile transition in TiAl intermetallic. *Intermetallics* 8:1–6.
9. Mitra, R. 2006. Mechanical behaviour and oxidation resistance of structural silicides. *Inter. Mater. Rev.* 5:13–64.
10. Aikin Jr, R. M. 1992. On the ductile-to-brittle transition temperature in $MoSi_2$. *Scr. Metall. Mater.* 26:1025–1030.
11. Stoloff, N. S., C. T. Liu, and S. C. Deevi. 2000. Emerging applications of intermetallics. *Intermetallics* 8:1313–1320.
12. Petrovic, J. J. 1997. High temperature structural silicides. *Ceram. Eng. Sci. Proc.* 18(3):3–17.
13. Petrovic, J. J. and A. K. Vasudevan. 1999. Key developments in high temperature structural silicides. *Mater. Sci. Eng. A* 261:1–5.
14. Yao, Z., J. Stiglich, and T. S. Sudarshan. 1999. Molybdenum silicide based materials and their properties. *J. Mater. Eng. Perf.* 8:291–304.
15. Lipsitt, H. A., M. J. Blackburn, and D. M. Dimiduk. 2002. High-temperature structural applications. In *Intermetallic Compounds, Principles and Practice, Vol. 3: Progress*, eds. J. H. Westbrook and R. L. Fleischer, 471–499. Chichester, UK: Wiley.
16. Miracle, D. B. and R. Darolia. 1995. NiAl and its alloys. In *Intermetallic Compounds, Principles and Practice, Vol. 2*, eds. J. H. Westbrook and R. L. Fleischer, 53–72. Chichester, UK: Wiley.

17. Huang, S. C. and Chesnutt, J. C. 1995. Gamma TiAl and its alloys. In *Intermetallic Compounds, Principles and Practice, Vol. 2*, eds. J. H. Westbrook and R. L. Fleischer, 73–90. Chichester, UK: Wiley.

18. Guo, H., L. Sun, H. Li, and S. Gong. 2008. High temperature oxidation behaviour of hafnium modified NiAl bond coat in EB-PVD thermal barrier coating system. *Thin Solid Films* 516:5732–5735.

19. Ito, K., H. Inui, Y. Shirai, and M. Yamaguchi. 1995. Plastic-deformation of $MoSi_2$ single-crystals. *Philos. Mag. A* 72:1075–1097.

20. Inui, H., Y. Toda, and M. Yamaguchi. 1993. Plastic-deformation of single-crystals of a Do(19) compound with an off-stoichiometric composition (Ti-36.5 at percent Al) at room-temperature. *Philos. Mag. A* 67:1315–1332.

21. Nakano, T., Y. Maeda, and Y. Umakoshi. 1996. Cold-rolling and recovery behaviors of Ti_3Al single crystals with D0(19) structure. *ISIJ Inter.* 36:111–120.

22. Blackburn, M. J. and M. P. Smith. 1978. Research to conduct an exploratory and analytical investigation of alloys, AFML-TR-78-18. US Air Force Materials Laboratory.

23. Banerjee, D. 1994. Ti_3Al and its alloys. In *Intermetallic Compounds, Vol. 2, Practice*, eds. J. H. Westbrook and R. L. Fleischer, 91–131. Chichester, UK: Wiley.

24. Blackburn, M. J. and M. P. Smith. 1981. US Patent 4294615.

25. Chan, K. S. 1992. Understanding fracture toughness in gamma TiAl. *J. Met.* 44(5):30–38.

26. Huang, S. C. 1992. Microstructures and property tradeoffs in wrought TiAl-base alloys. *Metall. Trans. A* 23:375–377.

27. Bewlay, B. P., M. R. Jackson, and H. A. Lipsitt. 1996. The balance of mechanical and environmental properties of a multielement niobium-niobium silicide-based *in situ* composite. *Metall. Mater. Trans. A* 27:3801–3808.

28. Pope, D. P. and R. Darolia. 1996. High-temperature applications of intermetallic compounds. *MRS Bull.* 21:30–36.

29. Menon, E. S. K., M. G. Mendiratta, and D. M. Dimiduk. 2001. High temperature oxidation mechanisms in Nb-silicide bearing multicomponent alloys. In *Structural Intermetallics 2001*, eds. K. J. Hemker, D. M. Dimiduk, H. Clemens, R. Darolia, H. Inui, J. M. Larsen, V. K. Sikka, M. Thomas, and J. D. Whittenberger, 591. Warrendale, PA: TMS.

30. Dimiduk, D. M. and J. H. Perepezko. 2003. Mo–Si–B alloys: Developing a revolutionary turbine-engine material. *MRS Bull.* 28:639–645.

31. Hebsur, M. G. 1999. Development and characterization of $SiC(f)/MoSi_2–Si_3N_4(p)$ hybrid composites. *Mater. Sci. Eng. A* 261:24–37.

32. Bewlay, B. P., M. R. Jackson, J. C. Zhao, P. R. Subramanian, M. G. Mendiratta, and J. J. Lewandowski. 2003. Ultrahigh-temperature Nb-silicide-based composites. *MRS Bull.* 28:646–653.

33. Bartlett, A. H., R. G. Castro, D. P. Butt, H. Kung, and J. J. Petrovic. 1996. Plasma sprayed $MoSi_2/Al_2O_3$ laminate composite tubes as lances in pyrometallurgical operations. *Ind. Heat.* 63:33–36.

34. Peters, M. I., R. U. Vaidya, R. G. Castro, J. J. Petrovic, K. J. Hollis, and D. E. Gallegos. 2001. Functionally graded $MoSi_2–Al_2O_3$ tubes for temperature-sensor applications. *Ind. Heat.* October:105–110.

35. Sundaram, S. K., J. Y. Hsu, and R. F. Speyer. 1994. Molten glass corrosion-resistance of immersed combustion-heating tube materials in soda lime silicate glass. *J. Am. Ceram. Soc.* 77:1613–1623.

36. Sundaram, S. K., J. Y. Hsu, and R. F. Speyer. 1995. Molten glass corrosion-resistance of immersed combustion-heating tube materials in e-glass. *J. Am. Ceram. Soc.* 78:1940–1946.

37. Sundaram, S. K. and R. F. Speyer. 1996. Electrochemical corrosion and protection of molybdenum and molybdenum disilicide in a molten soda-lime silicate glass environment. *J. Am. Ceram. Soc.* 79:1851–1856.

38. *Kanthal Super Bubbling Tubes in Glass Melting Tanks.* 1995. The Kanthal Corporation, Furnace Products.

39. Lin, W.-Y. and R. F. Speyer. 1994. Surface oxidation mechanisms of molybdenum disilicide in high-temperature combustion environments. *High Temperature Silicides and Refractory Alloys,* Materials Research Society Symposium Proceedings, Vol. 322, 267–272.

40. Castro, R. G., J. R. Hellmann, A. E. Segall, and D. L. Shelleman. 1994. Fabrication and testing of plasma-spray formed $MoSi_2$, and $MoSi_2$, composite tubes. *High Temperature Silicides and Refractory Alloys,* Materials Research Society Symposium Proceedings, Vol. 322, 81–86.

41. Nikkei Business Publications. 1994. *Nikkei Materials and Technology,* 94.6, 142. Nikkei Business Publications.

42. Yamada, K. and N. Kamiya. 1999. High temperature mechanical properties of Si_3N_4–$MoSi_2$ and Si_3N_4–SiC composites with network structures of second phases. *Mater. Sci. Eng. A* 261:270–277.

43. Liu, C. T. 1991. Intergranular fracture and boron effects in Ni_3Al and other intermetallics: Introductory paper. *Scr. Metall. Mater.* 25:1231–1236.

44. Liu, C. T. and D. P. Pope, 1994. Ni_3Al and its alloys. In *Intermetallic Compounds: Vol. 2, Practice,* eds. J. H. Westbrook and R. L. Fleischer, 17–51. Chichester, UK: Wiley.

45. Sikka, V. K., J. T. Mavity, and K. Anderson. 1992. Processing of nickel aluminides and their industrial applications. In *High Temperature Aluminides and Intermetallics,* eds. S. H. Whang, C. T. Liu, D. P. Pope, and J. O. Stiegler, 712–721. Essex, UK: Elsevier.

46. Sikka, V. K., J. T. Mavity, and K. Anderson. 1992. Processing of nickel aluminides and their industrial applications. *Mater. Sci. Eng. A* 153:712–721.

47. Sikka, V. K., M. L. Santella, P. Angellina, J. Mengel, R. Petrusha, A. P. Martocci, and R. I. Pankiw. 2004. Large-scale manufacturing of nickel aluminide transfer rolls for steel austenitizing furnaces. *Intermetallics* 12:837–844.

48. Han, Y. F., Z. P. Xing, and M. C. Chaturvedi. 1997. Development and engineering application of a DS cast Ni_3Al alloy IC6. In *Structural Intermetallics 1997,* eds. M. V. Nathal, R. Darolia, C. T. Liu, P. L. Martin, D. B. Miracle, R. Wagner, and M. Yamaguchi, 713–719. Warrendale, PA: TMS.

49. Austin, C. M., T. J. Kelly, K. G. McAllister, and J. C. Chestnut. 1997. Aircraft engine applications for gamma titanium aluminide. In *Structural Intermetallics 1997,* eds. M. V. Nathal, R. Darolia, C. T. Liu, P. L. Martin, D. B. Miracle, R. Wagner, and M. Yamaguchi, 413–425. Warrendale, PA: TMS.

50. Nazmy, M., C. Noseda, M. Staubli, and B. Phillipsen. 1997. In *Processing and Design Issues in High Temperature Materials,* eds. N. S. Stoloff and R. H. Jones, 159. Warrendale, PA: TMS.

51. Tetsui, T., K. Shindo, S. Kobyashi, and M. Takeyama. 2002. A newly developed hot worked TiAl alloy for blades and structural components. *Scr. Mater.* 47(6):399–403.

52. Schwaighofer, E., H. Clemens, S. Mayer, J. Lindemann, J. Klose, W. Smarsly, and V. Güther. 2014. Microstructural design and mechanical properties of a cast and heat treated intermetallic multi-phase γ-TiAl based alloy. *Intermetallics* 44:128–140.

53. Bewlay, B., M. Weimer, T. Kelly, A. Suzuki, and P. R. Subramanian. 2013. The science, technology, and implementation of TiAl alloys in commercial aircraft engines. In *Intermetallic Based Alloys: Science, Technology, and Applications*, eds. I. Baker, M. Heilmaier, K. S. Kumar, and K. Yoshimi, Materials Research Society Symposium Proceedings, Vol. 1516, 49. Cambridge: Cambridge University Press.

54. Isobe, S. and T. Noda. 1997. Automotive application of TiAl intermetallics. In *Structural Intermetallics 1997*, eds. M. V. Nathal, R. Darolia, C. T. Liu, P. L. Martin, D. B. Miracle, R. Wagner, and M. Yamaguchi, 427–433. Warrendale, PA: TMS.

55. Postans, P. J., M. T. Cope, S. Moorehouse, and A. B. Thakker. 1993. Application of titanium aluminides in gas turbine components. In *Titanium 92, Science and Technology Vol. III*, eds. F. H. Froes and I. L. Caplan, 2907. Warrendale, PA: TMS.

56. Nieh, T. G. and J. Wadsworth. 1999. Fine-structure superplastic intermetallics. *Int. Mater. Rev.* 44:59–75.

57. Sun, F. and D. L. Lin. 2001. Superplastic phenomenon in a large-grained TiAl alloy. *Scr. Mater.* 44:665–670.

58. Schimansky, F. P., K. W. Liu, and R. Gerling. 1999. Spray forming of gamma titanium aluminides. *Intermetallics* 7:1275–1282.

59. Vedula, K. 1994. FeAl and Fe₃Al. In *Intermetallic Compounds: Vol. 2, Practice*, eds. J. H. Westbrook and R. L. Fleischer, 199–209. Chichester, UK: Wiley.

60. Zheng, M., Y. He, and R. A. Rapp. 1997. *Proceedings of the 11th Annual Conference on Fossil Energy Materials*, ORNL/FMP-97/1, May 1997. Oak Ridge, TN: Oak Ridge National Laboratory.

6

Silicides: Processing and Mechanical Behavior

6.1 Introduction

The research and development effort on refractory metal-silicides is driven by the requirement for higher operating temperatures in the range of 1100°C–1500°C for the gas turbine components in jet engines and other aerospace applications. In addition, there are several other existing or potential applications, including heating elements in high-temperature furnaces, molten metal lances, and diesel engine glow plugs. The primary requirements of these materials are high melting points, elevated-temperature strength retention, and resistance to creep and oxidation, as well as reasonable ductility and fracture toughness in the temperature range of application. Of course, it is desirable for the material to possess sufficient ductility and fracture toughness at room temperature as well.

In general, the silicides of Mo, Nb, W, Ti, and Cr have melting points near or above 2000°C. A survey of structure and properties by Meschter and Schwartz has shown $MoSi_2$ and Ti_5Si_3 to be the most attractive among the binary silicides for elevated-temperature structural applications.[1] Both these materials are electrically conductive. Furthermore, $MoSi_2$ has outstanding oxidation resistance in the range of 700°C–1700°C, and its electrical resistivity increases with temperature, making it the perfect heating element material for use in air or oxidizing furnaces.[2,3] The chronology of all historical events related to research and development of $MoSi_2$, starting from its invention in 1907, may be found in an earlier review by Vasudévan and Petrovic.[4] The idea of developing $MoSi_2$ as a structural material appears to have originated in the early 1950s.[5] For load-bearing applications, structural integrity and strength retention at elevated temperatures are as important as resistance to environmental degradation. The initial studies by Maxwell on mechanical properties identified intrinsic room-temperature brittleness and poor fracture toughness as the major shortcomings of $MoSi_2$.[5] Although interest in brittle materials was insignificant prior to the 1970s, it is possible to find the work of Fitzer on coatings of $MoSi_2$ and Mo_5Si_3,[6] as well as that of

Nowotny on phase-diagram calculations.[7] Interest in $MoSi_2$-based materials surged in the late 1980s after significant improvement in fracture toughness by dispersion of the ductile Nb-wire reinforcements was reported.[8] Besides $MoSi_2$, Mo_5Si_3 and multiphase ternary or quaternary alloys involving boron as alloying element have generated interest over the past decade.[9–12] The interest in development of Mo–Si–B alloys as a high-temperature structural material grew significantly after Berczik was awarded a patent on a family of refractory Mo–Si(1.6–15.2 at.%)–B(0–39.4 at.%) alloys, which showed impressive oxidation resistance.[11] There are a few reviews[12–15] on development of Mo–Si–B alloys with an emphasis on compositions and processing, microstructural evolution, physical and mechanical properties, and oxidation behavior.

Information on structural silicides other than Mo-silicides can be found in a review article by Shah et al., where crystal structures, coefficient of thermal expansion (CTE) anisotropy, mechanical behavior, and oxidation resistance have been dealt with comprehensively.[16] Furthermore, information on processing, structure, and physical, mechanical and functional properties of different silicides can be found in the chapter by Kumar.[13] The development of refractory metal intermetallic composites (RMIC) for elevated temperature aero-engine applications has also been reviewed.[17,18] Some of the physical and mechanical properties, along with qualitative remarks regarding oxidation resistance of the silicides, are shown in Table 6.1.

The mechanical behavior of individual silicides depends on a variety of factors: (a) crystal structure, which influences types of nearest-neighbor atoms and their bonding as well as probable slip systems; (b) processing methods; (c) microstructural features, including grain size, structural defects, and oxide and impurity-based inclusions; (d) alloying additions, which influence crystal structure, nature of interatomic bonding, and microstructure; and (e) volume fraction, distribution, and morphology of brittle or ductile, continuous or discontinuous reinforcements. It is well known that five independent slip systems are required for plastic deformation of polycrystalline materials. However, even the single crystals of most of the silicides are brittle at room or intermediate temperature in at least some orientations. The poor symmetry of the crystal causes the magnitude of possible dislocation Burgers vectors to be large. As a result, formation of such dislocations is energetically unfavorable, and larger force is required for their movement. The brittle-to-ductile transition temperature (BDTT) of the single crystals depends on crystal structure, composition of the unit cell, and type of nearest neighbors along close-packed directions, as well as the nature of metal–silicon bonds, which are directly related to the Peierls stress. Interestingly, in spite of the brittleness observed in polycrystalline samples, single crystals of $MoSi_2$ with stress axis along selected soft orientations have shown evidence of plastic deformation at subzero or room temperatures.[19]

Besides the silicides of Mo, multiphase Nb-silicides have received considerable attention since the early 1990s. $Nb–Nb_5Si_3$ composites possess

TABLE 6.1

Property Data on Silicides of Mo, Nb, Ti, and Cr That Are Critical for Applications

Silicides	Melting Points (°C)	Density (g cm⁻³)	Young's Modulus (GPa)	BDTT (°C)	Fracture Toughness (MPa m$^{1/2}$)	Remarks on Oxidation Resistance
$MoSi_2$	2020	6.24	439.7[16,19,20]	SC: RT[21]	SC: 1.9–4[23]	Reliable: 700°C–1700°C[29]
				Poly: 1100–1300[22]	Poly: 2.5–4[24–28]	Pesting[30]
Mo_5Si_3	2180	8.24	323[31]	SC: 1250[32]	SC: 2–2.5[33] Poly: 2.9[34]	Poor pesting[35]
Mo_3Si	2025	8.9	295[36]	Poly: 1400[37]	Poly: 3.0[37]	Expected to be poor
Mo_5SiB_2	2160–2200	8.8	383[31]	SC: 1500[38]	SC: 2.0[38]	Good[39]
$Mo(Si,Al)_2$		6.2	370.7[40]	SC: 1100[41]	Not available	Good[42]
WSi_2	2160	9.86	467.9[16,20]	SC: 1100[23,43,44]	SC: 3.7[45]	Good, inferior to $MoSi_2$[4]
Ti_5Si_3	2130	4.32	156[46]	SC: 1200[47]	Poly: 2.1–3.2[46,48]	1200°C[49]
$NbSi_2$	1920	5.62	362.8[20]	SC: 400[50]	Not available	Pesting[51]
Nb_5Si_3	2484	7.16	188[52]		Poly: 1–3[53,54]	Pesting[55]
$CrSi_2$	1477	4.6	354.6[56]	SC: 800[57]	Not available	Good up to 1200°C[58,59]
Cr_3Si	1770	6.46	350[60]	Poly: 1200[61,62]	Poly: 1.863	No pesting Good up to 1200°C[64,65]

Note: SC, single crystal; poly, polycrystal.

impressive fracture toughness at room temperature, and are amenable to thermomechanical processing.[17,18] However, the resistance to oxidation is unsatisfactory, and many of the alloying additions are intended to improve the high-temperature oxidation resistance, along with optimization of the creep strength. The research on development of dual-phase and multiphase Nb-silicides has been reviewed periodically.[66–76] One also comes across a series of patents[77–80] related to the incremental developments on improvement of room-temperature fracture toughness, yield and fracture strengths, as well as oxidation resistance. The crystal structure of WSi_2 is similar to that of $MoSi_2$ and properties are only slightly different,[4] but it has attracted much less attention. Limited literature exists on the silicides of Ti[46–49] and Cr,[60–65] in which the elevated-temperature oxidation properties are particularly inferior to those of Mo-silicides.

This chapter illustrates the fundamentals regarding mechanical properties, including fracture toughness, flexural and compressive yield strengths, and compressive and tensile creep properties. A comparative assessment of properties, structure–property relationships, and mechanisms of deformation and oxidation is presented in this chapter.

6.2 Crystal Structure–Related Properties

Crystal structures, space groups, and lattice parameters of the molybdenum silicides and other silicides are shown in Tables 6.2 and 6.3, respectively. The atomic positions in most of these crystal structures may be found at

TABLE 6.2

Critical Data of Molybdenum Silicides, Based on Crystal Structure and Bonding

Molybdenum Silicides	Crystal Structure	Lattice Parameters (nm)	CTE at 27°C (K⁻¹)	Slip Systems
$MoSi_2$	Body-centered tetragonal $C11_b$ (tI6)	$a=0.3202$ $c=0.7845$	$\alpha_a=8.2\times10^{-6}$ $\alpha_c=9.4\times10^{-6}$ Anisotropy[33,83]: 1.15 Poly[4]: 7–10×10^{-6}	$\{010\}\langle100\rangle\{011\}\langle100\rangle$ $\{023\}\langle100\rangle\ \{1\bar{1}0\}\langle111\rangle$ $\{013\}\langle3\bar{3}1\rangle$[21,23,43,45,84,85]
Mo_5Si_3	Body-centered tetragonal $D8_m$ (tI32)	$a=0.959$ $c=0.487$	$\alpha_a=5.2\times10^{-6}$ $\alpha_c=11.5\times10^{-6}$ Anisotropy[33,86]: 2.2	$\{001\}\langle110\rangle\ \{100\}[001]$ $\{110\}[001]$[33,87]
Mo_3Si	Cubic A15 (cP8)	0.4892	3×10^{-6} [37]	$\{001\}\langle100\rangle$[36]
Mo_5SiB_2	Body-centered tetragonal $D8_1$ (tI32)	$a=0.6013$ $c=1.103$	$\alpha_a=7.1\times10^{-6}$ $\alpha_c=8.6\times10^{-6}$ Anisotropy[31]: 1.2 (RT) – 1.6 (1000°C) Poly[31]: 6×10^{-6}	$\{001\}\langle100\rangle\ \{112\}\frac{1}{2}\langle111\rangle$ $(010)[001]$[31,38,88]
$Mo(Si,Al)_2$	Hexagonal C40 (hP9)	$a=4.644$ $c=6.548$	Poly[89]: 10.4×10^{-6}	$(0001)\langle1\bar{2}10\rangle$[41]

Source: The information on structure has been taken from Powder Diffraction File, Inorganic Phases. International Centre for Diffraction Data, Swarthmore, PA; Villars, P. and L. D. Calvert, *Pearson's Handbook of Crystallographic Data for Intermetallic Phases*, 2nd edn, ASM International, Materials Park, OH, 1991.

Note: Poly, polycrystalline.

TABLE 6.3

Critical Data of Silicides Other than Those of Molybdenum, Based on Crystal Structure and Bonding

Silicides	Crystal Structure	Lattice Parameters (nm)	CTE at 27°C (/K)	Slip Systems
WSi_2	Body-centered tetragonal	$a = 0.3211$		$\{011\}\langle100\rangle$ $\{023\}\langle100\rangle$
	$C11_b$ (tI6)	$c = 0.7868$		$\{110\}\langle111\rangle$ $(001)\langle100\rangle^{90}$
Ti_5Si_3	Hexagonal	$a = 0.7444$	α_a: 5.1×10^{-6}	
	$D8_8$ (hP16)	$c = 0.5143$	α_c: 22.2×10^{-6} Anisotropy[52]: 4.1	
$NbSi_2$	Hexagonal	$a = 4.7971$		$\{0001\}\langle\bar{2}110\rangle^{50}$
	$C40$ (hP9)	$c = 6.592$		
Nb_5Si_3	Body-centered tetragonal	α phase:	α_a: 8.7×10^{-6}	
	α: $D8_1$ (tI32)	$a = 0.656$	α_c: 20.4×10^{-6}	
		$b = 1.187$	Anisotropy[52]: 2.5	
	β: $D8_m$	β phase: $a = 1.0$		
		$b = 0.507$		
$CrSi_2$	Hexagonal	$a = 0.4428$		$\{0001\}\langle\bar{2}110\rangle^{57}$
	$C40$ (hP9)	$c = 0.6363$		
Cr_3Si	Cubic	$a = 0.4556$	9×10^{-6} [60]	$\{001\}\langle010\rangle^{61}$
	$A15$ (cP8)			

Source: The information on structure has been taken from Powder Diffraction File, Inorganic Phases. International Centre for Diffraction Data, Swarthmore, PA; Villars, P. and L. D. Calvert, *Pearson's Handbook of Crystallographic Data for Intermetallic Phases*, 2nd edn, ASM International, Materials Park, OH, 1991.

http:/cst-www.nrl.navy.mil/lattice/struk/atype.html and http:/cst-www. nrl.navy.mil/lattice/struk/ctype.html. In most cases, single crystals of silicides have noncubic structure, are highly anisotropic, and show mechanical behavior differing significantly from that of polycrystalline silicides. The mechanical behavior of silicides is strongly dependent on the crystal structures, symmetries, and character of the interatomic bonds. The directionality of bonding changes with the type of crystal structure and the refractory metals involved.[91] In general, the metal–silicon electronic bonds are more directional in $C11_b$-structured silicides, compared with those having C40 and C54 structures. Between $MoSi_2$ and WSi_2, both of which have $C11_b$ structure, the directionality of metal–silicon bonds is stronger in the latter. Among the C40-structured silicides, the directionality of the metal–silicon bond is stronger in $CrSi_2$ when compared with that in $NbSi_2$. Tables 6.2 and 6.3 also present the CTEs of different silicides in single-crystalline as well as polycrystalline forms. As a result of strong directionality of chemical bonds, single crystals of silicides also possess varying degrees of anisotropy of CTE (Tables 6.2 and 6.3), which in turn leads to generation of thermal residual stresses during heating and cooling cycles. Microcracks generated by thermal stresses are also responsible for lower strength and fracture toughness (Table 6.1) of intermetallics with high CTE anisotropy.

6.3 Processing

The methods of processing of silicides have been reviewed in the past by Jeng and Lavernia,[92] Stoloff and coworkers,[93] and Yao et al.[94] The silicides are usually processed by various powder metallurgy or ingot metallurgy-based techniques. The methods of processing have a significant role to play in determination of impurity content and evolution of microstructure, which in turn strongly influence physical, mechanical, and oxidation properties. Therefore, special techniques have been devised to tailor the level of purity and control matrix grain size or sizes of dispersed phases.

Silicides and silicide-based composites can be synthesized by a variety of processes, including self-propagating high-temperature syntheses,[95–100] shock wave syntheses,[101] mechanical alloying,[102–106] plasma spray processing,[107–111] or *in situ* solid-state displacement reactions,[112–117] which give rise to powders or porous or dense compacts as finished products. The methods used for consolidation of powders include conventional hot pressing of mixtures of commercially available powders,[24–27] reactive hot pressing of elemental powders,[28,48,118–121] hot isostatic pressing,[122,123] the Ceracon process,[124] sintering in the absence of pressure,[125,126] liquid-phase sintering under pressure,[28,48] and spark plasma sintering,[127,128] as well as extrusion[129,130] and injection molding.[110] Multiphase Mo–Si–B alloys processed through an industrial technique used by Plansee[131] for refractory alloys, involving cold pressing of an intimate mixture of elemental powders, followed by pressureless sintering in a hydrogen environment and hot isostatic pressing, have demonstrated promising results.[132]

Ingot metallurgy processing methods include arc-melting,[133,134] induction skull melting,[70] XD™ (exothermic formation of dispersoid, trademark of Martin Marietta Corporation, Baltimore, MD) processing,[135–138] directional solidification,[139] single-crystal growth,[44,93,140] and investment casting.[75,76] Conventional methods of single-crystal growth such as Bridgman or Czochralski techniques cannot be used because of the high melting point of silicides. Hence, alternative processes involving electron-beam float zone, inductively coupled plasma pedestal growth, and induction float zone in atmospheric and pressurized chambers have been developed for producing single crystals of the refractory metal silicides.[44,93,140] As-cast or powder-processed silicides may be subjected to subsequent heat treatments, or deformation processing for microstructural refinement, which improve the mechanical properties further, particularly in multiphase Mo-silicides[130] or Nb-silicides.[54,139] In powder metallurgy processing of $MoSi_2$, the grain size increases with increase in the temperature of hot compaction,[24,141] as expected. It has been shown by Sadananda et al. that the activation energy for grain growth at low temperatures is 46 kJ mol^{-1}, which is much less than the 348 kJ mol^{-1} required at high temperature.[141] Such differences in the activation energies suggest surface or grain-boundary diffusion to be the dominant operating mechanism for mass transport at lower temperatures, and a lattice diffusion-controlled mechanism at higher temperatures. It will be seen in the course of this review that grain size strongly affects most of the mechanical properties.

Another factor affecting the mechanical properties of many of the silicides is the impurity content, particularly oxygen content, leading to formation of SiO_2 in the microstructure. It has been observed that *in situ* reaction sintering or reactive hot pressing of elemental powders is self-purifying, leading to a much lower oxygen content in the product than in the elemental powders used as raw materials.[28,97,98] The oxygen content of the *in situ* reaction-processed $MoSi_2$ has been found to be less than that of the elemental powders. Processes involving the *in situ* precipitation of reinforcement phases through a displacement reaction have been developed,[112-117] through which high-temperature creep strength[141,142] and fracture toughness[112,113,115,116,143,144] may be improved by simultaneous reduction of the SiO_2 content. Some of the examples of the displacement reactions resulting in reduced SiO_2 content are[112-115,141-144]:

$$SiO_2 \text{ (s)} + 3C \text{ (s)} \rightarrow SiC \text{ (s)} + 2CO \text{ (g)} \tag{6.1a}$$

$$SiO_2 \text{ (s)} + Al \text{ (l)} \rightarrow Al_2O_3 \text{ (s)} + Si \text{ (s or l)} \tag{6.1b}$$

The Si may remain in elemental form, or it may react with a small fraction of Mo_5Si_3 present in $MoSi_2$ powder as

$$Mo_5Si_3 \text{ (s)} + 7Si \text{ (l or s)} \rightarrow 5MoSi_2 \text{ (s)} \tag{6.2}$$

Excellent mechanical properties have been observed in the $MoSi_2$–SiC composites processed by displacement reaction of Mo_2C and Si[113]:

$$Mo_2C + Si \rightarrow 2MoSi_2 + SiC \tag{6.3}$$

In a similar manner, Ti_5Si_3 can be processed by starting from a mixture of Si and TiH_2 powders. Ti-sponge is less expensive than Ti powder and hence may be hydrided to form TiH_2, which can be then easily ground to a fine powder. The disadvantage of this process is that the impurity content of the

Ti-sponge is retained in the final product. The reactions involved in this process are[48]:

$$TiH_2 \rightarrow Ti + H_2 \qquad (6.4a)$$

$$Ti + Si \rightarrow Ti_5Si_3 \qquad (6.4b)$$

A large number of innovative powder metallurgy techniques have been developed to fabricate hybrid composites and functionally graded materials with optimized properties, the details of which may be found in a review by Stoloff.[93] Tubular components of laminated composites of $MoSi_2$ and Al_2O_3 have been fabricated by plasma spray processing.[111] The disadvantage of plasma spray processing is that it usually leads to the formation of $MoSi_2$ powders with high oxygen content, and this can be reduced by carrying it out in vacuum or an inert environment.[92]

6.4 Mechanical Behavior

The mechanical properties of the silicides, which are evaluated at room temperature, include fracture toughness and flexural strengths. At elevated temperatures, the capacity of strength retention is of interest, and this is evaluated in terms of yield strengths from flexural or compression tests as well as tensile or compressive creep rates. Moreover, high-temperature fracture toughness has also been reported for some of the selected silicides.

6.4.1 Fracture Toughness

Fracture behavior of silicides has been studied widely with the objectives of understanding the mechanisms of failure and devising possible ways of tailoring the microstructure to further improve the fracture toughness of some of the most brittle materials. Poor fracture toughness, which is also responsible for poor fatigue properties, is one of the key handicaps in the dynamic load-bearing applications of silicides.

6.4.1.1 Molybdenum Silicides

The room-temperature fracture toughness of single crystals of $MoSi_2$ and WSi_2 as well as their alloys has been reported to be in the range of 2.5–3.5 MPa $m^{1/2}$, respectively.[23] Wade and Petrovic studied the effect of hot pressing temperature on the fracture toughness of $MoSi_2$, which was found to vary within a range of 2.0–3.5 MPa $m^{1/2}$.[24] However, with improvement in processing techniques, polycrystalline $MoSi_2$ has exhibited slightly higher fracture toughness, ranging between 3 and 5 MPa $m^{1/2}$ at room temperature.[10,14,24–28,145] The lower fracture toughness of $MoSi_2$ has been attributed to its inherent brittleness due to the covalent character and directionality[91] of the Mo–Si bonds and the presence of a large number of cleavage planes in this material. In coarse-grained $MoSi_2$, the indentation cracks[146] and fracture surfaces[28] are found to be largely transgranular as a result of abundance of cleavage planes. Due to the prevalence of this type of fracture mechanism, the room-temperature fracture toughness shows a modest increase with decreasing grain size.[28,141] Moreover, the fracture surfaces of fine-grained $MoSi_2$ are found to be mostly intergranular, which involves a crack path more tortuous than in the transgranular fracture observed in the case of coarse-grained $MoSi_2$. Generally, the room-temperature fracture toughness of the brittle materials can be moderately improved by increasing the tortuosity of crack paths, which in turn raises the energy of fracture. Petrovic has reviewed the strategies of improving the fracture toughness of $MoSi_2$-based materials by the use of reinforcements or alloying.[147] Improvement in fracture toughness has been achieved in $MoSi_2$-based composites with brittle ceramic[9,10,24–28,145,146–161] or refractory metal reinforcements.[162–173] The mechanisms of toughening, commonly observed in the composites, include crack closure, bridging, branching, and deflection due to brittle reinforcements (Figure 6.1) as well as crack-tip blunting and ligament bridging by ductile reinforcements. The work hardening of ductile reinforcements prior to their fracture has been found to be responsible for nonlinear load–displacement plots.

Reinforcing the $MoSi_2$ matrix with brittle reinforcements, such as SiC particles and whiskers, Si_3N_4 particles, or Mo_5Si_3, is found to improve the fracture toughness to approximately 5–7 MPa $m^{1/2}$.[9,10,22–26,145–161] Selected results indicating increase in fracture toughness and flexural strength on addition of brittle reinforcements are shown in Table 6.4. It has been reported that

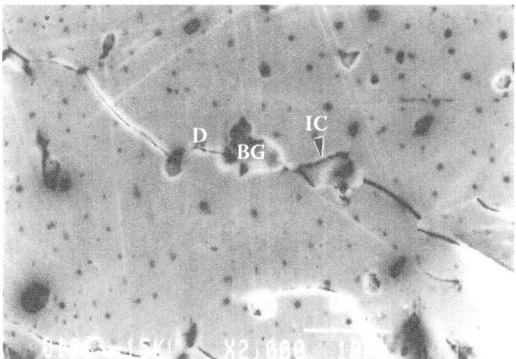

FIGURE 6.1
SEM image depicting the toughening mechanism in the *in situ* $MoSi_2$–Al_2O_3 composite ($MoSi_2$–5.5 at.% Al): crack deflection (D), crack bridging (BG), and interface cracking (IC). (Reprinted from *Intermetallics*, 7, Mitra, R., et al., Effect of small aluminum additions on microstructure and mechanical properties of molybdenum di-silicide, 213–232, Copyright (1999), with permission from Elsevier.)

TABLE 6.4

Room-Temperature Fracture Toughness Data and Methods of Testing of $MoSi_2$ and $MoSi_2$ Matrix Composites with Brittle and Ductile Reinforcements

Material	Processing	Fracture Toughness ($MPa\ m^{1/2}$)	Fracture Toughness Test	Flexural Strength (MPa)	References
$MoSi_2$	Hot pressed	5.3	Chevron-	140–160	Gac and
$MoSi_2$–20SiC_w	(at 1625°C)	8.20	notched	310	Petrovic[25]
$MoSi_2$	Hot pressed	4.8	3-pt. single		Yang and
$MoSi_2$–20TiC	(at 1700°C)	5.0	edge notch		Jeng[148]
$MoSi_2$-20SiCw		6.4	bend	263	
$MoSi_2$	Hot pressed	0.6–4.8	4-pt. bend	300–475	Henager
$MoSi_2$–30 SiC	displacement reaction (Mo_2C+Si)	6.7	Chevron- notched		et al.[113,152]
$\rho=5.366\ g/cc$					
$\rho=5.499\ g/cc$		7.9			
$MoSi_2$	Hot pressed (1700°C)	2.6	Indentation		Petrovic et al.[149,150]
$MoSi_2$–30PSZ		6.6			
$MoSi_2$– 20ZrO_2 (U-YSZ)	Hot isostatic pressed (1700°C)	7.9			
$MoSi_2$– 10ZrO_2(S,U)– 10SiC		3.8–6.8		250–600	
$MoSi_2$–30–50 Si_3N_4	Hot pressed and HIPed	5.2	CNB (RT)		Hebsur and Nathal[156]
		35	(1400°C/Ar)		
$MoSi_2$–30 Si_3N_4– 30SCS–6		65	(1400°C/Ar)		
$MoSi_2$	HIPed	$W_R=57\ J\ m^{-2}$	Tensile		Lu et al.[162]
$MoSi_2$–Nb (laminated)		$W_R=264\ J\ m^{-2}$	testing		
$MoSi_2$–Nb (laminated)	Hot pressed (1400°C)		4 pt. bend Chevron		Xiao and Abbaschian[163]
Uncoated Nb		15.2±1.3	notched		
Al_2O_3 coated		14.0±1.5			
ZrO_2 coated		12.8±1.5			
$MoSi_2$	Low-pressure plasma spray	4.5±0.173	Chevron notched		Castro et al.[107,164]
$MoSi_2$–20 wt.% Ta		9.97±0.351	4 pt. bend		
$MoSi_2$	HIPed	$W=1.4\ J\ cm^{-2}$	3 pt. bend flexural tests		Alman and Stoloff[165]
$MoSi_2$–20%		$W=1.2\ Jcm^{-2}$			
Nb particles		2.4–12.3			
Short fibers		29.7–44.1			

optimum improvement in fracture toughness and R-curve behavior characterized by rise in fracture resistance with increasing crack length occurs in the case of the $MoSi_2$ matrix composites for a volume fraction of 20% SiC reinforcements.[159] Fracture toughness can also be remarkably improved by reinforcing with unstabilized or partially stabilized tetragonal ZrO_2.[149,150] The stress-induced tetragonal-to-monoclinic transformation of ZrO_2 contributes to toughening by localized volume expansion, which in turn leads to: (a) generation of compressive stresses enforcing crack closure, (b) formation of microcracks accompanied by energy dissipation, and (c) lowering of elastic modulus ahead of the crack tip. Moreover, partial transformation of ZrO_2 is known to occur at ≈1175°C during cooling from the temperature of processing. As this temperature is around the BDTT of $MoSi_2$, the compressive strains due to volume expansion are accommodated by a significant amount of dislocation generation and multiplication. A higher density of dislocations aids in reducing the BDTT. Significant R-curve behavior was obtained in the hybrid $MoSi_2$–SiC–ZrO_2 composites, and the fracture toughness was about three times higher than that of $MoSi_2$.[145]

Reinforcement of the composites with randomly oriented whiskers of 30 or 50 vol.% β-Si_3N_4 has improved the fracture toughness by a crack deflection mechanism to ≈5.8 MPa $m^{1/2}$ at ambient temperature.[156] More significant improvement has been obtained at high temperatures (1400°C in argon atmosphere), where fracture toughness of $MoSi_2$–30 vol.% Si_3N_4 and $MoSi_2$–30 SiC_f–30 Si_3N_{4p} hybrid composites has been found to be 35 and 65 MPa $m^{1/2}$, respectively. The plots of load against displacement have been found to be nonlinear, indicating the occurrence of noncatastrophic failure, with absorption of higher energy.

Alloying[45,142–144,174–177] and *in situ* formation of reinforcements[113–115,143,144] have been used as strategies for improvement of both room- and high-temperature fracture toughness. Waghmare et al. have calculated surface energies and generalized stacking fault energies using the quantum mechanics–based first-principles density-functional approach to give an idea of ductility or brittleness,[174] and a disembrittlement factor was evaluated for different alloying elements. Such theoretical calculations have predicted that the substitution of Mo sites by alloying elements such as Nb and V, and Si sites by Al, is expected to improve the ductility of $MoSi_2$, while alloying with Re is expected to enhance the brittleness. The results of calculations regarding the Nb–Si system agree well with those of Tanaka et al., showing a less directional character of the Nb–Si bonds in C40 $NbSi_2$.[91] Furthermore, alloying with Al[41,128,143,144,174–176] is expected to enhance the ductility of $MoSi_2$, as the covalent Mo–Si bonds are replaced by more metallic Mo–Al bonds. Alloying of polycrystalline or single-crystalline $MoSi_2$ by Al has resulted in reduction of hardness, implying solid-solution softening. However the improvement in fracture toughness has been found to be relatively modest, with the maximum value of 6.4 MPa $m^{1/2}$ being achieved on alloying with 9 at.% Al.[144] Moreover, single crystals of $Mo(Si,Al)_2$ have shown a BDTT higher than that of MoSi$_2$ (Table 6.1) and only one slip system being operative (Table 6.2). Alloying with W has also not helped to increase fracture toughness or ductility, even though the stacking faults were found to be more abundant.[45] Polycrystalline MoSi$_2$ alloyed with 1 at.% Nb has exhibited limited compressive plastic strain prior to failure at room temperature, which can be considered to be encouraging.[177] The improved ductility on alloying of MoSi$_2$ with Nb could be due to reduced directionality of interatomic bonding, as discussed above.

At elevated temperatures, the purity of grain boundaries plays an important role in the fracture behavior of polycrystalline specimens, as well as the grain size. For example, the fracture toughness of polycrystalline MoSi$_2$ decreases with increasing temperature if the intergranular SiO$_2$ content is high. As the SiO$_2$ is converted to SiC reinforcements by *in situ* reaction with carbon, alloying with 2 wt.% C leads to a significant increase in the high-temperature fracture toughness, whereas the extent of improvement at room temperature is limited.[112,161]

The fracture-toughness values of MoSi$_2$-based composites with ductile reinforcements, which include discontinuously distributed lamellae, particles, or random short fibers[162–167] of Nb or Ta or continuous fibers, wires, or filaments[166–171] of Nb, Mo, W, and W–1% ThO$_2$, are shown in Table 6.4. Use of Nb foils in MoSi$_2$–Nb laminated composites has led to a significant increase in damage tolerance compared with that of monolithic MoSi$_2$.[166,167,172,173] It has been observed that the energy of fracture increases significantly if unconstrained plastic deformation of the ductile phase is permitted.[172,173] This observation is consistent with that of Ashby, who has earlier shown that the extent of ductility of the phase incorporated into a brittle matrix and the ease of decohesion at the interface play an important role in enhancement of fracture toughness.[178] The chief obstacle to improvement of fracture toughness arises from the embrittling effect of the interfacial phases formed by either chemical reactions between matrix and reinforcement or interdiffusion at interfaces. An appropriate coating or barrier layer such as Y$_2$O$_3$, ZrO$_2$, or Al$_2$O$_3$ on the ductile laminates has been used by a few investigators[172,173] to minimize the occurrence of interfacial reactions between matrix and reinforcement. The fracture toughness, measured on the basis of the peak stress of the linear elastic part of the load–displacement plot (termed damage tolerance), for MoSi$_2$–Nb composites with uncoated or coated lamellae has been found to be in the range of 12–18 MPa m$^{1/2}$, which is a remarkable improvement compared with that of unreinforced MoSi$_2$. Furthermore, the work of rupture estimated from the area included within the plots of stress against displacement is reported to be significantly higher in composites having Al$_2$O$_3$ or ZrO$_2$ coatings on Nb lamellae.

The fracture toughness values of Mo$_5$Si$_3$, Mo$_3$Si, and Mo$_5$SiB$_2$ have been reported to lie between 2 and 3 MPa m$^{1/2}$ (Table 6.1),[33,34,37,38] primarily due to low symmetry of the crystal structures of these phases, which does not allow plastic deformation through dislocation formation and movement. The low fracture toughness of Mo$_5$Si$_3$ and Mo$_5$SiB$_2$ is also due to anisotropy of the

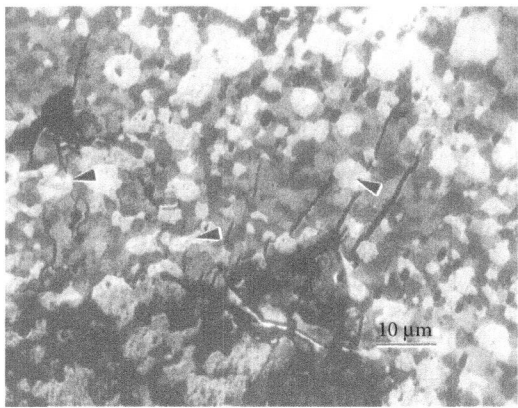

FIGURE 6.2
SEM image depicting an indentation crack path in 76Mo–14Si–10B alloy, where crack arrest and bridging due to the presence of α-Mo particles are observed. *Arrows* show α-Mo particles involved in crack bridging. (Reprinted from *Key Eng. Mater.*, 395, Mitra, R., et al., Effect of ductile and brittle phases on deformation and fracture behaviour of molybdenum and niobium silicide based composites, 179–192, Copyright (2009), with permission from Elsevier.)

CTE along the *c* and *a* axes, which contributes to internal residual stress in these materials.

In contrast to the behavior observed in the case of composites with single-phase intermetallic matrices, multiphase Mo–Si–B alloys having a dispersion of the ductile α-Mo phase have been found to be promising for achieving a significant gain in fracture toughness. Presence of a significant volume fraction (≥30%) of the α-Mo phase leads to toughening by crack arrest and bridging, because the plastic deformation of the ductile phase consumes significantly more energy (Figure 6.2). In general, fracture toughness of ≈7–15 MPa m$^{1/2}$ has been observed at room temperature in Mo–Si–B alloys.[12,15,179–188] Moreover, alloys with coarse α-Mo particles forming a more or less continuous network have shown a considerable improvement in fracture toughness to a range of 15–21 MPa m$^{1/2}$,[184] whereas R-curve behavior becomes predominant due to crack arrest by the α-Mo phase. It has also been observed that an alloy with 30 vol.% coarse α-Mo particles shows much higher fracture toughness compared with an alloy with about 50 vol.% fine α-Mo, suggesting the importance of the size of the dispersed ductile phase. The load–displacement plot obtained from a fracture-toughness test on single-edge notch-bend specimens of a 76Mo14Si10B alloy tested using a three-point bend fixture, as shown in Figure 6.3, has been found to be nonlinear with distinct evidence of work hardening of the ductile α-Mo phase, which in turn increases the total energy absorbed during failure.[185] Whereas solid-solution strengthening of α-Mo is achieved by minor alloying with Si, embrittlement also occurs due to its segregation at dislocation cores and grain boundaries.[15] Further decrease in ductility of Mo occurs due to interstitial alloying with

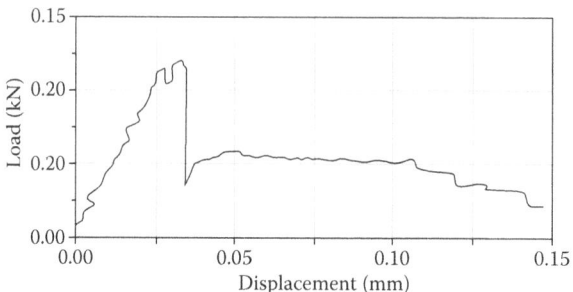

FIGURE 6.3

Load–displacement plot obtained on fracture toughness testing of 76Mo14Si10B alloy. (Reprinted from *Intermetallics*, 14, Mitra, R., et al., Microstructure and mechanical behaviour of reaction hot pressed multiphase Mo–Si–B and Mo–Si–B–Al intermetallic alloys, 1461–1471, Copyright (2006), with permission from Elsevier.)

O and C. Recently, some studies have shown that improvement in ductility of the α-Mo phase is obtained by either alloying with Zr or addition of $MgAl_2O_4$ particles.[15,186–188] It has been suggested that such additions lead to gettering of oxygen. Interestingly, the fracture toughness shows a marked improvement with increasing temperature, almost by 50% at 1300°C, because the ductility of the bcc-Mo solid solution increases and its flow stress decreases rapidly due to enhanced dislocation movement by thermal activation.

Table 6.5 shows the room-temperature flexural strength and fracture toughness of some of the Mo–Si–B multiphase alloys. The fracture toughness of the Mo–Si–B alloys exceeds that of the average $MoSi_2$–ceramic matrix composites, but it is found to be less than that of the $MoSi_2$–$Si_3N_{4(p)}$–SiC_f composite. One more observation to be noted is the consistently much higher flexural strength of the Mo–Si–B alloys compared with that of $MoSi_2$ and most other $MoSi_2$-based composites. The higher flexural strength is again attributed to the constrained plastic deformation involving higher-than-normal yield strength and work hardening of the α-Mo phase. However, further research is required for the purpose of improving both impact properties and fatigue strength of the Mo–Si–B alloys.[12] Choe et al. have found that the Mo–12Si–8.5B alloy shows a high crack-initiation toughness, and the threshold stress intensity factor is increased with temperature in a manner similar to that under static conditions.[36,183] On the other hand, the resistance to fatigue-crack growth has been found to be only modest in this composite.[183]

6.4.1.2 Niobium Silicides

The intermetallic Nb_5Si_3 is known to be brittle at both ambient and elevated temperatures. However, *in situ* composites with Nb solid solution (Nb_{ss}) and Nb_5Si_3 as the constituent phases have shown remarkable improvement in fracture toughness with respect to that of the latter phase, while retaining

TABLE 6.5

Room-Temperature Fracture Toughness and Flexural Strength of Mo–Si–B Multiphase Alloys Processed by Different Methods

Material	Processing	Fracture Toughness (MPa m$^{1/2}$)	Fracture-Toughness Test	Flexural Strength (MPa)	References
Mo9.4Si13.8B	Hot pressed	10	SENB	570	Liu et al.[179]
Mo–Si–B alloys	Arc melted			322–597	Schneibel et al.[180]
Mo14Si10B	Annealed at 1600°C for 24 h			630	
Mo10Si18B				604	
Mo12Si8.5B	Powder metallurgy	14.0–15.6	Chevron notched	376–412	Schneibel et al.[181]
Mo12Si8.5B	Ingot metallurgy, annealed at 1600°C	9.1–9.8		484–539	
Mo20Si10B	Arc melted, HIPed, 1600°C	13.5–15.1	Chevron notched		Schneibel et al.[182]
Mo12Si8.5B	Arc melted	7.2	Disk-shaped CT		Choe et al.[36,183]
Mo16.8Si8.4B	Powder metallurgy	4.1			
Mo12Si10Nb8.5B	Arc melted	6.3			
Mo20Si10B with coarse α-Mo	HIPed, 1600°C	21	Disk-shaped CT		Kruzic et al.[184]
Mo14Si10B	Hot pressed at 1600°C	5.0±0.8	SENB	404±10	Mitra et al.[185]

Note: Methods of fracture-toughness tests are also shown.
SENB, single edge notch bend; CT, compact tension.

high strength at elevated temperatures.[65–78,189–197] The flexural-strength and fracture-toughness values of selected Nb-silicides, along with the bulk alloy composition as well as the method of experimentation, are shown in Table 6.6. Typical indentation cracks in the alloys containing Nb_{ss} and Nb_5Si_3 phases as constituents are shown in Figure 6.4.[197,198] The Nb_5Si_3 particle shown in Figure 6.4a is cracked, while the neighboring primary Nb_{ss} phase does not show any evidence of cracking, because this crack is blunted and arrested at the Nb_{ss}/Nb_5Si_3 interface. On the other hand, Figure 6.4b shows bridging of the indentation cracks by a secondary Nb_{ss} phase present as a constituent of a nonlamellar eutectic.

As expected, the fracture toughness increases with increasing volume fraction, tensile yield strength, and size of the ductile particles. It has also been reported[76] that the primary Nb_{ss} shows almost three times higher microhardness than the nominally pure Nb, which is attributed to the constrained

TABLE 6.6

Room-Temperature Mechanical Properties of Nb-Silicides Processed by Different Methods

Material	Processing	Fracture Toughness (MPa m$^{1/2}$)	Mechanical Tests	Flexural Strength (MPa)	References
Nb–10 Si	As cast	8.9–9.5	SENB		Mendiratta et al.[54]
	As cast, 1500°C: 100 h	9.6–9.7	3 pt. bend tests	480	
	Extruded	16.4–16.5			
	Extruded, 1500°C: 100h	20.3–21.1			
Nb–16 Si	As cast	5.4			
	As cast 1500°C: 100h	6.4–8.1		500	
Nb–16.5 Si	As cast, hot extruded	12.6	SENB	670	Mendiratta, Dimiduk[189]
Nb–Ti–Hf–Cr–Al–Si	As cast	18.2–23.3	SENB, tensile	820	Bewlay et al.[139]
Nb–10 Si	As cast, hot extruded	24.1±2.2	SENB	638±104	Rigney, Lewandowski[190]
Nb(10–22)Si	Arc-melting or directional solidification	4–12	SENB		Kim et al.[196]
Nb(5–22)Si5Mo		10–15			
Nb(10–22)Si15Mo		6–8			
Nb19.1Si5.2Mo	Arc-melting	8.8±0.8	SENB		Chattopadhyay et al.[197]
Nb17.9Si26.3Mo		5.5±0.2			
Nb12.8Si4.1Mo		6.6±0.3			
Nb12.3Si14.8Mo		5.6±0.5			
Nb–Si–Al–Ti	Arc-melted, (Nb$_{ss}$, Nb$_3$Al, Nb$_5$Si$_3$)	10	SENB		Murayama and Hanada[191]
Nb10W10SixB	Arc melted, 1597°C, 100 h	8–9	SENB, tensile σ_{YS} σ_{UTS}	840–935 1300– 1450	Sha et al.[192]
Layered: Nb–Nb$_5$Si$_3$	Field-activated reaction and densification	21–30	Microindentation		Heian et al.[193]

plastic deformation of the former phase. The constraint to plastic deformation is imposed by the surrounding undeformable Nb$_5$Si$_3$ phase. Fracture begins by microcrack initiation in the brittle Nb$_5$Si$_3$ phase, followed by partial decohesion, stretching, and work hardening of the ductile Nb$_{ss}$ phase involved in crack bridging. However, due to incomplete decohesion and plastic constraint,

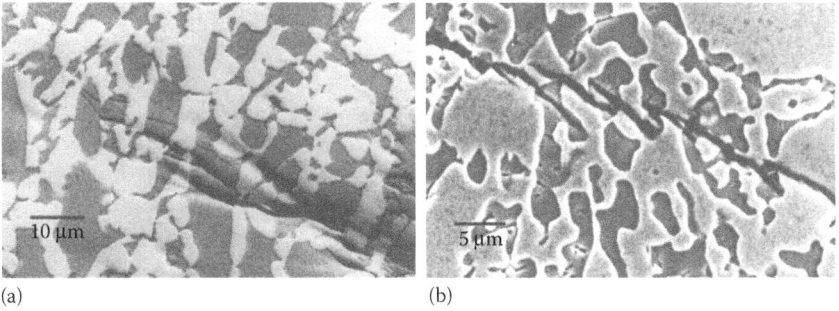

(a)   (b)

FIGURE 6.4
SEM (BSE) images depicting paths of indentation cracks through Nb–Si–Mo alloy, where (a) crack arrest and (b) bridging by the ductile Nb_{ss} phase ligaments are observed. BSE, backscattered electrons; SEM, scanning electron microscopy. (From Mitra, R., K. Chattopadhyay, A. K. Srivastava, K. K. Ray, and N. Eswara Prasad., *Key Eng. Mater.*, 395, 179–192, 2009. Reprinted with permission from Trans Tech. Publications.)

the majority of the small Nb_{ss} particles show cleavage fracture, while the large primary Nb_{ss} phase shows evidence of extensive plastic deformation. Therefore, it is possible to achieve a more significant rise in fracture toughness, peaking at around 30 MPa $m^{1/2}$ in the Nb-Nb_5Si_3 laminated composites,[193] compared with that in the case of the discontinuously reinforced composites.

Nb-based RMIC have a few distinct advantages when compared with those based on Mo_{ss}, particularly because the former material has been found to possess lower densities and higher damage tolerance at room temperature. The densities of the Nb-based RMICs lie in the range of 6.6–7.2 g cm^{-3}, while those of the Mo-based RMICs are in the range of 8.6–9.4 g cm^{-3}. The difference in fracture toughness manifests itself in a ductile metal-type fatigue behavior in Nb-RMICs and a more brittle fatigue behavior of Mo-RMICs.[72]

It has been noticed that the fracture toughness of Nb–Si alloys can be improved significantly by selecting suitable alloying elements and processing conditions. For example, the Nb–10 Si (at.%) alloy has shown fracture toughness of ≈9.6 MPa $m^{1/2}$ in the arc-melted condition, but in the range of ≈14–22 MPa $m^{1/2}$ on being processed by directional solidification followed by extrusion and heat treatment at 1500°C for 100 h.[76] Furthermore, Rigney and Lewandowski have shown R-curve behavior for this material, which is characterized by a rise in stress intensity factors from 8 to 36 MPa $m^{1/2}$ with increasing crack length.[190] The fracture toughness of the as-cast Nb–Si alloy is lowered to ≈5.4 MPa $m^{1/2}$ on increasing the Si content to about 16 at.%, because the volume fraction of Nb_5Si_3 is raised. However, on annealing at 1500°C for 100 h, the fracture toughness of this alloy has been found to exhibit a modest increase to ≈7.5 MPa $m^{1/2}$.[192] The fracture toughness can be further increased to 9 MPa $m^{1/2}$ by alloying with W and B,[73,134] whereas addition of Ti, Cr, and Al improves the fracture toughness significantly to 13–16 MPa $m^{1/2}$.[74] Further improvement in fracture toughness to the range of 18.2–23.3 MPa $m^{1/2}$ has been reported for alloys containing Hf.[134] Therefore, multicomponent Nb–Si-based alloys are considered to be promising for engineering applications.

The solid solubility of Mo is higher in Nb_{ss} than in Nb_5Si_3, and leads to significant solid-solution strengthening.[197,198] Interestingly, the thickness of the eutectic Nb_{ss} phase in Nb–xSi–5Mo alloy has been found to be more than that in the corresponding binary Nb–xSi alloy, which leads to higher fracture toughness in the former material. On the other hand, alloying with 15 at.% Mo has led to reduction in the fracture toughness. The improvement in fracture toughness on addition of 5 at.% Mo to near-eutectic compositions is most obvious after an extended annealing treatment of the as-cast specimens, which probably allows the state of equilibrium to be reached. For alloys with 15 at.% Mo, higher fracture toughness could be achieved in arc-melted samples having a maze-like microstructure with thick Nb_{ss}, compared with that in the directionally solidified eutectics.[196]

6.4.1.3 Silicides of W, Ti, and Cr

The fracture-toughness values of monolithic WSi_2,[45] Ti_5Si_3,[46,48] and Nb_5Si_3[53] are very poor, in the range of $\approx$1.5–4.0 MPa m$^{1/2}$. Single crystals of WSi_2 tested with the [001] direction parallel or perpendicular to the notch during four-point bend tests have shown fracture toughness of $\approx$3.1 and 3.7 MPa m$^{1/2}$, respectively.[45] This study showed a negligible effect of the presence of stacking faults in the (001) plane on the fracture toughness of WSi_2. Similar fracture-toughness values could be estimated from tests on single crystals of the $(Mo,W)Si_2$ alloys as well. Fracture toughness does not increase much on alloying of the aforementioned intermetallics, because the directionality of the metal–silicon covalent bond is hardly altered.

The room-temperature fracture toughness of Ti_5Si_3 has been found to depend on grain size, which in turn depends on processing conditions. It is found to vary from 2.1 MPa m$^{1/2}$ for the coarse-grained (20–50 μm size) microstructure[46] to 2.7 MPa m$^{1/2}$ or 3.2 MPa m$^{1/2}$ for relatively fine-grained (5–6 μm size) samples.[48,199] Furthermore, R-curve-type behavior is also observed, suggesting improvement in fracture resistance with reduction in the grain size. While the fracture toughness of the coarse-grained (20–50 μm) microstructure has been reported to increase to 7.5 MPa m$^{1/2}$ at 800°C,[46] it increases to about 12 MPa m$^{1/2}$ for grain size in the range of 5–6 MPa m$^{1/2}$.[200] Therefore, it can be inferred that reduction in grain size is more helpful for increasing fracture toughness at elevated temperatures than at ambient temperature.

The major problem with Ti_5Si_3 is its high CTE anisotropy along the *c* and *a* axes (as shown in Table 6.3), which results in significant internal stresses inside the grains during cooling from the temperature of processing, and microcracking.[16,34,48,52] It has been conclusively shown that reduction of grain size in Ti_5Si_3 significantly lowers the adverse impact of CTE anisotropy,[34] as the possibility of microcrack formation due to thermal stresses is reduced, and therefore both hardness and fracture toughness are found to increase.

Moreover, both hardness and fracture toughness of polycrystalline Ti_5Si_3 could be noticeably improved either by alloying or by adding reinforcements.[34,48,96] The positive effect of alloying on reduction of both directionality of interatomic bonding and CTE anisotropy has been demonstrated. For example, alloying of Ti_5Si_3 with Nb[52,201] has resulted in a significant reduction in the CTE anisotropy, along with a reduced propensity for microcrack formation. Furthermore, the presence of ceramic reinforcements leads to R-curve behavior through crack deflection and bridging.[48] Figure 6.5a and b compare the indentation cracking pattern observed in Ti_5Si_3 and Ti_5Si_3–TiC composite. The composite material shows greater frequency of crack deflection at higher angles due to the presence of TiC reinforcement, and therefore the crack path appears to be highly tortuous (Figure 6.5c). Further improvement in fracture toughness can be achieved in this material by

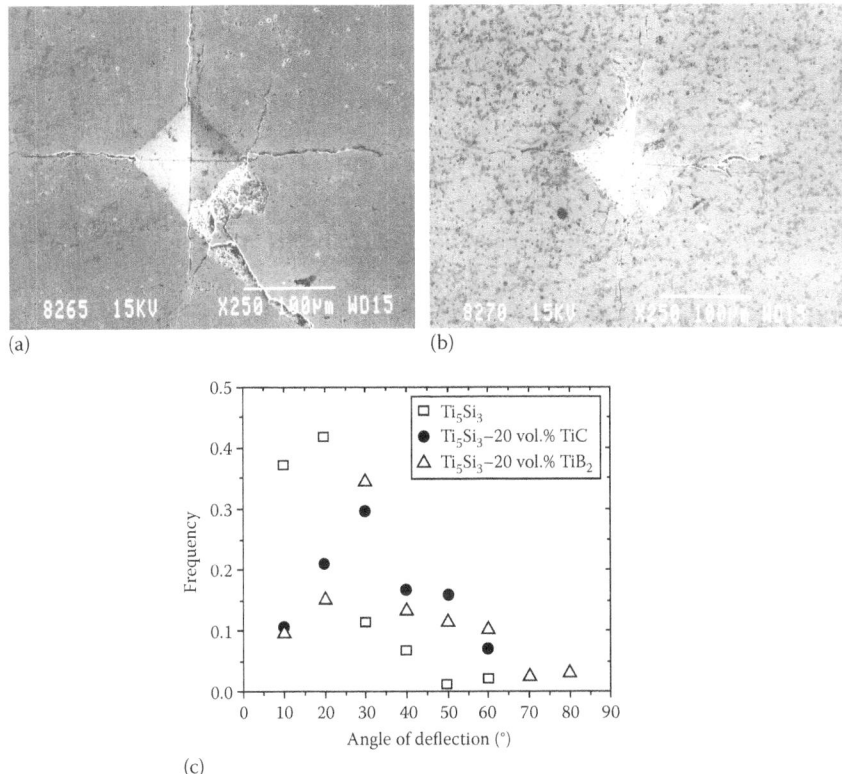

(a) (b)

(c)

FIGURE 6.5
Indentation cracking: (a) Ti_5Si_3, (b) Ti_5Si_3–20 vol.% TiC composite, and (c) variation of crack deflection frequency with angle. (With kind permission from Springer Science+Business Media: *Metall. Mater. Trans. A*, Elevated-temperature oxidation behavior of titanium silicide and titanium silicide based alloy and composite, 29, 1998, 1665–1675, Mitra, R. and V. V. R. Rao, copyright notice displayed with material.)

suitable design of microstructure, processing, and addition of ductile phase. For example, unidirectionally solidified Ti–Ti$_5$Si$_3$ composite produced by the electron-beam-zone melting technique has been reported to possess fracture toughness of $\approx$11 MPa m$^{1/2}$ at room temperature.[202]

Fracture toughness of Cr$_3$Si–Mo (Cr$_{40}$Mo$_{30}$Si$_{30}$) alloy containing a mixture of (Cr,Mo)$_3$Si and (Cr,Mo)$_5$Si$_3$ phases has been found to be between 2 and 3 MPa m$^{1/2}$,[64] which is only a modest improvement compared with that (1.8 MPa m$^{1/2}$) of the unalloyed Cr$_3$Si.[63] The fracture toughness of Cr$_3$Si–Al$_2$O$_3$ fiber composites has also been reported to be poor,[203] probably due to poor densification during processing. However, the Cr–Cr$_3$Si two-phase alloy has shown relatively higher fracture toughness (indentation) of 5.5–8.0 MPa m$^{1/2}$,[63] due to the positive contribution of the ductile metallic phase.

6.4.2 High-Temperature Deformation Behavior of Molybdenum Silicides at Constant Strain Rates

The high-temperature deformation behavior of MoSi$_2$, Mo$_3$Si, Mo$_5$Si$_3$, Mo$_5$SiB$_2$, and multiphase Mo–Si–B alloys has been discussed, considering the different mechanisms involved and mechanical properties observed.

6.4.2.1 Dislocation-Based Deformation Mechanisms in MoSi$_2$

The slip systems operating during deformation of single and polycrystalline MoSi$_2$ are shown in Tables 6.7 and 6.8, respectively, with the references in chronological order. Moreover, the Burgers vectors of the dislocations and the interplanar spacings of the observed slip systems are listed and analyzed in Table 6.9. The shortest lattice vectors, $\langle 100 \rangle$, ½$\langle 111 \rangle$, and $\langle 110 \rangle$, are the preferred Burgers vectors (*b*), as such dislocations have lower self-energies, while the Peierls stress for dislocation movement is reduced with increase in interplanar spacing (*d*) of the slip planes or the *d/b* ratio.

Single crystals of MoSi$_2$ exhibit plastic deformation when tested in compression even at room temperature,[21] while polycrystalline specimens show ductile behavior only at 1100°C or above and in some cases at 1300°C.[22] Such a sharp difference in the deformation behaviors of single and polycrystalline MoSi$_2$ is attributed to the absence of five independent slip systems in the latter type of material. In the case of anisotropic single crystals, the deformation behavior and the types of operating slip systems depend on the crystal orientation. The first systematic study of the slip systems operating in single crystals of MoSi$_2$ by Umakoshi and coworkers[84] has reported plastic deformation at temperatures greater than 1000°C, irrespective of crystal orientation. The crystal with [001] orientation, undergoing slip on the $\{01\bar{3}\}\langle 331 \rangle$ system, has been found to possess the highest critically resolved shear stress (CRSS), while other orientations involving slip on $\{1\bar{1}0\}\langle 331 \rangle$ have a lower

TABLE 6.7

Results of Experimental Research on Slip Systems in Single Crystals of Molybdenum Silicides in Chronological Order, with the Conditions of Mechanical Testing

Year	Operative Slip Systems	Test Temperature	Method of Testing and Orientations	References
1968	{100}⟨001⟩, {110}⟨001⟩	RT	Indentation	Vahldiek et al.[204]
1989	{013} and {110}½⟨3$\bar{3}$1⟩	900°C	Compression, near ⟨001⟩ and ⟨100⟩	Umakoshi et al.[205]
1990	{013} and {110}½⟨3$\bar{3}$1⟩	1100°C–1500°C	Compression, near ⟨001⟩ and ⟨100⟩	Kimura et al.[43]
1990	{013} and {110}½⟨3$\bar{3}$1⟩ ⟨100⟩ and ⟨110⟩ (glide, climb)	900°C–1200°C >1300°C	Compression, near ⟨001⟩ and ⟨100⟩	Umakoshi et al.[84]
1992	{110}⟨100⟩	RT	Indentation	Boldt et al.[206]
1993	{$\bar{1}$03}⟨331⟩ (screw), mixed: decompose into ½⟨111⟩, ⟨110⟩	1000°C	Compression	Maloy et al.[207]
1995	{013}⟨100⟩, {011}⟨100⟩, {1$\bar{1}$0}½⟨111⟩, {0$\bar{1}$3}½⟨331⟩	900°C–1600°C	Compression [001], [021], and [771]	Maloy et al.[208]
1995	{1$\bar{1}$0}⟨111⟩, {011}⟨100⟩, {010}⟨100⟩, {023}⟨100⟩, and {0$\bar{1}$3}⟨331⟩	−196°C–1500°C	Compression [001], [110], [221], [0 15 1]	Ito et al.[21]
1996	⟨100⟩ and ⟨111⟩, climb	1400°C	Compression [112]	Evans et al.[209]
2002	Up to 1200°C: {1$\bar{1}$0}⟨111⟩ Above 1200°C: {011}⟨100⟩	500°C–1500°C	Tensile [131]	Nakano et al.[210]
2002	Planar faults on (001) plane	400°C–1000°C	*In situ* tensile straining in TEM along [201]	Guder et al.[211]

value of CRSS. Based on an exhaustive study of the deformation behavior of MoSi$_2$ single crystals between −196°C and 1500°C, Ito et al. have reported that [001]-oriented crystals involving the activation of {0$\bar{1}$3}⟨331⟩ slip system deform only above 1300°C, while deformation could be observed at room temperature in crystals with other orientations.[21] It should be noted that the unit cell of MoSi$_2$ contains Mo–Si–Si–Mo covalent bonds along the [001] direction, which makes it harder to deform than those with the stress axis in other directions. The five slip systems are {1$\bar{1}$0}⟨111⟩, {011}⟨100⟩, {010}⟨100⟩, {023}⟨100⟩, and {0$\bar{1}$3}⟨331⟩ (Figure 6.6a–e),[21,23] and their operation depends on

TABLE 6.8

Results of Experimental Research on Slip Systems in Polycrystalline Molybdenum Silicides in Chronological Order, with the Conditions of Mechanical Testing

Year	Operative Slip Systems	Test Temperature	Experimental Method	References
1990	{010}, {011} and {013} $\langle100\rangle$	1200°C	4 pt. bending	Unal et al.[212]
	{1$\bar{1}$0}½$\langle111\rangle$, {1$\bar{1}$1}$\langle110\rangle$			
1992	Primary: {011}$\langle100\rangle$	RT–1000°C	Indentation	Maloy et al.[85]
	Secondary: {1$\bar{1}$0}½$\langle111\rangle$	>600°C		
1992	{0kl}$\langle100\rangle$, {1$\bar{1}$0}½$\langle111\rangle$	1200°C–1500°C	4 pt. bending, hot hardness	Mitchell et al.[213]
1993	{1$\bar{1}$0}½$\langle111\rangle$ dissociated into two ¼$\langle111\rangle$ and SISF	1400°C	Compression	Evans et al.[214]
1993	Dislocation networks of $\langle100\rangle$, ½$\langle111\rangle$, and $\langle110\rangle$	1400°C	Compression	Evans et al.[215]
1995	Mainly $\langle100\rangle$ dislocations on {013}, {011}, {001}, and {010}	900°C–1300°C	Compression	Campbell et al.[216]
1997	$\langle100\rangle$, ½$\langle111\rangle$, and $\langle110\rangle$ Unusual planes at 1400°C: climb.	1200°C, 1400°C, 1600°C 10^{-3}, 10^{-4}, 10^{-5} s^{-1}	Compression	Evans et al.[217]

the crystal orientation during loading. Moreover, the types of dislocations and slip systems involved in plastic deformation have been found to change with both temperature and strain rate. Umakoshi et al.[84] and Maloy et al.[208,209] have reported that the $\langle331\rangle$ dislocations gliding on the {013} planes account for deformation in single crystals with specific orientations (Table 6.9), while this could not be observed in other crystal orientations. The preference for the dislocation with relatively larger Burgers vector has been justified by Maloy et al. on the basis of its ability to dissociate into collinear partial dislocations, separated by antiphase boundaries (APB) or superlattice intrinsic stacking fault (SISF)[208,209]:

$$\tfrac{1}{2}\langle331\rangle \rightarrow \tfrac{1}{6}\langle331\rangle + \tfrac{1}{6}\langle331\rangle + \tfrac{1}{6}\langle331\rangle \tag{6.5}$$

The 1/6$\langle331\rangle$ partial dislocation has the shortest Burgers vector, $b = 0.2616$ nm. At 1000°C, the ½$\langle331\rangle$ dislocations with screw character remain stable, whereas the mixed dislocations dissociate. Furthermore, based on experimental observations, it has been argued that at higher temperatures

TABLE 6.9

Slip Systems Observed in Single and Polycrystalline MoSi$_2$

Slip Systems	Interplanar Spacing, d (nm)	Burgers Vector, b (nm)	b^2/a^2	d/b
{013}⟨100⟩	0.202	0.3204	1	0.630
{010}⟨100⟩	0.160	0.3204	1	0.499
{1$\bar{1}$0}⟨110⟩	0.226	0.4531	2	0.499
{1$\bar{1}$0}½⟨111⟩	0.226	0.4531	2	0.499
{011}⟨100⟩	0.099	0.3204	1	0.309
{1$\bar{1}$0}½⟨331⟩	0.226	0.7848	6	0.288
{110}⟨001⟩	0.226	0.7850	6	0.288
{0$\bar{1}$3}½⟨331⟩	0.202	0.7848	6	0.257
{100}⟨001⟩	0.160	0.7850	6	0.204

Source: Ito, K., et al., *Phil. Mag. A*, 72, 1075–1097, 1995; Maloy, S. A., et al., *Acta Metall. Mater.*, 40, 3159–3165, 1992; Vahldiek, F. W. and S. A. Mersol, *J. Less-common Met.*, 15, 2, 165–176, 1968; Umakoshi, Y., et al., *Phil. Mag. Lett.*, 59, 159–164, 1989; Boldt, P. H., et al., *Mater. Sci. Eng. A*, 155, 251–258, 1992; Maloy, S. A., et al., *Phil. Mag. Lett.*, 67, 313–321, 1993; Maloy, S. A., et al., *Acta Metall. Mater.*, 43, 657–668, 1995; Evans, D. J., et al., *Phil. Mag. A*, 75, 17–30, 1997; Nakano, T., et al., *Acta Mater.*, 50, 3731–3742, 2002; Guder, S., et al., *Philos. Mag.*, 82, 14, 2737–2754, 2002; Unal, O., et al., *J. Am. Ceram. Soc.*, 73, 1752–1757, 1990; Mitchell, T. E., et al., *Mater. Sci. Eng. A*, 155, 241–249, 1992; Evans, D. J., et al. In *High Temperature Ordered Intermetallics V*, Materials Research Society Symposium Proceedings, Vol. 288, 567–572, MRS, Pittsburgh, PA, 1993; Evans, D. J., et al., *Phil. Mag. Lett.*, 67, 331–341, 1993; Campbell, J. P., et al. In *High Temperature Ordered Intermetallics VI*, Materials Research Society Symposium Proceedings, Vol. 364, 893–898, MRS, Pittsburgh, PA, 1995; Evans, D. J., et al., *Phil. Mag. A*, 75, 1–15, 1997.

Note: Data on Burgers vector and interplanar spacings of different slip systems are shown.

(≥1000°C) the ½⟨331⟩ dislocation is unstable and dissociates into the following perfect dislocations[208,209]:

$$½⟨331⟩ \rightarrow ½⟨111⟩ + ⟨110⟩ \tag{6.6}$$

At temperatures ≥1200°C, ⟨100⟩ or ½⟨111⟩ dislocations have been found to predominate.[21,23,209] In contrast to the aforementioned observations, Ito et al. have reported the presence of [100] dislocations on (011), (010), and (023) slip planes even at room temperature.[21,23] The basis of such contradictory observations could be the fact that the character of operating slip systems is extremely orientation specific. At temperatures of ≥1300°C, edge dislocations with Burgers vector along ⟨100⟩, ⟨110⟩, and ⟨111⟩ have been observed in planes

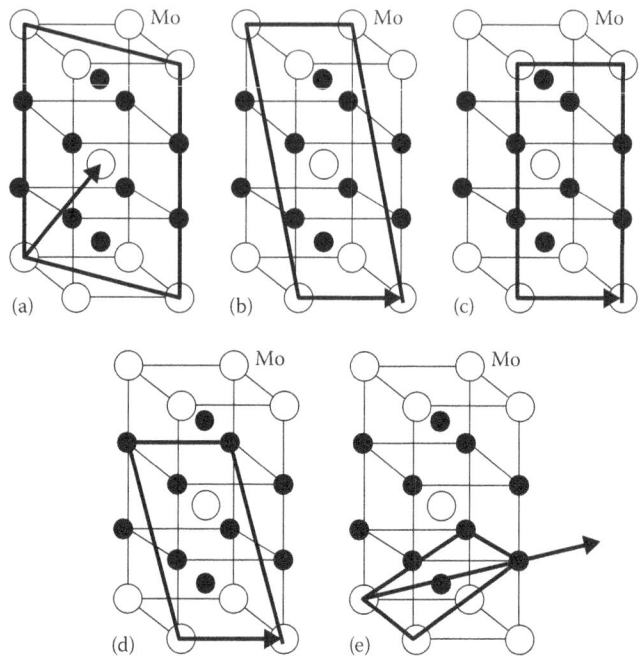

FIGURE 6.6

Slip systems observed in MoSi$_2$ single crystals: (a) $\{1\bar{1}0\}\frac{1}{2}\langle111\rangle$, (b) $\{011\}\langle100\rangle$, (c) $\{010\}\langle100\rangle$, (d) $\{023\}\langle100\rangle$, and (e) $\{0\bar{1}3\}\langle331\rangle$. (From Ito, K., et al., *Phil. Mag. A*, 72, 1075–1097, 1995.)

different from the regular slip planes, which is suggestive of the predominance of climb.[210]

The major slip systems operating during deformation of polycrystalline MoSi$_2$[214-217] are more or less the same as those for the single-crystalline samples, as discussed above. The only exception is that the presence of $\langle331\rangle$ dislocations in either dissociated or undissociated form has been reported quite rarely for polycrystalline MoSi$_2$.[209,218] One of the explanations for the rare observation of $\langle331\rangle$ dislocations in the polycrystalline MoSi$_2$ is the fact that the magnitude of $b=\frac{1}{2}\langle331\rangle$ is larger than that of other commonly found types of dislocations. The large magnitude of the Burgers vector of this dislocation is not only responsible for the high stress required for its activation in most grains, but also for its high self-energy, which in turn could make it unstable at temperatures $\geq1000°C$. As 1000°C is below the BDTT of polycrystalline MoSi$_2$, it is not possible to observe dislocations with $b=\frac{1}{2}\langle331\rangle$ in the microstructures of the deformed samples. At low temperatures, the primary operating slip system has been found to be $\{011\}\langle100\rangle$, while $\{1\bar{1}0\}\langle111\rangle$ becomes operative as the secondary slip system at temperatures $\geq600°C$.[85] Although the dissociation of dislocations has not been reported in the specimens deformed at 1200°C, the $\frac{1}{2}\langle111\rangle$ dislocations have been found to be dissociated in the specimens compression tested at 1400°C using a nominal strain rate of $10^{-3}\,s^{-1}$ following the reaction[213-215]

$$\frac{1}{2}\langle 111\rangle \rightarrow \frac{1}{4}\langle 111\rangle + SISF + \frac{1}{4}\langle 111\rangle \qquad (6.7)$$

Although the Von Mises criterion requires the operation of five independent slip systems, deformation involving slip on $\{0kl\}\langle 100\rangle$ and $\{1\bar{1}0\}\frac{1}{2}\langle 111\rangle$ systems accounts for only four independent slip systems up to 1000°C.[85] The fifth slip system, $\{0\bar{1}3\}\langle 331\rangle$, is activated in the grains oriented with the [001] direction along the stress axis only at temperatures ≥1300°C.[21] Hence, it has been argued that, in the absence of five independent slip systems, the compatibility of deformation of different grains during deformation is ensured by a combination of dislocation climb and grain-boundary diffusion.[216,217] TEM studies[218] on compression-tested MoSi$_2$ specimens have shown that the majority of the dislocations have a Burgers vector of $\langle 100\rangle$. Evans et al. have also shown that deformation at 1200°C involves dislocation glide at all strain rates, whereas at 1400°C or above, the contribution of climb is higher, which is supported by observation of the arrays of $\langle 100\rangle$ dislocations of edge character.[217] Even at 1400°C, glide has a role to play at higher strain rates, such as 10^{-3} s^{-1}. However, arrangement of the dislocations in low angle boundaries has been observed by Mitra et al. on compressive deformation of fine-grained MoSi$_2$ at 1200°C using a strain rate of 10^{-3} s^{-1} (Figure 6.7).[218] It is intuitive that

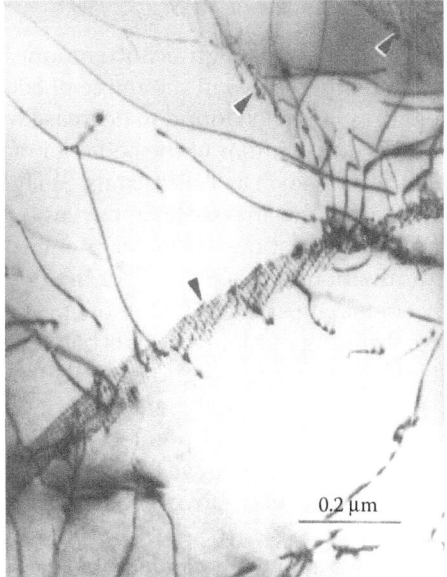

FIGURE 6.7
Dislocations arranged in low angle boundaries in MoSi$_2$ with an average grain size of 5 μm after compressive deformation at 1200°C. (With kind permission from Springer Science+Business Media: *Metall. Mater. Trans. A*, High-temperature deformation behavior of coarse- and fine-grained MoSi$_2$ with different silica contents, 34, 2003, 1069–1088, Mitra, R., et al., copyright notice displayed with material.)

occurrence of climb can occur at even lower temperatures in fine-grained $MoSi_2$, because of the role of grain boundaries as sources of vacancies.

Apparently, the BDTT of the polycrystalline $MoSi_2$ has also been found to depend on the oxygen content of the specimens, which in turn depends on the purity of raw materials and method of processing.[219-221] Patankar and Lewandowski have reported the BDTT of $MoSi_2$ processed by mechanical alloying and hot pressing to be $\approx1100°C$.[221] On the other hand, Aikin,[22] as well as Srinivasan, Schwarz and coworkers,[104,222,224] has reported a reduced BDTT of 1000°C in high oxygen-containing $MoSi_2$ and 1300°C in the case of high-purity $MoSi_2$. The variation in the BDTT with oxygen content as well as grain size is attributed to premature softening of the SiO_2 particles at the grain boundaries during high-temperature deformation.

The BDTT of $MoSi_2$ can be lowered with an increase in the amount of plasticity by using TiC particles as reinforcements in its composites.[222,223] Although the TiC is known to be brittle at room temperature, its BDTT is known to be in the range of 600°C–800°C, above which appreciable compressive deformation is possible.[222,223] As the TiC particles are deformed, stress concentration builds up at the $MoSi_2$–TiC interfaces, which act as sources for generation of dislocations inside the $MoSi_2$ grains.

6.4.2.2 Yielding and Flow Behavior of MoSi$_2$

$MoSi_2$ single crystals have shown a significant amount of yield-point phenomena at intermediate temperatures in the range of 800°C–1000°C,[21,23] and the magnitude of yield drop has been found to decrease with increasing temperature. This behavior is quite similar to the behavior of materials with low density of mobile dislocations, such as LiF crystals.[224] The sharp yield points are followed by serrations in the stress–strain curves, which are attributed to low density of mobile dislocations or lower mobility of dislocations in the $MoSi_2$ grains. Examination of the surfaces of the deformed crystals has shown formation of coarse slip bands.[21] Moreover, *in situ* dynamic TEM studies[225,226] involving straining of $MoSi_2$ crystals between 800°C and 1000°C have shown distinct evidence of viscous glide of ½⟨110⟩ dislocations on {1$\bar{1}$0} slip planes. It has been proposed that the energy of the dislocation core is reduced by the formation of clusters of point defects such as Si vacancies, the concentration of which increases above 800°C.[227] The point-defect atmosphere is dragged as the dislocation moves, so that its movement becomes discontinuous and involves jerks.

$MoSi_2$ single crystals also show an anomalous increase of CRSS with increasing temperature and a relatively higher rate of work hardening. Moreover, it is interesting to note that the temperature range for anomalous increase in yield stress and serrated plastic flow depends on the crystal orientations, as the nature of operating slip systems also varies.[21] The process of interaction of mobile dislocations with interstitial impurities or point-defect clusters is extrinsic in character, and is somewhat inadequate to explain why the temperature range of the yield stress anomaly would vary with the nature of operating slip systems. Hence, an intrinsic process involving the alteration of the dislocation core structure, accompanied by decomposition of perfect dislocations into partials, which occur in the different slip systems at different temperatures, has been proposed by Mitchell et al. Yield-point phenomena and serrated plastic flow have also been observed in polycrystalline specimens with interstitial impurities (C, O, N),[218] but only under specified conditions of temperatures and strain rates (Figure 6.8a).[228]

The elevated-temperature yield strengths of polycrystalline $MoSi_2$ depend on microstructural parameters such as grain size,[218,229–231] SiO_2 content,[28,218–220,230,231] type of alloying elements,[232–236] and presence of reinforcements,[26–28,237,238] as well as specific test conditions[218] including both temperature and strain rates. The high-temperature hardness has also been found to be

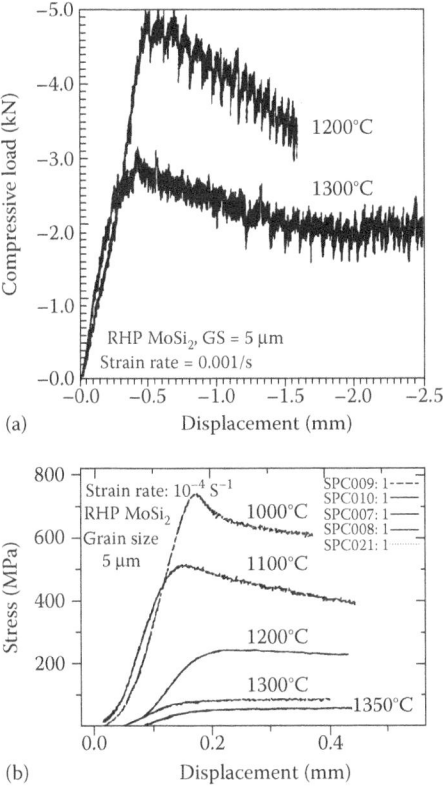

(a)

(b)

FIGURE 6.8

Stress–strain curves showing yield-point phenomena and serrated plastic flow in polycrystalline $MoSi_2$ with average grain sizes of 5 μm, tested at strain rate of (a) $10^{-3}\,s^{-1}$ and (b) $10^{-4}\,s^{-1}$. (With kind permission from Springer Science+Business Media: *Metall. Mater. Trans. A*, High-temperature deformation behavior of coarse- and fine-grained $MoSi_2$ with different silica contents, 34, 2003, 1069–1088, Mitra, R., et al., copyright notice displayed with material.)

influenced by grain size, SiO_2 content, and methods of processing.[28,131,206,230,231] Compression tests[218] in the temperature range of 1000°C–1350°C at nominal strain rates in the range of 10^{-3} to $10^{-4} s^{-1}$ on reaction hot-pressed (RHP) $MoSi_2$ with significantly low SiO_2 content and fine average grain size of 5 μm have shown lower yield stress compared with that processed by hot pressing of commercial-grade $MoSi_2$ powders, with a higher SiO_2 content and coarse average grain size of 27 μm at higher temperatures or reduced strain rates, and vice versa (Figure 6.8b). At lower temperatures or higher strain rates, the $MoSi_2$ follows Hall–Petch-type behavior because the grain boundaries act as barriers to dislocation motion. On the other hand, at higher temperatures or reduced strain rates, the effect of the thermally activated diffusion process is more predominant in the fine-grained specimens, probably because the grain boundaries act as sources and sinks for vacancies and dislocations.

In the course of high-temperature deformation, the relatively fine-grained RHP $MoSi_2$ has shown evidence of dynamic recovery, the extent of which is found to increase with increasing temperature (Figure 6.8a). This behavior is attributed to increased mobility of edge dislocations by climb, leading to the formation of low angle boundaries (Figure 6.7).[218] The thermally activated deformation of the fine-grained ($\approx$5 μm) RHP $MoSi_2$ is more obvious from the sharp increase in the strain-rate sensitivity from 0.2 at 1100°C to 0.63 at 1350°C and demonstration of increasing compressive ductility with increasing temperature (Figure 6.9). The coarse-grained ($\approx$27 μm) Starck $MoSi_2$ has shown a more brittle character, with the strain-rate sensitivity being $\approx$0.03 at 1100°C and $\approx$0.34 at 1350°C (Figure 6.9).

High-temperature tension tests[229] on $MoSi_2$ with grain sizes of 1 and 10 μm in the temperature range of 1127°C–1327°C and strain rates ranging between

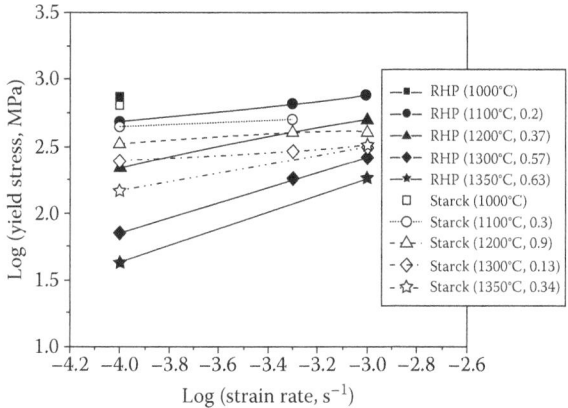

FIGURE 6.9
Variation of the logarithm of yield stress with logarithm of strain-rate sensitivity for RHP $MoSi_2$ (5 μm grain size) and Starck $MoSi_2$ (27 μm grain size). (Reprinted from *Intermetallics*, 12, Mitra, R., et al., Effect of microstructural parameters and Al alloying on creep behavior, 827–836, Copyright (2004), with permission from Elsevier.)

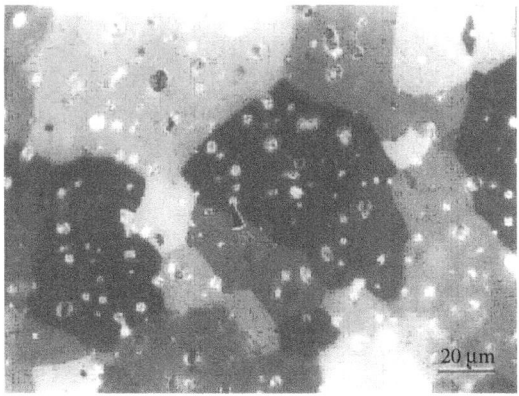

FIGURE 6.10
Polarized light optical micrograph of $MoSi_2$, showing the location of globular SiO_2 particles at grain interior and grain boundaries. (With kind permission from Springer Science+Business Media: *Mater. Sci. Eng. A*, Processing–microstructure–property relationships in reaction hot-pressed $MoSi_2$ and $MoSi_2/SiC_p$ composites, 225, 1997, 105–117, Mitra, R., et al., copyright notice displayed with material.)

10^{-5} and $10^{-3}\,s^{-1}$ have shown a trend similar to that observed in the compression tests on RHP and Starck $MoSi_2$.[218] $MoSi_2$ with 1 µm grain size has shown a high strain-rate sensitivity of 0.55, while coarse-grained samples showed a sensitivity of 0.12 at higher strain rates and lower temperatures, and 0.35 at lower strain rates at higher temperatures. Also, the elongation of $MoSi_2$ with 10 µm grain size is increased from 3% at 1127°C to 10% at 1327°C, while that of the sample with 1 µm grain size is increased from 5% at 1127°C to 33% at 1327°C, indicating the more ductile character of fine-grained specimens with increasing temperature. In spite of the high strain-rate sensitivity of the fine-grained specimens, superplasticity could not be observed in this material due to intergranular cavitation.

Intergranular cavitation in $MoSi_2$ is usually attributed to the presence of amorphous or glassy SiO_2 at the grain boundaries and triple points. The SiO_2 particles are globular to begin with, and do not wet the grain boundaries (Figure 6.10).[28,218–221,230,231,239] With increasing temperature, the SiO_2 particles soften, thereby leading to flow localization at grain boundaries and triple points. The SiO_2 undergoes viscous flow along the grain boundaries by shear and eventually crack (Figure 6.11). With increasing temperature, strain softening is initiated at relatively smaller plastic strains than at lower temperatures. It is intuitive that the presence of intergranular SiO_2 is more harmful in the case of fine-grained $MoSi_2$, because of the larger grain-boundary area in this material.

6.4.2.3 Effect of Alloying

Effects of alloying with W, Re, Cr, Nb, Zr, Ta, V, and Al (=M) on hardness and yield strengths of both single and polycrystalline $MoSi_2$ have been reported in the

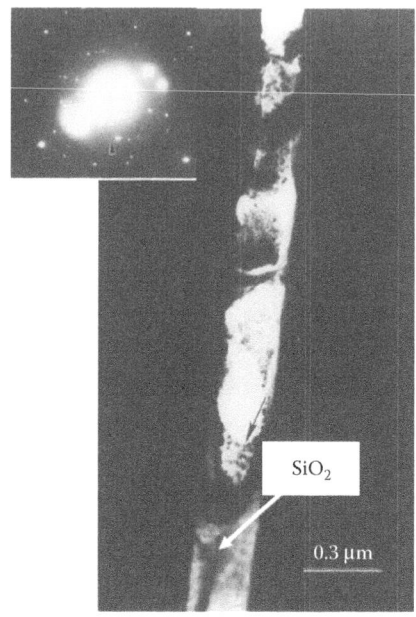

FIGURE 6.11
Intergranular cracking on softening of amorphous SiO_2 layer during deformation at 1300°C. A diffraction pattern depicting a diffuse halo (arrowed) as evidence of amorphous intergranular film is shown as an inset. (From Mitra, R., et al., *Metall. Mater. Trans. A*, 34, 1069–1088, 2003.)

literature.[114,175–177,232–236,240,241] With the exception of Al substituting in the Si site, all the aforementioned elements (=M) substitute in the Mo site to form a compound of the type $(Mo_{(1-x)}M_x)Si_2$. Interestingly, microhardness measurements carried out on samples of polycrystalline $MoSi_2$ and its ternary alloys by Harada et al. have demonstrated the hardness of only $C11_b$-structured $MoSi_2$–W or Re alloy to be higher than that of unalloyed $MoSi_2$.[234] In contrast, a decrease in microhardness has been observed at temperatures <700°C on alloying with elements responsible for the formation of either C49 structure (e.g., Zr) or C40 structure (e.g., Cr, Nb, Ta, V, and Al). Studies by Mitchell and Misra[235] on the effect of alloying with Re, and subsequently by Sharif et al.[236] on the effects of alloying with Al, Re, and Nb, have shown results similar to those of Harada et al.,[234] that is, increase of hardness on alloying with Re and softening on alloying with Nb or Al.

High-temperature compressive yield stress follows a trend mostly similar to that observed during microhardness tests at similar temperatures to those described above.[177,178,240,241] In compression tests of [0 15 1]-oriented single crystals between room temperature and 1500°C, Inui et al. have shown an anomalous increase in the yield stress with increasing temperature between 800°C and 1100°C.[240] Furthermore, at temperatures ≥1300°C, the yield strengths of all ternary Mo–Si–X alloys (X = V, Cr, or Nb) have been found to be higher than that of the unalloyed $MoSi_2$, with the exception of X = Al.

A particularly significant increase in the high-temperature yield strength has been noticed on alloying $MoSi_2$ with Cr and Nb.[217] However, in the case of [001]-oriented crystals tested at 1400°C and 1500°C, the maximum stress in the stress–strain curves of all the ternary $MoSi_2$-based alloys, including that containing Al as the alloying element, is greater than that obtained from the flow curve of the unalloyed binary $MoSi_2$.[176]

Minor alloying additions of Al to polycrystalline $MoSi_2$ have been found to increase the elevated-temperature yield strength to 1300°C, as the SiO_2 present in the raw materials is converted to Al_2O_3 by an *in situ* reaction during processing.[144,241] The strength is high until the matrix phase has an Al content of less than 3 at.%, which ensures that its $C11_b$ structure is retained, and conversion to C40 structure does not occur. Figure 6.12 shows the comparison of yield strengths of polycrystalline $MoSi_2$–Al alloys with those of $MoSi_2$ and single-crystal specimens.[241] The flow curves of $MoSi_2$–Al alloys have shown serrations, which is suggestive of dynamic strain-aging behavior. Furthermore, excess addition of Al leads to the formation of a high volume fraction of the C40 phase, $MoSi_{2-x}Al_x$ ($0.2 \leq x \leq 2$), which in turn has been reported to lower the yield strength significantly.

The plastic deformation behavior of C40-structured $Mo(Si,Al)_2$ single crystals has been studied by Inui et al.[41] The $Mo(Si,Al)_2$ phase showed plastic deformation only at temperatures $\geq$1100°C for orientations in which the $(0001)\langle11\bar{2}0\rangle$ basal slip with dislocations having $b=\frac{1}{3}\langle11\bar{2}0\rangle$ is operative. However, if the Al concentration is higher, as in the case of $Mo(Si_{0.85},Al_{0.20})_2$ crystals, plastic deformation is initiated at 1200°C instead of 1100°C, which is

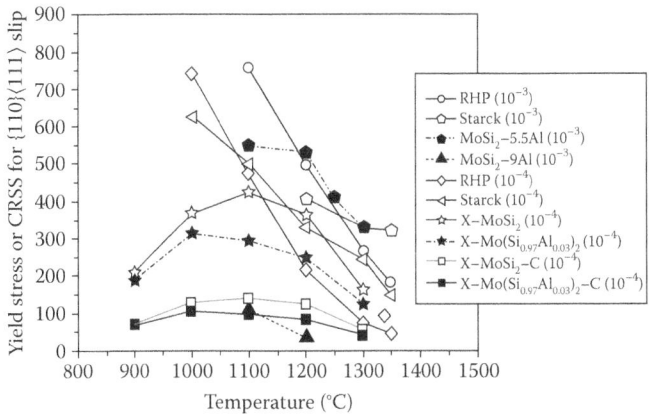

FIGURE 6.12
Variation of yield stresses of $MoSi_2$ and $MoSi_2$–Al alloys with temperature at strain rates of 10^{-3} and $10^{-4}\,s^{-1}$. Strain rates are shown in parentheses. The yield stresses of single crystals (marked as X in the legend) were obtained in the [0 15 1] orientation. The CRSS corresponds to the {1 10}⟨111⟩ slip system. (Reprinted from *Intermetallics*, 12, Mitra, R., et al., Effect of microstructural parameters and Al alloying on creep behavior, threshold stress and activation volumes of molybdenum disilicides, 827–836, Copyright (2004), with permission from Elsevier.)

followed by a decrease in the yield strength with increasing temperature. It has been proposed that the basal slip occurs through a *synchroshear* mechanism, involving the dissociation of the perfect dislocations into partials with Burgers vector of $1/6\langle 11\bar{2}0\rangle$, separated by a stacking fault. The spreading of the core of the perfect dislocation in the basal planes makes their cross-slip into the prismatic planes difficult. The synchroshear mechanism of deformation, found also in $CrSi_2$,[54,242] has been attributed to noncentral or directional bonding and low stability of the C40 phase compared with that of the $C11_b$ phase. C40 silicides deforming by the synchroshear mechanism show higher values of CRSS than those deforming by a more regular dislocation mechanism. Therefore, the yield stress has been found to decrease very sharply as the test temperature is raised from 1100°C to 1200°C.

Softening observed at low temperatures in C40-structured ternary $MoSi_2$ single crystals alloyed with V, Cr, and Nb, and strengthening found at high temperatures, can be explained on the basis of dislocation structures and stacking fault energies.[176–178,235–237] Transmission electron microscopy (TEM) studies have shown that dislocations with $b=\frac{1}{2}\langle 111\rangle$ dissociate into two identical partial dislocations with $b=\frac{1}{4}\langle 111\rangle$, separated by a stacking fault [Reaction 6.7]. The stacking fault energies on the {110} planes are much lower in the C40-structured ternary $MoSi_2$ alloy compared with those of the $C11_b$ alloy, which in turn promotes the dissociation of dislocations with $b=\frac{1}{2}\langle 111\rangle$. In the $C11_b$ $MoSi_2$ with [0 15 1] orientation, transition from $\{01\bar{3}\}\langle 331\rangle$ slip to $\{1\bar{1}0\}\langle 111\rangle$ occurs at around 400°C–500°C. However, such an alteration in the slip system occurs at a much lower temperature in C40-structured $MoSi_2$-based ternary alloys, which may explain the cause of softening at room temperature. Polycrystalline $MoSi_2$ alloyed with 1 at.% Nb has shown yielding and noticeable plastic strain prior to fracture in compression at room temperature.[236] This observation can be considered a significant development, because plastic deformation is absent in unalloyed, polycrystalline $MoSi_2$ at temperatures lower than 900°C. Observations of deformed $MoSi_2$–Nb alloy specimens using the TEM have shown an increase in the stacking fault width in the $\frac{1}{2}\langle 111\rangle$ dislocations (see the dislocation Reaction 6.7), suggesting that the stacking fault energy is lowered. Higher width of the stacking faults reduces the barrier for nucleation of kinks and their migration on the $\frac{1}{4}\langle 111\rangle$ partial dislocations and thereby promotes slip. The improvement in ductility on alloying with Nb agrees well with an earlier theoretical prediction by Waghmare et al.[174]

Inui et al. have reported an increase in yield strength of the $(Mo,W)Si_2$ alloy with increasing W content in solid solution through compression tests between room temperature and 1500°C.[240] The observations regarding the flow behavior of single crystals of $(Mo_{(1-x)}W_x)Si_2$ (where $x=0.25, 0.5, 0.75$) at different orientations may be summarized as follows: (i) the minimum temperature for the onset of plastic deformation in $(Mo_{0.5}W_{0.5})Si_2$ is 400°C–600°C, while that of $MoSi_2$ is room temperature; and (ii) the CRSS for slip on $\{011\}\langle 100\rangle$ and $\{1\bar{1}0\}\langle 111\rangle$ slip systems increases with increasing x in the $(Mo_{(1-x)}W_x)Si_2$ alloys.

For example, in [110]-oriented crystals, $(011)\langle 100 \rangle$ slip is operative at room temperature in $MoSi_2$, while the same slip system is not activated at temperatures less than 400°C in $(Mo_{0.5}W_{0.5})Si_2$ and $(Mo_{0.25}W_{0.75})Si_2$ alloys. The effect of W on the CRSS has been attributed to the increased interaction between W atoms and lattice dislocations, which in turn increases the Peierls stress at the dislocation cores. Also, the directionality[91] of the W–Si bonds has been found to be greater than that of the Mo–Si bonds, and hence substitution of the Mo sites in $MoSi_2$ by W atoms increases the BDTT significantly. Further information is provided during the discussion on deformation behavior of WSi_2 in Section 6.4.4.1.

6.4.2.4 Effect of Reinforcements

The addition of ceramic reinforcements to $MoSi_2$ is helpful for retention of strength at elevated temperatures (Figure 6.13). Aikin has reported significant improvement in compressive yield strengths of XD™ $MoSi_2$-based composites with 15–45 vol.% SiC, TiB_2, ZrB_2, and HfB_2 reinforcements with respect to those of $MoSi_2$.[237,238] The yield strength increases with increasing volume fraction of ceramic reinforcements. The compressive yield strength has been found to be a function of particle size, volume fraction, and distribution of particles, which could be reduced to only one variable of importance, that is, interparticle spacing. Studies using the TEM and correlation of interparticle spacing with yield stress have shown that the dispersed ceramic particles act as obstacles to the motion of dislocations, and smaller interparticle spacings reduce the mean free path for dislocation movement,[238] which results

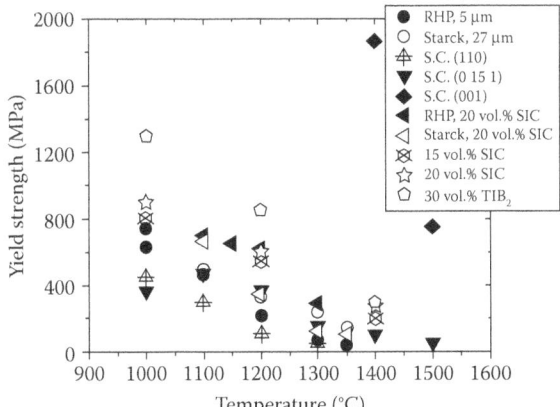

FIGURE 6.13
Comparison of the elevated-temperature yield stress of the $MoSi_2$-based composites with that of $MoSi_2$. (From Mitra, R., et al., *Mater. Sci. Eng. A*, 225, 105–117, 1997; Miller, M. K., et al., *Scr. Mater.*, 46, 299–303, 2002; Mitra, R., et al., *Metall. Mater. Trans. A*, 34, 1069–1088, 2003; Aikin Jr, R. M., *Mater. Sci. Eng. A*, 155, 121–133, 1992; Inui, H., et al., *Mat. Sci. Eng. A*, 261, 131–138, 1999.)

in strengthening. Addition of 2–4 wt.% C,[112] which leads to the formation of a high volume fraction of SiC by displacement reaction with SiO_2, enhances the high-temperature hardness as well as the yield strength. Significant improvement in high-temperature strength retention has also been observed in RHP $MoSi_2$–SiC composites.[26,154] Moreover, composites having equimolar $MoSi_2$–WSi_2 polycrystalline alloy as matrix and 20 vol.% SiC whiskers as reinforcements have shown yield strength approximately 8–10 times greater than those of $MoSi_2$.[243]

The flexural strength of $MoSi_2$ matrix composites with 15 vol.% SiC particles of 200–300 nm size has been found to be 760±47 MPa at room temperature, which is about five times more than that of unreinforced $MoSi_2$ (160±8 MPa).[156] The impressive gain in room-temperature strength and its retention at high temperature in the $MoSi_2$–SiC nanocomposites can be explained on the basis of uniform and small interparticle spacings. Furthermore, it has been possible to retain flexural strengths of 1.0 GPa until 1100°C by reinforcing $MoSi_2$ and $Mo(Si,Al)_2$ with 15 vol.% 50–200 nm size SiC particles, which restrict grain growth. Such a strengthening mechanism involving Hall–Petch-type behavior is operative at lower temperatures or higher strain rates in unreinforced $MoSi_2$ as well.[244]

6.4.3 Creep Behavior of Molybdenum Silicides

6.4.3.1 MoSi₂

The creep behavior of molybdenum silicides has normally been studied in the range of 1000°C–1500°C, bearing in mind the probable temperatures of structural application. The resistance to creep deformation of $MoSi_2$-based composites has been found to be superior to that of Ni-base superalloys, and comparable to that of structural ceramic materials, such as Si_3N_4.[141,245] However, the deformation mechanisms of $MoSi_2$-based composites are significantly different from those of Si_3N_4, and also vary with both temperature and stress.

In a review on the creep behavior of intermetallics and composites, Sadananda and Feng have shown that at 1100°C the creep strain rates of $MoSi_2$-based composites are lower than those of either TiAl- or NiAl-based composites and the MAR-M 240 superalloy.[245] For a given stress, the creep rates of aluminides are six orders of magnitude higher, while those of the superalloy are inferior by two orders of magnitude, compared with those of $MoSi_2$. On suitable alloying or on addition of suitable reinforcement, the creep resistance of $MoSi_2$ can be further improved. Hence, there is a strong motivation to understand the creep mechanisms of $MoSi_2$-based materials. The mechanisms of creep deformation in single- and polycrystalline $MoSi_2$ are, of course, different. The stress exponents and activation energies found in different creep studies of single- and polycrystalline $MoSi_2$ are shown in Table 6.10, while those of the $MoSi_2$-matrix composites are presented in Table 6.11.

TABLE 6.10

Stress Exponents and Activation Energies for Creep Deformation of Single and Polycrystalline $MoSi_2$

MoSi$_2$ Processing (Grain Size, µm)	Test Type	Temperature (°C)	Stress Exponent, n	Q$_c$ (kJ [mol.K]$^{-1}$)	References
Single crystal	Compression load axis [210]	1200	3	251	Bose[246]
Single crystal	Compression load axis [771]	1100	7		Maloy et al.[247]
Single crystal	Compression load axis [021]	1100	20		Maloy et al.[247]
Single crystal	Compression load axis [010]	1200–1400	7	900	Maloy et al.[247]
Single crystal	Compression load axis [1 4 1]	1200–1400	3	520	Umakoshi et al.[248]
Single crystal	Compression	1200–1400			Inui et al.[249]
	[0 15 1]: 75 MPa	1300	2.19	372	
	[001]: 400 MPa	1300	2.43	327	
Poly, HP	Compression	1200	3	306	Bose[246]
Poly, HP	Tension	1200	3	348	Bose[246]
Poly, HP (10 µm)	Compression	1100–1450	1.8	380–450 (10–40 MPa)	Sadananda et al.[250]
Poly, HP + HIP (26 µm)	Compression (step strain rate)	1200–1400	3–4.3		Ghosh and Basu[251]
Poly, HP + HIP (20 µm)	Compression	1050–1300	3.5	430	Suzuki et al.[138]
Polycrystalline, pressureless sintered (7 µm)	Bending	1100 1150 1200	2.4 1.9 1.2	159–634	Dusza and Hvizdos[126]
Polycrystalline, HP (4–25 µm)	Compression	1200	1 and 4	160–430	Sadananda et al.[141]
Polycrystalline, HP: 5 µm HP: 27 µm	Compression (incremental load change tests)	1200	2 4.7		Mitra et al.[241]

Note: The conditions of creep tests are also shown.

The results depicted in Table 6.10 indicate that orientation of the crystal with respect to the stress axis has a strong influence on the creep behavior of $MoSi_2$.[246-249] Similarly to the trend observed for high-temperature yield strength, the single-crystalline specimens with stress axis along the hard [001] orientation have exhibited much higher resistance to creep compared with the specimens with other orientations. Dislocation glide and climb have

TABLE 6.11

Stress Exponents and Activation Energies for Creep Deformation of $MoSi_2$-Based Alloys and Composites

MoSi₂-Based Alloys and Composites	Test Type	Temperature (°C)	Stress Exponent, n	Q_c (kJ [mol.K]⁻¹)	References
50/50 $MoSi_2 + WSi_2$, HP	Compression	1100–1400 1100–1300	2.4	540	Sadananda et al.[250]
$MoSi_2$–20 v% SiC, HP		1400 1100–1400	3.6 3.3–5.2	460	
$MoSi_2$–CaO, Si_3N_4, SiC	Compression	1200	3		Ghosh et al.[251]
$MoSi_2$–30 vol.% SiC, XD, HP+HIP	Compression	1050–1300	3.5	430	Suzuki et al.[138]
$MoSi_2$–10 to 40 vol.% SiC, HP	Tension	1050–1200	2.6–6.0		French et al.[252]
$MoSi_2$–9 vol.% SiC, plasma spray	Compression	1100–1300 1300–1500	2.5 1.5	300 190	Jeng et al.[253]
$(Mo,W)Si_2$–20 vol.% SiC_W	Tension Compression	1100–1225	3.2 2.3	557 312	Wiederhorn et al.[254]
$MoSi_2$+1, 2, and 4 wt.% C	Compression	1200	1.4–2.1	460	Sadananda and Feng[142]
$MoSi_2$–Mo_5Si_3	Step strain rate Compression	1200–1400	4.1	355	Mason and Van Aken[86]
$MoSi_2$–50–100 vol.% Si_3N_4	Compression	1200	1–5		Sadananda et al.[141]
$MoSi_2$–Al $(C11_b)$ alloy + Al_2O_3	Compression	1200–1400	1.9–3.7	271–415	Sadananda et al.[141]
$MoSi_2$–SiC composite	Compression	1200	4.2		Mitra et al.[244]

been found to be the mechanism of creep at 1300°C for the tests carried out under the stress range of 50–100 MPa on the samples with [0 15 1] orientation and 300–500 MPa in the case of the [001] orientation.[249] However, in the creep-tested crystals with [001] orientation, the dislocations reported are the same as those observed during compression testing at the strain rate of $10^{-4} s^{-1}$, that is, ⟨331⟩, ⟨100⟩, ⟨111⟩, and ⟨110⟩. Similarly to the observations for the tests carried out at constant strain rates, TEM examination of the creep-tested specimens has shown evidence for decomposition of the ⟨331⟩ dislocations into ⟨110⟩- and ⟨111⟩-type perfect dislocations. Interestingly, the

⟨110⟩ dislocations are fewer than the other aforementioned types of dislocations because of their propensity to decompose into ⟨100⟩-type dislocations.
Moreover, <100>-type dislocations having nearly edge orientations have
been found to constitute the low angle boundaries. The crystals with [0 15 1]
orientations have shown dislocations with a Burgers vector similar to that
with [001] orientation, except for the absence of ⟨331⟩ dislocations.

In polycrystalline $MoSi_2$,[86,141,142,250–254] the factors found to affect the creep
behavior include grain size, SiO_2 content, alloying, and presence of hard and
nondeforming reinforcements, as well as the test conditions of temperature
and stress. The stress exponents range between 1 and 4 depending on either
temperature or stress used for the test, grain size, SiO_2 content, and alloying additions.[141,241] At 1200°C, $MoSi_2$ specimens with grain sizes between 4
and 20 μm have shown diffusion-controlled creep (i.e., stress exponent = 1)
for applied stress <70 MPa and dislocation-creep regime (i.e., stress exponent = 4–5) for stresses ≥80 MPa (Figure 6.14a and b). Furthermore, $MoSi_2$
specimens with grain size ≥25 μm undergo only dislocation creep. In both
diffusion and dislocation-creep regimes, the grain-size exponent has been
found to be about 4.3, which is higher than the grain-size exponent of 3.0
expected for Coble creep. Such strong dependence of creep rates on grain size
in the dislocation-creep regime is unusual. Moreover, the transition from the
diffusion to dislocation-creep regime is observed to take place at a smaller
stress with increasing grain size. Such high dependence of creep behavior
on grain size may be explained by the operation of only four independent
slip systems, as a result of which climb enhanced by vacancy generation
and grain-boundary diffusion is essential for the accommodation of strain
incompatibility between neighboring grains. It is also interesting to note
that, for specimens predeformed in the Newtonian viscous-flow regime (or
diffusion-creep regime) until the transition to the dislocation-creep regime,
the steady-state creep rate at any stress beyond the transition has been found
to be the same for all specimens, showing grain-size-independent behavior
(Figure 6.14c).

Tensile creep is associated with cavitation at SiO_2-containing $MoSi_2$ grain
boundaries, similar to the observations recorded by Suzuki et al. for compression creep.[138] In tune with the abovementioned observation, the creep
rates are found to be significantly higher in specimens with high SiO_2 content, and the effect is more obvious in fine-grained specimens with a large
grain-boundary area, as has been depicted in Figure 6.15. For the same grain
size of 5–6 μm, the stress exponent of high SiO_2-containing $MoSi_2$ has been
found to be ≈1, while that of a low SiO_2-containing $MoSi_2$ has been reported
as ≈2. It is well accepted that a stress exponent of 1 is suggestive of Newtonian
viscous flow, while a stress exponent of 2 is suggestive of grain-boundary
sliding, accommodated by slip. The weakening effects of SiO_2 and fine grain
size at high temperatures may camouflage the strengthening effect of SiC
reinforcements (Figure 6.15) in the case of $MoSi_2$–SiC composites.[141,142,241] For
example, the stress exponent has been found to be ≈1.4 for a $MoSi_2$–20 vol.%

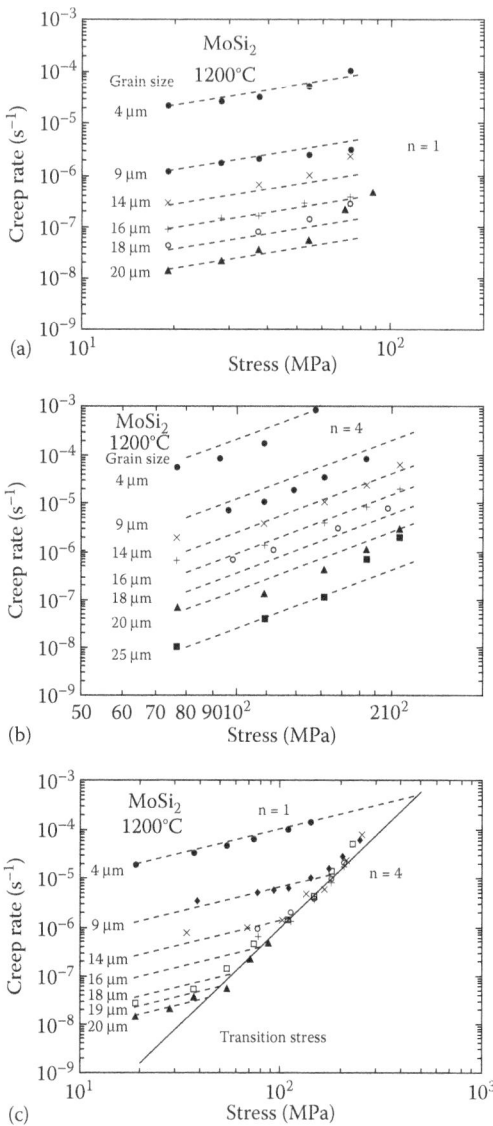

FIGURE 6.14
Creep behavior of MoSi$_2$ with different grain sizes: (a) diffusion creep, (b) dislocation creep, and (c) transition from diffusion to dislocation-creep regime when subjected to incremental loading. (Reprinted from *Mater. Sci. Eng. A*, 261, Sadananda, K., et al., Creep and fatigue properties of high temperature silicides and their composites, 223–238, Copyright (1999), with permission from Elsevier.)

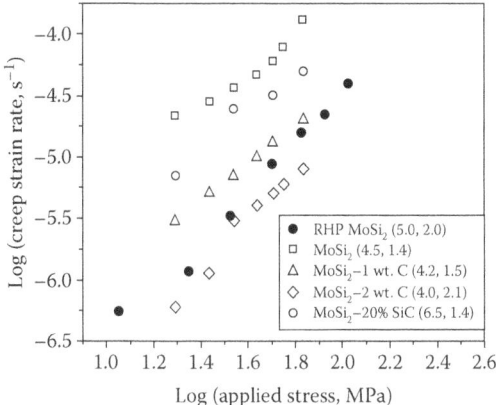

FIGURE 6.15

Comparison of creep rates of MoSi$_2$ with varying SiO$_2$ content and SiC reinforcements added externally or formed by *in situ* reaction. (Reprinted from *Intermetallics*, 12, Mitra, R., et al., Effect of microstructural parameters and Al alloying on creep behavior, threshold stress and activation volumes of molybdenum disilicides, 827–836, Copyright (2004), with permission from Elsevier.)

SiC composite with a matrix grain size of 6–7 μm and oxygen content of 2.5 wt.%, whereas it has been found to be ≈4.2 for matrix grain size of 18 μm and oxygen content of 0.24 wt.%.[241] In coarse-grained MoSi$_2$ and MoSi$_2$–SiC composites with coarse grain size, the effect of SiO$_2$ is less obvious because the grain-boundary area is smaller.

For MoSi$_2$ with 2 wt.% C, in which the SiO$_2$ is converted to SiC through an *in situ* reaction, the SiC volume fraction in the microstructure has been found to be ≈30%–40%, which aids in improving the creep properties significantly.[142] The creep rates of polycrystalline MoSi$_2$ alloyed with C are compared with those of specimens containing different amounts of SiO$_2$ in Figure 6.15. Tensile creep and stress–rupture studies on MoSi$_2$–SiC (containing 10–40 vol.% SiC) composites[252] processed by mechanical mixing of MoSi$_2$ and SiC powders, as well as MoSi$_2$–SiC (with 0 to 25 vol.% SiC) composites processed by addition of C, have shown significant improvement in their properties due to removal of SiO$_2$. The formation of Mo$_5$Si$_3$C as a dispersed phase through *in situ* reactions leads to further improvement in the creep resistance.

Creep studies[126] in bending mode on pressureless sintered MoSi$_2$ having an average grain size of 7 μm have shown behavior similar to that observed in compression creep. The stress exponents obtained at the temperatures of 1100°C, 1150°C, and 1200°C have been found to be 2.4, 1.9, and 1.2, respectively. Similar stress exponents have also been obtained by Jeng et al. from creep experiments conducted on fine-grained MoSi$_2$–9 vol.% SiC composites fabricated by plasma spray processing.[253] It has been proposed that creep takes place by grain-boundary sliding accommodated by lattice diffusion of Si atoms at 1200°C and by grain-boundary sliding involving shearing of intergranular SiO_2 films at 1300°C.

Creep studies on $MoSi_2$–SiC composites have shown improvement in creep properties only when the reinforcement volume fraction is ≥20 vol.%.[141] If the volume fraction is <20%, the effect of SiC on creep strengthening is counterbalanced by an accompanying decrease in the grain size due to refinement in the course of processing. In an $MoSi_2$–SiC composite with the matrix grain sizes varying between 18 and 35 μm, the stress exponent has been found to be ≈5.2 at 1100°C and ≈3.3 at higher temperatures.[241] Figure 6.16 shows plots of steady-state creep rate against stress for $MoSi_2$ with grain sizes of 5 μm (RHP $MoSi_2$), 18 μm, and 27 μm, RHP $MoSi_2$–20 vol.% SiC composite (18 μm matrix grain size), and $MoSi_2$–5.5 and 9 at.% Al alloys with matrix grain sizes of 20 μm and 25 μm, respectively.[241,250] Based on the results in this figure, it is obvious that the RHP $MoSi_2$–20 vol.% SiC composite with matrix grain size of 18 μm has more creep resistance and a higher stress exponent compared with $MoSi_2$ with a similar grain size. This observation confirms that the presence of SiC particles as reinforcement is effective in strengthening against creep, provided the matrix grain size is similar to that of the unreinforced $MoSi_2$. Moreover, a significant increment in creep resistance has also been observed in the case of $MoSi_2$–$Si_3N_{4(p)}$ composites.[141]

Sadananda and Feng have analyzed the different strengthening mechanisms in creep of $MoSi_2$–matrix composites and inferred that the rule of mixtures based on isostrain and isostress provides the two limiting bounds within which the other predictions fall.[255] Nondeforming SiC or Si_3N_4 particles deform only elastically, and impede the motion of dislocations, while

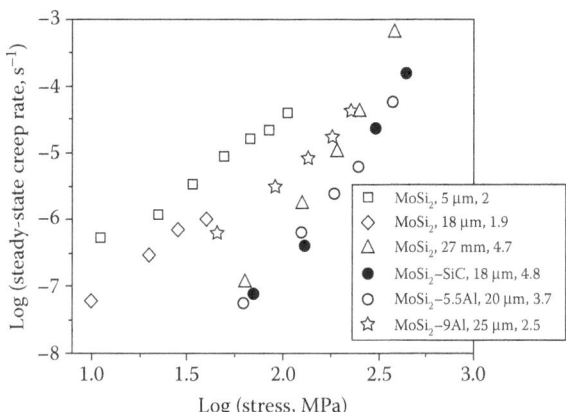

FIGURE 6.16
Comparison of steady-state creep rates of polycrystalline $MoSi_2$ with grain sizes of 5, 18, and 27 μm, $MoSi_2$–20 vol.% SiC composite with matrix grain size of 18 μm, and $MoSi_2$–5.5 and 9 at.% Al alloys with matrix grain sizes of 20 μm and 25 μm, respectively. The stress exponents are shown next to the grain size. (The data have been taken from Mitra, R., et al., *Intermetallics*, 12, 827–836, 2004; Sadananda, K., et al., *Mater. Sci. Eng. A*, 155, 227–239, 1992.)

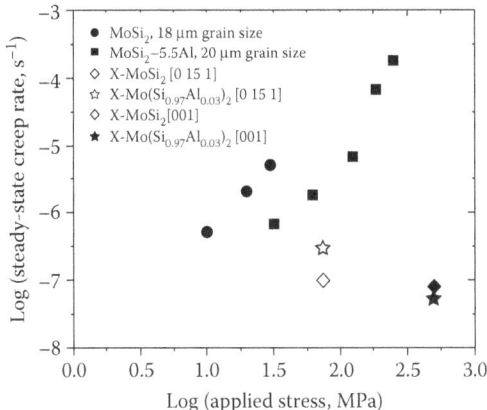

FIGURE 6.17
Comparison of steady-state creep rates of polycrystalline $MoSi_2$ and $MoSi_2$–5.5 at.% Al alloy, as well as single crystals of $MoSi_2$ and $Mo(Si_{0.97}Al_{0.03})_2$ tested with [0 15 1] and [001] orientations at 1300°C. Single crystals are marked as X. (From Mitra, R., et al., *Intermetallics*, 12, 827–836, 2004.)

load transfer from the matrix to the reinforcement takes place by the shear-lag mechanism.[256] It is creep of the matrix that governs the creep rate of the composite. However, if the reinforcement phase undergoes a higher creep rate than the matrix, as in the case of $MoSi_2$–Nb_f composites, the creep rates are found to exceed that of the unreinforced $MoSi_2$.[123]

Figure 6.17 presents the plots showing the comparison of creep rates of single- and polycrystalline $MoSi_2$ with those of $MoSi_2$ alloyed with Al. On alloying with Al,[241] the creep properties are improved with respect to that of polycrystalline $MoSi_2$, as long as the SiO_2 present at grain boundaries is converted to Al_2O_3 and the concentration of Al in $MoSi_2$ does not alter the structure from $C11_b$ to C40. However, the creep rates of $MoSi_2$–5.5 at.% Al alloy are found to be less than those of the $MoSi_2$–9 at.% Al alloy because of the presence of C40 $Mo(Si,Al)_2$ phase in the latter alloy, and the observation is consistent with the trend observed for variation of yield strength with Al content of the alloys (Figure 6.13).[241] Inui et al. have shown earlier that, on alloying with those elements in the case of which the tetragonal ($C11_b$) structure is retained, solid-solution strengthening is noticed during compression tests at constant strain rates.[176,249] In contrast, softening is observed on alloying with the elements contributing to the C40 structure. Single crystals of $Mo(Si_{0.97}Al_{0.03})_2$ with hexagonal C40 structure have shown higher creep rates than those of $MoSi_2$ single crystals for the orientation of the stress axis along the [0 15 1] direction. Interestingly, $Mo(Si_{0.97}Al_{0.03})_2$ has been found to be slightly more creep resistant than $MoSi_2$ with change of the stress axis to the [001] direction (Figure 6.13).

Compression creep studies in the temperature range of 1200°C–1400°C have shown that alloying of single-crystalline $MoSi_2$ with W decreases the creep strength of the samples with [0 15 1] orientation, while improved creep resistance has been observed in those with [001] orientation.[249] The anisotropy of creep strength is certainly related to the dependence of the operating slip system and types of dislocations on orientation. For example, dislocations with Burgers vector along $\langle 331 \rangle$ have been noticed only in samples deformed with [001] orientation. It has been reported that alloying of polycrystalline $MoSi_2$ with W leads to a pronounced increment in creep resistance at 1100°C, with the difference decreasing with increasing temperature until it becomes quite negligible at 1300°C.[249] The combined effect of alloying with W and reinforcing with SiC particles leads to even higher creep strength, as expected. Moreover, the stress exponents have been found to increase on alloying with W or addition of SiC reinforcements or both simultaneously.

Consistently with the yielding behavior in the high-temperature regime, alloying of $MoSi_2$ with Nb and Re has led to a remarkable improvement in creep resistance at 1300°C for a stress axis along both [0 15 1] and [001] orientations.[249] Furthermore, TEM observations of Nb- and Re-alloyed single-crystalline $MoSi_2$ specimens deformed in creep have shown a high density of $\langle 110 \rangle$ dislocations in crystals with [0 15 1] orientations, matching the observations reported for the specimens tested at constant strain rates.

Analyses of creep data obtained from various published reports[241,257] show that different parameters, such as stress exponent, threshold stress, activation energies, and activation volumes, depend on grain size, alloying, and the presence of nondeforming reinforcements. Threshold stress represents the level of stress below which the creep strains are zero or do not exceed the anelastic limit. In other words, it is the minimum stress required to set the dislocations in motion in the active slip systems. It is found by estimating the value of stress corresponding to the steady-state strain rate equal to zero by the method of extrapolation, and may be taken as an index of creep resistance. The values of threshold stress obtained as a result of analyses of the creep data of $MoSi_2$, $MoSi_2$–SiC composites, and $MoSi_2$–Al alloys appear to be higher in coarse-grained specimens or those reinforced with SiC or Al_2O_3 particles. In coarse-grained $MoSi_2$, the higher values of threshold stress are believed to be due to the Peierls barrier and nucleation of glide dislocations at preexisting dislocation networks. The Peierls barrier may be overcome in specimens with smaller grain sizes because of the larger area of grain boundaries, providing a higher flux of vacancies, which aid the thermal activation process and consequently reduce the threshold stress. In $MoSi_2$, the presence of point defects and solute atoms also affects the magnitude of the Peierls stress. On the other hand, the threshold stress observed in the composites is probably due to load transfer by a shear-lag mechanism from the matrix to the elastically deforming reinforcement, or the back stress experienced by mobile dislocations in the matrix due to higher dislocation density at particle–matrix interfaces. The values of activation volumes[241] have been found to vary from $27b^3$ at low stresses to $2b^3$ at high stresses. The activation-volume analyses imply that deformation by creep requires overcoming glide-type barriers, such as Peierls stress at low stresses and dislocation climb at high stresses.

The creep–fatigue–environment interactions of $MoSi_2$–50 mol.% WSi_2–30 vol.% SiC composites at 1200°C in air have been studied by Ramamurty et al.[258] Samples have been observed to undergo catastrophic failure at room temperature if the difference between the upper and lower stress intensity factors (ΔK) exceeds the mode-I plane strain fracture toughness (K_{IC}) of this material. At higher temperatures, the composite is found to demonstrate a higher fatigue threshold, followed by considerable subcritical crack growth, when compared with that of the unreinforced $(Mo,W)Si_2$ alloy. The slower crack growth in the composite is due to R-curve-type behavior, characterized by crack deflection and bridging due to the presence of SiC reinforcements. It has been observed that a glassy film is formed at the matrix grain boundaries and particle–matrix interfaces due to the oxidation of $(Mo,W)Si_2$ and SiC phases. The failure mechanism involves viscous flow of glassy films along grain boundaries and interfaces, accompanied by nucleation, growth, and coalescence of cavities and microcracking at the crack tips. The crack growth under cyclic loading conditions is slower than that under static loading, even when the maximum stress intensity of a fatigue cycle equals the stress intensity factor corresponding to the static loading. It has been proposed that the slower crack growth in cyclic loading could be due to resintering of intergranular cavities by viscous flow of glassy film in the compressive part of the cycles.

6.4.3.2 Mo_3Si

The compressive yield strengths of Mo_3Si, as experimentally determined at 1400°C,[37] have been found to decrease sharply with decreasing strain rate from 10^{-3} to 10^{-5} s^{-1}, implying very high strain-rate sensitivity and thermally activated deformation. The yield strength of Mo_3Si has also shown a strong dependence on its stoichiometry, decreasing as the Si concentration is increased from 22 to 25 at.%. At higher concentrations of Si, precipitation of brittle Mo_5Si_3 is expected, which might enhance the propensity for internal cracking.

Studies[87] using the TEM have shown a high density of dislocations inside the Mo_3Si phase in the multiphase Mo–Si–B alloy, subjected to creep between 1220°C and 1320°C. Here, the operative slip system has been found to be (001) $\langle 100 \rangle$, with dislocations of mixed character.

6.4.3.3 Mo_5Si_3

Mo_5Si_3 single crystals have been compression tested with a stress axis near [101] and [111] and along [100] and [001] crystallographic directions.[32] These tests have shown the BDTT to be $\approx$1250°C. Moreover, the flow curves have been found to possess upper and lower yield points in single-crystal

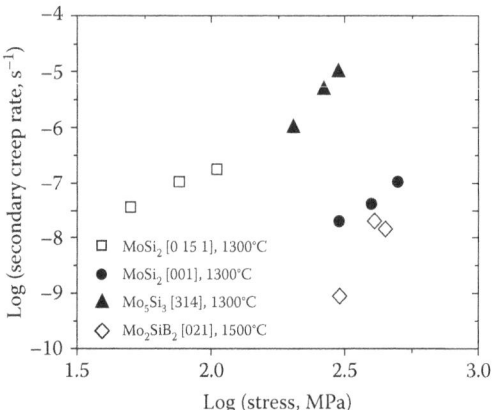

FIGURE 6.18
Log–log plots of secondary creep rates against stress for [0 15 1] and [001]-oriented $MoSi_2$, [314]-oriented Mo_5Si_3, and [021] Mo_5SiB_2. (From Ito, K., et al., *Intermetallics*, 9, 591–602, 2001; Mason, D. P. and D. C. Van Aken, *Acta Metall. Mater.*, 43, 1201–1210, 1995; Inui, H., et al., *Intermetallics*, 8, 1159–1168, 2000.)

specimens with either [101] or [100] orientation, as is expected when the density of mobile dislocations is low. Figure 6.18 shows a sharp decrease in yield stress of Mo_5Si_3 single crystals with increasing temperature,[32] indicating thermally activated deformation behavior. The average value of strain-rate sensitivity has been found to be 0.16, which is suggestive of reasonable ductility. Experimental observations by Yoshimi et al. of the slip traces in [101]- and [100]-oriented crystals deformed in compression in the temperature range of 1250°C–1300°C have shown the slip planes to be (110), (101), and (201).[32] On the other hand, study of the microstructure of the multiphase Mo_3Si–Mo_5Si_3–Mo_5SiB_2 alloy after creep deformation between 1220°C and 1320°C has shown no evidence of dislocations in the Mo_5Si_3 phase.[87]

Very few data on creep deformation of monolithic Mo_5Si_3 are available (see Table 6.12). For a creep test carried out under 69 MPa at 1200°C, Anton et al. have reported a steady-state strain rate $\approx 4 \times 10^{-8}\,s^{-1}$, which is about one-fifth of the creep rate reported for $MoSi_2$ under identical test conditions.[259] Studies by Mason and Van Aken on the creep behavior of the directionally solidified $MoSi_2$–Mo_5Si_3 eutectic composite have shown the creep rates to depend primarily on the interlamellar spacing of the Mo_5Si_3 phase.[86] The climb of (001) ½⟨110⟩ partial dislocations with edge character has been found to control the creep rates of the Mo_5Si_3 phase in the eutectic. Creep studies between 1200°C and 1400°C in Mo_5Si_3 single crystals with a stress axis along the [314] direction have shown stress exponent ≈ 6 and activation energy ≈ 510 kJ mol^{-1},[86] which are higher than those of $MoSi_2$ in a similar temperature range.

TABLE 6.12

Types of Tests, Stress Exponents, and Activation Energies for Creep Deformation of Multiphase Mo–Si–B Alloys

Mo–Si-Based Intermetallics (Other Than $MoSi_2$)	Type of Creep Tests	Temperature (°C)	Stress Exponent, n	Activation Energy, Q_c (kJ [mol.K]$^{-1}$)	References
[314] Mo_5Si_3	Compression	1200–1400	6	510	Mason and Van Aken[86]
Mo_5Si_3,Mo_3Si, $Mo_5(Si,B)_3$	Step strain rate Compression	1220–1320	4.3	396	Meyer et al.[87]
[021] Mo_5SiB_2	Compression	1500	6.8	740	Ito et al.[31]
Mo–9.4Si–13.8B	Compression	1400–1450	2.8	740	Nieh et al.[261]
Mo–12Si–8.5B	Compression	1300	3.2	295	Schneibel[262]
Mo–16.8Si–8.4B		1300	2.1–2.8	327–376	
Mo–13.5Si–10.8B		1400	3.9	328	
[021] Mo_5SiB_2	Compression	1500–1600	6.8	740	Hayashi et al.[260]
[001] Mo_5SiB_2			4.3	400	
			4.6		

6.4.3.4 Mo_5SiB_2

Ito et al. have reported the stress–strain curves obtained from the tests carried out at 1500°C on a single crystal of Mo_5SiB_2 with stress axes along the [001], [010], [110], [443], and [021] directions.[31] Analysis of these stress–strain curves indicates that the specimens with [021] and [443] orientations are ductile, while others are brittle. The Mo_5SiB_2 crystal with the stress axis oriented along the [021] direction had a yield strength of ≈650 MPa, accompanied by a high rate of strain hardening. On the other hand, a maximum stress of >2.0 GPa has been recorded for a stress axis along the [110] direction, indicating that this orientation is much harder and more brittle than the [021] orientation. Furthermore, the preferred slip system for compressive deformation along the [021] direction has been found to be [001](010).[38]

Table 6.12 shows the results of creep studies carried out on Mo_5SiB_2. Ito et al.'s studies on the creep of [021]-oriented single crystals of Mo_5SiB_2 under 432 MPa at 1500°C have shown the steady-state creep rate to be ≈3.1 × 10^{-8} s^{-1}.[31] If this result is extrapolated to 1300°C, the obtained value appears to be three orders of magnitude lower than that of $MoSi_2$ in its hardest orientation of [001] and also less than those of [314]-oriented Mo_5Si_3 crystals and AS800–Si_3N_4. Figure 6.18 shows the creep rates of single crystals of [001] $MoSi_2$ and [314] Mo_5Si_3 at 1300°C along with the results obtained from tests on Mo_5SiB_2

at 1500°C. Based on the results in this figure, it is inferred that the creep rates of single-crystal Mo_5SiB_2 at 1500°C are less than those of both $MoSi_2$ and Mo_5Si_3 at 1300°C, and, as expected, Mo_5Si_3 is more creep resistant than $MoSi_2$. Moreover, the activation energy for creep of single-crystal Mo_5SiB_2 ($\approx$740 kJ mol^{-1}) is greater than that of [001]-oriented $MoSi_2$ (327 kJ mol^{-1}) and [314]-oriented Mo_5Si_3 (510 kJ mol^{-1}). Hayashi et al. have studied the creep-deformation behavior of Mo_5SiB_2 single crystals with [021] and [001] orientations and polycrystalline specimens.[260] Data on the steady-state creep rates and Larson–Miller parameters have shown that the highest creep resistance is noticed in the case of single crystals with the stress axis oriented along the [001] direction, followed by those with [021] orientation and polycrystalline specimens. [021]-oriented Mo_5SiB_2 specimens, after deformation in creep, have shown long, straight edge dislocations with Burgers vector along [001] on (010) planes. On the other hand, crystals with [001] orientation have not shown evidence of much dislocation activity.

During creep testing of the multiphase Mo–Si–B alloy with Mo_5SiB_2 as the main constituent in the temperature range of 1220°C–1320°C,[87] it has been observed that {001}⟨010⟩ and {001}⟨110⟩ slip systems are activated, which can be justified on the basis of the fact that ⟨010⟩ is the shortest lattice vector (measuring 0.6 nm) in the Mo_5SiB_2 phase, followed by ⟨110⟩, while the {001} plane is ranked second in density of packing. TEM studies on crept specimens by Field et al.[88] have shown evidence of dislocations with Burgers vectors of either [100] or ½[111], which differ from those observed in the multiphase composites with Mo_5SiB_2 as a constituent.[87]

6.4.3.5 Mo–Si–B Multiphase Alloys

Limited information is available on the high-temperature yield and flow behavior of Mo–Si–B alloys containing Mo_5SiB_2 (T2), Mo_3Si, and α-Mo phases.[261,262] Mo–Si–B alloys containing 40–50 vol.% of the α-Mo phase have shown tensile or compressive yield strengths in the range of 480–580 MPa when tested between 1200°C and 1400°C at strain rates of 10^{-4} or 10^{-5} s^{-1}. Apparently, the high-temperature strength and ductility of Mo–Si–B alloys have been found to be sensitive to both composition and volume fraction of the constituent phases in the alloy. Tensile ductility of $\approx$100% could be observed on testing at 1400°C with a strain rate of 10^{-4} s^{-1}, probably because of the high volume fraction of α-Mo in the alloy with a composition of Mo–9.4Si–13.8 B. On the other hand, the tensile ductility of the Mo–8.9Si–7.7B alloy containing approximately 55 vol.% α-Mo has been found to be 5% at 1200°C and nil at 1100°C.[132] The method of processing has also been found to play an important role in controlling the morphology of α-Mo particles and the resulting tensile behavior.[130] A near-continuous network of α-Mo formed in the Mo–3Si–1B alloy through extrusion has been found to reduce the BDTT by 200°C with respect to that of a hot isostatically pressed alloy of similar composition but with discontinuously distributed globular α-Mo.

For an alloy with composition of 76Mo–14Si–10B, yield strength and maximum stress are found to decrease by 45% and 40%, respectively, whereas the strain-hardening exponent is halved on increase of the compression test temperature from 1100°C to 1200°C.[185] However, the strength is not found to vary much on further increasing the temperature to 1300°C. These observations suggest that the thermally activated deformation of the α-Mo phase primarily controls the mechanical behavior of these multiphase alloys, and its ductility is significantly increased as the test temperature is raised from 1100°C ($0.48T_m$) to 1200°C ($0.51T_m$).

In a study on a two-phase (α-Mo + T2) alloy and a three-phase alloy having compositions of Mo–6.1Si–7.9B and Mo–8.6Si–8.7B, respectively, Alur et al. have shown that the compressive flow behavior of these alloys tested in the temperature range of 1000°C–1400°C at strain rates varying from 10^{-3} to 10^{-6} s^{-1} is controlled by deformation of the α-Mo phase.[263] Whereas α-Mo undergoes either dynamic recovery or recrystallization, the T2 phase has been found to either crack or show subgrain formation, depending on the temperature and strain rate used for tests.[263] Dynamic recovery has been observed in the two-phase alloy for tests carried out at relatively slower strain rates. Moreover, both the aforementioned alloys have shown values of strength superior to that of the commercial TZM (99Mo–0.5Ti–0.08Zr–C) alloy.

The methods and some of the results of creep studies on the multiphase Mo–Si–B alloys can be seen at a glance in Table 6.12. The first report[87] on creep testing of the Mo–Si–B multiphase alloys is a study on a material with B-doped Mo_5Si_3 (T1) as the matrix with dispersion of Mo_3Si and Mo_5SiB_2 (T2) phases. For compression creep at temperatures in the range of 1220°C–1320°C and stresses in the range of 140–180 MPa, stress exponents varying from 3.8 to 5.0 and activation energy of ≈400 kJ mol^{-1} have been found. The values of stress exponents are suggestive of a dislocation glide and climb-based mechanism, which is supported by observation of a high density of dislocations in the Mo_3Si phase. However, little dislocation activity could be observed in the T1 matrix phase (B-doped Mo_5Si_3), which is believed to undergo grain-boundary or interphase sliding. Moreover, the samples subjected to 13% creep strain have shown a high density of cracks in the T1 phase.

Tensile creep tests carried out by Jain and Kumar[264] on an Mo solid-solution (Mo–3Si–1.3B), two-phase alloy with α-Mo + 35 vol.% T2 (Mo-6Si-8B) and a three-phase alloy (Mo–8.6Si–8.7B) with 50 vol.% Mo_3Si + T2 have shown that the creep resistance scales with Si concentration of the α-Mo phase. This observation is consistent with that of another study by Jain et al., which has shown that both strength and work-hardening rates of Mo solid solution at elevated temperatures (1000°C and 1200°C) increase with increasing concentration of Si and B.[265] The values of activation energies from various elevated-temperature tensile creep and compressive tests have been found to be in the range of 360–460 kJ mol^{-1}, which are reasonably close to that of Mo self-diffusion (≈405 kJ mol^{-1}). It is therefore inferred that atomic diffusion of Mo in α-Mo plays a significant role in high-temperature deformation. Studies of postcreep specimens have shown evidence for precipitation of T2 phase at grain boundaries during the test, most likely due to segregation of Si and B at these preferred locations.[264]

Schneibel has used high-temperature compression tests at a slow strain rate of $10^{-5}\,\text{s}^{-1}$, defining the flow stress at 2% plastic strain as the creep stress.[262] Engineering stress–strain curves with flow stress remaining more or less constant after yielding due to a dynamic recovery-type mechanism have been used to derive creep data, instead of actual tests under constant load or stress. Both arc-melted and powder metallurgy-processed alloys with or without Nb or W, and containing α-Mo, Mo_3Si, and T2 as constituent phases, have been investigated in this manner. These alloys have shown stress exponents varying between 2 and 4 and activation energies in the range of 300–500 kJ mol⁻¹. Creep resistance has been found to improve remarkably on alloying with Nb and only modestly on alloying with W. Correlation of microstructures with creep strengths shows that the presence of discontinuously dispersed α-Mo particles is more desirable than a continuous network-type structure for improvement in creep strength. This inference suggests that the type of α-Mo phase morphology required for high creep strength is in sharp contrast to that required for achieving high room-temperature fracture resistance, and hence microstructural optimization is required for achieving a desirable combination of mechanical properties at a given temperature.

In recent years, a breakthrough has been achieved through the development of superplastic Mo–Si–B alloys.[266,267] An ultrafine (submicrometer phase size) three-phase Mo–2.7Nb–8.9Si–7.7B alloy, processed in steps involving mechanical alloying of the elemental powders in a hydrogen environment, followed by cold isostatic pressing of these powders at 200 MPa, sintering at 1450°C in a hydrogen atmosphere, and hot isostatic pressing at 1500°C, has shown superplastic behavior, with 200%–400% tensile strain to failure in tests carried out using a strain rate of 10^{-3} at 1300°C. Analysis of the results of these tests has shown the values of the stress exponent and activation energy for creep to be 2.3 and 470 kJ mol⁻¹, respectively. The value of activation energy has been found to be reasonably close to that of Mo self-diffusion ($\approx$405 kJ mol⁻¹), which indicates that diffusion of Mo plays an important role in accommodating strains caused by grain-boundary or interface sliding. Furthermore, the results of compressive creep tests carried out by Jehanno et al. in the range of 1050°C–1315°C have shown the steady-state strain rates of the relatively coarse-grained annealed (such as 10 h at 1700°C) hot isostatically pressed (HIPed) samples samples to be an order of magnitude lower than those of ultra-fine-grained as-HIPed samples.[268] These observations imply that suitable heat treatment or thermomechanical processing should be carried out on these alloys after near-net shaping by superplastic forming in order to obtain a relatively coarser average grain size, so that desirable creep resistance is achieved.

Superplasticity has been reported by Li et al.[267] for the powder metallurgy-processed fine-grained (3 μm) Mo–9Si–8B–3Hf alloy when subjected to tests at a strain rate of $3 \times 10^{-4}\,s^{-1}$ in the temperature range of 1400°C–1560°C, and additionally at $1 \times 10^{-3}\,s^{-1}$ at 1530°C or 1560°C. The tensile strain to failure has been found to vary from 156% to 410%, with the maximum elongation being observed for the test with a strain rate of $3 \times 10^{-4}\,s^{-1}$ at 1560°C. Interestingly, both Li et al.[267] and Jehanno et al.[268] have reported the microstructure to be stable with negligible grain growth in the course of the creep tests. Furthermore, high-temperature tensile tests carried out using a strain rate of $2 \times 10^{-2}\,s^{-1}$ in vacuum by Yu et al. have shown Mo–10Si–14B–3Hf to exhibit impressive tensile strength of ≈560 MPa at 1400°C.[269]

Fatigue-crack-growth studies in tension-tension mode on an Mo–6Si–8B alloy with two phases (α-Mo + 38 vol.% T2) have shown the Paris law slope to be ≈20–30 at room temperature, ≈6 at 1200°C, and ≈3.5 at 1400°C.[270] The results of fracture-toughness tests carried out at elevated temperatures using monotonic loading have also been found to be strongly affected by the crosshead speeds. The fracture toughness of this alloy has been found to be ≈8 MPa m$^{1/2}$ at room temperature, rising to ≈18 MPa m$^{1/2}$ at 1200°C and ≈25 MPa m$^{1/2}$ at 1400°C.

6.4.4 Deformation Behavior of Silicides of W, Nb, and Cr

The slip systems of WSi_2, $NbSi_2$, and $CrSi_2$ found through different investigations are shown in Table 6.13. The slip systems in operation depend in most cases on the crystal structure (C11$_b$ or C40), the crystallographic orientation of the stress axis, and the temperature of compression testing.

6.4.4.1 *WSi$_2$*

Although the structure (Table 6.1), physical properties (melting point, CTE, and elastic constants), and electronic structure of WSi_2 are very similar to those of $MoSi_2$, the plastic deformation behavior of these materials differs quite significantly. The studies on deformation behavior of the WSi_2 single crystals by Kimura and coworkers[43] have indicated that, at temperatures ≥1300°C, single crystals of WSi_2 oriented near the ⟨001⟩ and ⟨100⟩ directions exhibit plastic flow with {013}⟨331⟩ and {1$\bar{1}$0}⟨331⟩ as the activated systems, respectively (Table 6.13). Moreover, stacking faults have been observed in the {110} planes. Interestingly, and contrary to the behavior exhibited by $MoSi_2$, the WSi_2 crystals with orientations close to ⟨100⟩ have shown higher yield strength than those with ⟨001⟩ orientations. This observation is attributed to a higher CRSS for slip on the {1$\bar{1}$0}⟨331⟩ system than that for {0$\bar{1}$3}⟨331⟩, probably due to higher density of stacking faults formed in the former slip system.

It has been shown by Ito et al. that the onset temperature for plastic deformation of WSi_2 single crystals depends on their orientation.[90] For example,

TABLE 6.13

Results of Experimental Research on Slip Systems in Single Crystals of WSi_2, $NbSi_2$, and $CrSi_2$ in Chronological Order

Year Published	Operative Slip Systems	Test Temperature	Test Method and Orientations	References
WSi_2 (BDTT = 1100°C)				
1990	1300°C and above: {0$\bar{1}$3} ⟨331⟩, {1$\bar{1}$0}⟨331⟩		Compression near ⟨001⟩, ⟨100⟩	Kimura et al.[43]
1991	1300°C and above: {0$\bar{1}$3} ⟨331⟩, {1$\bar{1}$0}⟨331⟩	1300°C–1500°C	Compression near ⟨001⟩, ⟨100⟩	Hirano et al.[44]
1996	{1$\bar{1}$0}⟨111⟩, {011} ⟨100⟩, {023}⟨100⟩, {001}⟨100⟩	1100°C–1500°C	Compression [$\bar{1}$13], [$\bar{1}$12], [$\bar{2}$21], and [011]	Ito et al.[23]
1999	{1$\bar{1}$0}⟨111⟩, {011} ⟨100⟩, {023}⟨100⟩, {001}⟨100⟩	RT–1500°C	Compression [0 15 1], [011], [$\bar{1}$10], [$\bar{2}$21], [$\bar{1}$12], [$\bar{1}$13], and [001]	Ito et al.[90]
$NbSi_2$ (BDTT = 500°C)				
1997	(0001)⟨1$\bar{2}$10⟩	RT–1500°C	Compression [0001], [1$\bar{2}$12], and [1$\bar{2}$10]	Moriwaki et al.[50]
2001	(0001)⟨1$\bar{2}$10⟩	RT–1500°C	Compression [1$\bar{2}$12], [7$\bar{7}$08]	Inui and Yamaguchi[242]
2002	{10$\bar{1}$0}⅓⟨1$\bar{2}$10⟩, {10$\bar{1}$1}⅓⟨1$\bar{2}$10⟩, {1010}⟨0001⟩, and {112Y}⟨112X⟩; X.Y = −6	500°C–1500°C	Tension [$\bar{8}$449]	Nakano et al.[271]
$CrSi_2$ (BDTT = 800°C)				
1997	(0001)⟨1$\bar{2}$10⟩	RT–1400°C	Compression [0001], [1$\bar{1}$00], and [7$\bar{7}$08]	Inui et al.[56]
2001	(0001)⟨1$\bar{2}$10⟩	RT–1500°C	Compression [1$\bar{2}$12], [7$\bar{7}$08]	Inui and Yamaguchi[242]

1100°C and 1400°C are the minimum temperatures required for plastic deformation caused by loading with the stress axis along the [$\bar{1}$10] and [001] directions, respectively. This observation confirms that a crystal with [001] orientation is harder to deform than one having the [$\bar{1}$10] orientation. The operating slip systems in the WSi_2 single crystals have been identified as {1$\bar{1}$0}⟨111⟩, {011}⟨100⟩, {023}⟨100⟩, and {001}⟨100⟩ (Table 6.13). Three of the four slip systems identified in WSi_2 by Ito et al. have also been reported for $MoSi_2$,[23] whereas $\{001\}\langle100\rangle$ is unique to the former material. Interestingly, slip on the $MoSi_2$ $\{0\bar{1}3\}\langle331\rangle$ system, as reported by Kimura and coworkers[43] for a stress axis along the [001] direction, has seldom been observed in the studies by Ito et al. on WSi_2. However, partial dislocations with $b=1/6\langle331\rangle$ formed by dissociation of $\langle100\rangle$ dislocations have been observed in crystals with [001] orientation. It has been proposed that plastic deformation of crystals with [001] orientation is supported by climb of $1/6\langle331\rangle$ partial dislocations on the (001) planes.

The deformation mechanisms of WSi_2 may be distinguished from those of $MoSi_2$ due to the more directional character of the metal–Si bonds in the former material,[91] as well as the presence of stacking or condensation faults on the (001) planes.[45] The Frank-type stacking faults found through TEM studies appear to form by removal or relocation of a pair of (001) layers of Si atoms from the lattice. Moreover, high-resolution transmission electron microscopy (HRTEM) observations have revealed the occurrence of dilation of the lattice perpendicular to the fault. When $MoSi_2$ is alloyed with WSi_2, W occupies the sites on the Mo sublattice and the density of stacking faults gradually increases. The stacking-fault density has been found to be most significant for compositions with W concentrations between those in $(Mo_{0.25}W_{0.75})Si_2$ and WSi_2.

In WSi_2 crystals deformed with the stress axis along the $[\bar{1}10]$ direction, the dissociation of the $\frac{1}{2}\langle111\rangle$ dislocations into pairs of $\frac{1}{4}\langle111\rangle$ partial dislocations, separated by a SISF, was recorded.[90] A much higher CRSS has been recorded for operation of the $\{1\bar{1}0\}\langle111\rangle$ slip system in WSi_2 when compared with that of $MoSi_2$, due to the different directions of alignment of dislocation line vectors and resultant variation in the magnitude of the Peierls barrier. Furthermore, TEM studies have shown that the line vectors of the $\frac{1}{2}\langle111\rangle$ dislocations in the $\{1\bar{1}0\}$ planes of WSi_2 tend to align along the $\langle110\rangle$ direction. As the $(1\bar{1}0)$ and (001) planes intersect, the $\frac{1}{2}[111]$ dislocations gliding on the former plane encounter the stacking faults mentioned above, and thereby their movement is impeded.

Like the CRSS, the flow behavior also depends on the orientation of the WSi_2 crystals. Yield-point phenomena can be observed in WSi_2 crystals with the stress axis along the $[\bar{1}10]$ direction between 1100°C and 1300°C and along the $[\bar{1}12]$ direction between 1300°C and 1500°C.[271] Moreover, the magnitude of yield drop is found to decrease with increasing temperature. This observation is consistent with the expected increase in the density of mobile dislocations, as their movement is made easier by thermal activation and by operation of additional slip systems with increase in temperature.

6.4.4.2 Niobium Silicides

6.4.4.2.1 NbSi₂

The only slip system in C40-structured $NbSi_2$ is $(0001)\langle1\bar{2}10\rangle$.[50,242] Although this intermetallic is structurally similar to $Mo(Si,Al)_2$, it is interesting to note that it exhibits very different plastic deformation behavior (Table 6.13). Whereas compression tests have shown that the onset of plastic deformation takes place in single crystals of Mo(Si,Al)$_2$ at 1100°C, it occurs at the much lower temperature of 500°C in the case of NbSi$_2$. Furthermore, single crystals of NbSi$_2$ predeformed at 1400°C have shown plastic deformation at temperatures as low as 200°C. It is intuitive that prior plastic deformation at 1400°C creates a reasonable density of potentially mobile dislocations, which enhances the ductility or reduces the BDTT and yield stress at lower temperatures. Prior to any kind of deformation, a single crystal of NbSi$_2$ has a lower density of dislocations, because creation of dislocations at lower temperatures requires higher stress. Flow curves obtained from tests on NbSi$_2$ at different temperatures exhibit a yield-point phenomenon, with the drop in stress decreasing with increase in temperature between 500°C and 800°C. This behavior is attributed to dependence of dislocation mobility on thermal activation. Moreover, an anomalous increase in yield strength has been observed between 1200°C and 1500°C, with the peak being observed at 1300°C. It has also been observed that the regime of serrated plastic flow observed between 900°C and 1300°C coincides with that of anomalous increase in yield strength. Moriwaki et al. have observed serrated plastic flow in NbSi$_2$ with an impurity level of 40 wppm and 250 wppm of interstitial and metallic (including W, Ta, and Mo) impurities, respectively,[50] and this behavior is attributed to interactions of solute atoms with the dislocation core. The dislocation structure of single crystals of MoSi$_2$ and NbSi$_2$ deformed in the temperature range of anomalous increase in yield strength can be well understood from the results of a comparative study carried out by Nakano et al.[271]

Alloying with Mo, Al, or Ti has been found to stabilize C11$_b$, C40, and C54 structures, respectively.[271] While Mo or Ti atoms substitute in the Nb sublattice sites, the Al atoms substitute in sites occupied by Si atoms. As the stacking faults in the C40 lattice have C11$_b$ structure, it is more likely that the Mo atoms will tend to segregate and stabilize the SISF splitting the dislocations, thereby causing an increase in both the CRSS and the temperature regime of the yield-stress anomaly. On the other hand, substitution of Si or Nb lattice sites with Al or Ti atoms, respectively, has a comparatively modest effect on the serrated plastic flow behavior.

6.4.4.2.2 Multiphase Nb-Silicides

6.4.4.2.2.1 Nb–Si-Based Binary Alloys Studies on the high temperature yield and creep behavior of the Nb$_{ss}$–Nb$_5$Si$_3$ two-phase alloys have been extensively reported in the literature.[66–74,272–284] Moreover, the flexural strength at room temperature has been found to be between 500 and 700 MPa, and a drastic fall in strength could be observed only above a temperature range of 1100°C–1450°C, depending on the composition of these composites.

The stress exponent of Nb or Nb$_{ss}$[275–277] and Nb$_5$Si$_3$[274] has been reported to be 5.8 and 1, respectively, while that of the *in situ* Nb$_{ss}$–Nb$_5$Si$_3$ composite has been found to be between 2 and 3.[272] Furthermore, the creep rates of the binary Nb_{ss}–Nb_5Si_3 dual-phase alloys have been found to be similar to those of the Ni-based superalloys. Through analyses of the experimental results, it has been shown that the creep behavior of monolithic Nb_5Si_3 is controlled by diffusion of Nb atoms in Nb_5Si_3.[274] Using suitable theoretical models, the cause for wide variation in the stress exponents has been explained on the basis of the presence of rigid or creeping silicides.[278] The composites with a creeping matrix and a rigid silicide phase show stress exponents similar to that of the matrix. On the other hand, the stress exponent of the composite with both matrix and silicide constituents creeping has been found to be similar to that of the latter phase.

6.4.4.2.2.2 Nb–Si-Based Multicomponent Alloys   Solid-solution strengthening plays a significant role in the high-temperature deformation of multicomponent Nb-silicides. Studies on the mechanical behavior of the Nb–Si–Mo alloys have shown significant solid-solution strengthening of the Nb_{ss} phase.[197,198] For a similar Mo content in these alloys, both yield strength and maximum compressive stress are found to scale with the volume fraction of $(Nb,Mo)_5Si_3$ phase, and the hypereutectic alloys are found to have higher strength than that of the hypoeutectic alloys. Moreover, the hypoeutectic alloys have been found to have a greater ability to retain strength at elevated temperatures than the hypereutectic alloys, probably due to the presence of silicide precipitates in the primary Nb_{ss} phase.[198] Studies have shown impressive strength retention by multicomponent Nb–Si-based alloys at temperatures exceeding 1200°C.[75] The compressive yield strength of Nb–W–Si–B alloys has been found to be in the range of 540–570 MPa at 1200°C and 320–340 MPa at 1400°C.[192] High-temperature tensile deformation of *in situ* processed multiphase Nb-silicides with the composition of Nb–18Si–5Mo–5Hf–2C,[279] which are constituted of Nb_{ss} and $(Nb,X)_5Si_3$ (X = Mo or Hf) phases, has shown a maximum yield stress of about 320 MPa at 1300°C. The amount of elongation increases from as low as 0.8% at 1200°C to 37% at 1300°C, with a further increase to 72% at 1400°C, suggesting that the BDTT of this alloy lies between 1200°C and 1300°C. Although Nb_5Si_3 is undeformable, the Nb_{ss} phase is ductile and it softens further with increasing temperature, such that it arrests cracks easily. However, plastic deformation of the Nb_{ss} phase is limited by the constraint due to its bonding with the neighboring Nb_5Si_3 phase. Based on the studies of fracture surfaces at higher temperatures of 1300°C and 1400°C, it has been proposed that partial debonding at the Nb_{ss}–Nb_5Si_3 interface permits enhanced plastic deformation of the ductile Nb_{ss} phase to the point of necking. The mechanisms of high-temperature tensile fracture involve: (i) cleavage of brittle Nb_5Si_3 at temperatures <1200°C; (ii) interfacial decohesion followed by ductile rupture of the Nb_{ss} phase at 1300°C and 1400°C; and (iii) ductile rupture of the Nb_{ss} phase followed by cleavage of Nb_5Si_3 at ≥1500°C. As the flow stress of the Nb_{ss} phase decreases significantly at 1500°C, it is less than the stress required for interfacial decohesion.

Nb-silicide-based composites have been developed, aiming at a target of <1% creep strain recorded after being subjected to stresses exceeding 170 MPa for 125 h at 1200°C.[75] Nb–8Hf–25Ti–18Si alloys have shown 0.1% creep strain at a stress of 140 MPa and 1200°C, which is not only close to the target but also superior to the second-generation Ni-base superalloys (2GSX).

6.4.4.3 *CrSi₂*

Inui et al. have reported that single crystals of $CrSi_2$ undergo plastic deformation only at temperatures ≥800°C through basal slip, involving dislocations with a Burgers vector of $\frac{1}{3}\langle 1\bar{2}10\rangle$ (Table 6.13).[56] Tests on single-crystal specimens oriented along the $[\bar{7}708]$ direction have shown the yield stress to decrease very drastically with increasing temperature between 800°C and 1000°C in a manner similar to what has been observed in the case of $Mo(Si,Al)_2$ above its BDTT. Another similarity of $CrSi_2$ to $Mo(Si,Al)_2$ is the characteristic yield drop, which decreases with increasing temperature and is almost negligible at temperatures ≥1100°C.

It has been observed that perfect dislocations with $b=\frac{1}{3}\langle 1\bar{2}10\rangle$ dissociate further into partial dislocations with $b=1/6\langle 1\bar{2}10\rangle$, which are separated by SISF. The dissociation width has been found to be independent of the orientation of the dislocations, similar to that observed in $Mo(Si,Al)_2$, and much larger than that in $NbSi_2$. Furthermore, HRTEM observations have shown that the stacking sequence of the faulted region occurs by a synchroshear mechanism, similar to that observed in the case of $Mo(Si,Al)_2$. The synchroshear mechanism involves deformation encountering a higher Peierls barrier. Moreover, the BDTT of $CrSi_2$ is reported to be higher than that of $NbSi_2$, which demonstrates conventional shear. Relatively higher Peierls stress of $CrSi_2$ is consistent with higher directionality of its chemical bonding than that of other C40 silicides such as $NbSi_2$ and $Mo(Si,Al)_2$.

The yield strengths of selected orientations of single crystals of $MoSi_2$, WSi_2, Ti_5Si_3, $NbSi_2$, and $CrSi_2$ are shown in Figure 6.19. Single crystals of $MoSi_2$ with [001] orientation, WSi_2 with $[\bar{1}10]$ and [0 15 1] orientations, and Ti_5Si_3 with [0001] orientation have demonstrated the highest strength retention at temperatures in the range of 1300°C–1500°C. The plots shown in Figure 6.19 are consistent with the fact that $MoSi_2$ [001] is harder than other investigated orientations of $MoSi_2$. This trend is also in tune with the conclusions of Tanaka et al. on the amount of bond directionality in silicides.[91] The directionality of bonding in the $C11_b$ unit cell of WSi_2 is expected to be greater than that in $MoSi_2$ based on theoretical predictions. However, it appears to be stronger along the [001] $MoSi_2$ due to the presence of stacking faults in the (001) planes of WSi_2. Figure 6.20 shows plots depicting the creep behavior of polycrystalline $MoSi_2$, Ti_5Si_3, Nb_5Si_3, and Cr_3Si at 1200°C. From the results in this figure, it is clear that Nb_5Si_3 has the highest creep resistance among the aforementioned intermetallics. Quite interestingly, the creep strengths of $MoSi_2$ and Ti_5Si_3 with similar grain sizes are very close. It may be noted that the melting

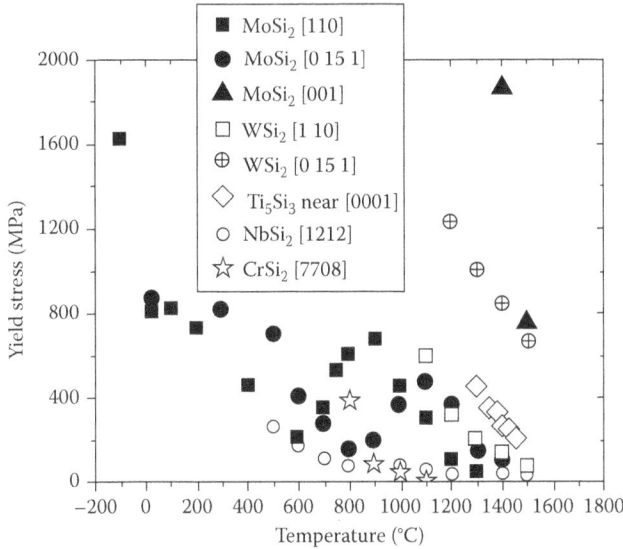

FIGURE 6.19

Comparison of yield strengths of selected orientations of single crystals of MoSi$_2$, WSi$_2$, Ti$_5$Si$_3$, NbSi$_2$, and CrSi$_2$. (From Ito, K., et al., *Phil. Mag. A*, 72, 1075–1097, 1995; Ito, K., et al., *Intermetallics*, 4, S119–S131, 1996; Umakoshi, Y. and T. Nakashima, *Scr. Metall. Mater.*, 30, 11, 1431–1436, 1994; Moriwaki, M., et al., *Mater. Sci. Eng. A*, 239–240, 69–74, 1997; Inui, H., et al., *Mater. Sci. Eng. A*, 239–240, 63–68, 1997.)

point of Ti$_5$Si$_3$ is slightly higher than that of MoSi$_2$, and grain-boundary diffusion plays an important role in creep deformation of both.

6.4.4.4 Ti$_5$Si$_3$

The deformation behavior of single crystals of Ti$_5$Si$_3$ has been examined by Umakoshi and Nakashima through compression tests at a strain rate of $2 \times 10^{-4}\,s^{-1}$ in the temperature range of 1300°C–1500°C.[47] In this study, a decrease in yield strength, accompanied by a sharp increase in ductility, with increasing temperature has been observed. The yield strengths of Ti$_5$Si$_3$ single crystals at a particular temperature strongly depend on the crystal orientation, similarly to the observations reported for other silicides. Serrated plastic flow has also been observed during the compressive deformation of Ti$_5$Si$_3$. Moreover, TEM studies of the postdeformed microstructures have shown the presence of dislocations and a large number of deformation twins; based on this, twinning in the $\{10\bar{1}2\}\langle10\bar{1}\bar{1}\rangle$ system has been proposed as the major mode of deformation.

Polycrystalline Ti$_5$Si$_3$ with coarse grain size ranging between 20 and 50 μm has been tested in compression between 1000°C and 1500°C by Rosenkranz et al. at a strain rate of $10^{-4}\,s^{-1}$.[46] In another study, Ti$_5$Si$_3$ with an average grain size of 5 μm has been tested in the temperature range of 1100°C–1200°C at

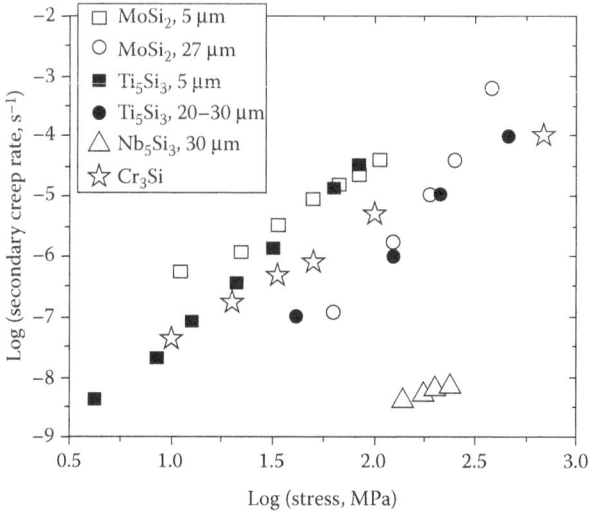

FIGURE 6.20

Comparison of the secondary creep rates of polycrystalline $MoSi_2$, Ti_5Si_3, Nb_5Si_3, and Cr_3Si at 1200°C. (Tanaka, K., et al., *Intermetallics*, 6, 607–611, 1998; Raj, S. V., *Mater. Sci. Eng. A*, 201, 229–241, 1995; Sadananda, K., et al., *Mater. Sci. Eng. A*, 261, 223–238, 1999; Mitra, R., et al., *Intermetallics*, 12, 827–836, 2004; Subramanian, P. R., et al. In *High Temperature Silicides and Refractory Alloys*, *Mat. Res. Soc. Symp. Proc.* MRS, Warrendale, PA, 322:491, 1994.)

a strain rate of 10^{-3} s^{-1} by Mitra.[48] The yield strengths of both studies are presented in Table 6.14. Serrated plastic flow has been observed in the case of polycrystalline Ti_5Si_3 as well (Figure 6.21). Both interaction of dislocations with solute atoms and twinning are possible mechanisms for serrations in the flow curves of single and polycrystalline Ti_5Si_3.

Similarly to the behavior of $MoSi_2$, the yield strength of fine-grained Ti_5Si_3 is greater than that of coarse-grained samples at higher strain rate and lower temperature, while the reverse is true at lower strain rates and higher temperatures. The dependence of yield strength on grain size, temperature, and strain rates suggests a strong role of grain boundaries in deformation behavior. Similarly to the behavior of single crystals, the flow curves show a sharp drop in the yield strength with increasing temperature.[48]

TABLE 6.14

Yield Strength of Ti_5Si_3 of Different Grain Sizes

Temperature (°C)	Yield Strength (MPa)	
	Strain Rate = 10^{-3} s^{-1}, Grain Size = 5 μm	Strain Rate = 10^{-4} s^{-1}, Grain Size = 25 μm
1100	1115	826
1200	550	609

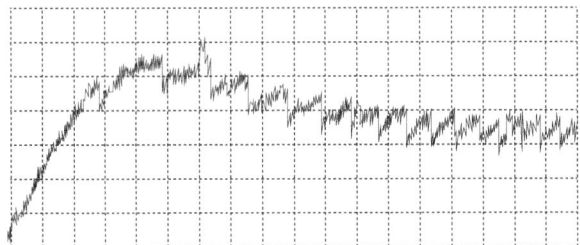

FIGURE 6.21
Typical serrations in the flow curve of polycrystalline Ti_5Si_3 compression tested at 1200°C with a strain rate of $10^{-3}\,s^{-1}$. (From Mitra, R., *Metall. Mater. Trans. A*, 29, 1629–1641, 1998.)

The creep data of Ti_5Si_3-based materials are shown in Table 6.15. Rosenkranz et al. have observed the stress exponent to be ≈ 3 in the temperature range of 1000°C–1200°C,[46] which is the same as that reported by Sadananda et al. in the temperature range of 1200°C–1400°C.[141] A similar stress exponent of 3.0 for both coarse and fine grain-size specimens suggests the occurrence of power-law creep for both. However, the secondary creep rates of Ti_5Si_3

TABLE 6.15

Type of Creep Tests, Stress Exponents, and Activation Energies of Ti_5Si_3 and Multiphase Nb-Silicides

Silicides	Type of Creep Test	Temperature (°C)	Stress Exponent, n	Activation Energy, Q_c (kJ [mol.K]$^{-1}$)	References
Ti_5Si_3 (20–50 µm)	Compression	1000–1200	3	350	Rosenkranz et al.[46]
Ti_5Si_3 (6 µm)	Compression	1200–1400	3	620–640	Sadananda et al.[141]
Ti_5Si_3– 20 vol.% TiB_2	Compression	1200–1400	2		Sadananda et al.[141]
Ti_5Si_3–8 wt.% Al	Compression	1200–1400	2		Sadananda et al.[141]
Nb_5Si_3	Compression	1100–1400 (143–174 MPa)	0.8–0.9		Subramanian et al.[272]
		1400 (174–418 MPa)	2.2		
Nb–10 Si (Nb–Nb_5Si_3)	Compression	1200	2.7		Subramanian et al.[273]
Nb–16Si	Compression	1200	3.1		Bewlay et al.[274]
Nb–Ti–Hf–Si			2–3, 11		

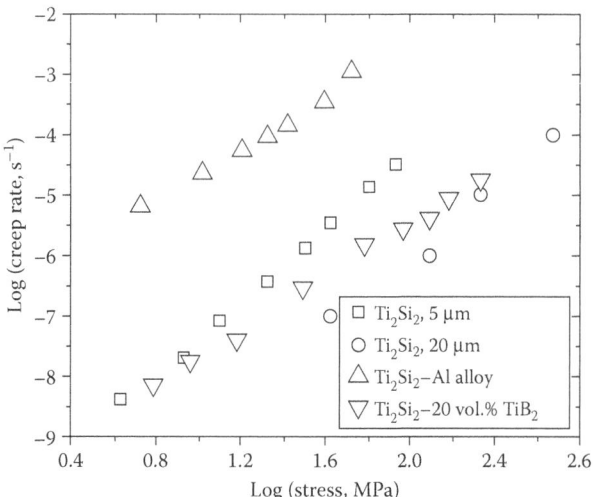

FIGURE 6.22
Comparison of the secondary creep rates of Ti_5Si_3 having average grain size of 5 μm and in the range of 20–30 μm and the effects of alloying with Al or dispersion of TiB_2 reinforcement particles. (From Rosenkranz, R., et al., *Mater. Sci. Eng. A*, 152, 288–294, 1992; Sadananda, K., et al., *Mater. Sci. Eng. A*, 261, 223–238, 1999.)

with fine grain size are higher than those of the coarse-grained specimens (Figure 6.22), indicating grain size dependence even in the dislocation-creep regime, and this behavior is somewhat similar to that observed in the case of $MoSi_2$.[141] Moreover, comparison of the creep behavior of Ti_5Si_3 with that of $MoSi_2$ having similar grain sizes has shown that power-law creep with a stress exponent of ≈3 takes place in the former material at lower stresses. The higher activation energy in the temperature regime of 1200°C–1400°C, compared with that in the range of 1000°C–1200°C, is probably due to the predominance of grain-boundary diffusion at lower temperatures and lattice diffusion at higher temperatures.

Figure 6.22 also shows a comparison of the steady-state creep rates of Ti_5Si_3 with those of Ti_5Si_3–8 wt.% Al alloy and Ti_5Si_3–20 vol.% TiB_2 composite, having an average matrix grain size of ≈5 μm at 1200°C. On alloying with 8 wt.% Al, the creep rates increase by three orders of magnitude. Alloying of Ti_5Si_3 with Al leads to the dispersion of $Al_{0.67}Si_{0.08}Ti_{0.25}$ phase and Al_2O_3 in the microstructure. The $Al_{0.67}Si_{0.08}Ti_{0.25}$ phase loses its strength rapidly with increasing temperature. On the other hand, improvement in creep resistance can be observed in the case of the Ti_5Si_3–20 vol.% TiB_2 composite, because of the strengthening effect of TiB_2 particles. The lower creep rate of Ti_5Si_3 with coarse grain size of 20 μm, when compared with that of Ti_5Si_3–20 vol.% TiB_2 composite with a finer matrix grain size, is also suggestive of a strong effect of grain size on creep, as has been found for the $MoSi_2$ matrix composites.[141]

6.5 Summary

The major challenges faced in the development of silicide-based composites are due to their inherent brittleness, poor fracture toughness, and inadequate high-temperature strength retention. Single crystals of silicides exhibit superior high-temperature yield and creep strengths, as expected. However, the fracture toughness of single crystals is found to be low because of covalent and directional bonding, thermal and elastic anisotropy, and the presence of low-energy cleavage planes. The factors affecting polycrystalline silicide specimens include all those affecting single crystals, as well as impurity concentration or segregation at grain boundaries, grain size, and methods of processing. Some breakthroughs have been possible in the case of polycrystalline silicides by the use of brittle or ductile reinforcements. Notable are the results on hybrid $MoSi_2$–SiC_f–Si_3N_4 composites and $MoSi_2$–Nb (coated) composites. Further improvement in fracture toughness has been possible in multiphase Mo–Si–B alloys and *in situ* Nb–Nb_5Si_3 composites. The mechanical behavior of the multiphase alloys or composites depends on the volume fraction, morphology, distribution, and properties of the constituent phases.

The high-temperature deformation behavior of single crystals is closely related to orientation and structure, as well as the character of the metal–Si bonds, which can be altered by suitable alloying. One can appreciate the role of bonding from the differences in behavior of silicides with similar crystal structures, particularly $MoSi_2$ and WSi_2 ($C11_b$) or $NbSi_2$ and $CrSi_2$ (C40). Yield-point phenomena, anomalous increase in yield strength, and serrated plastic flow are quite common in single crystals of silicides, indicating poor mobility of dislocations, high Peierls barrier, and strong interaction of substitutional alloying elements or interstitial impurities with the dislocation core. The insufficient number of slip systems calls for an important role played by grain-boundary diffusion and thermal activation in the elevated-temperature deformation of polycrystalline silicides. The creep behavior appears to be strongly dependent on grain size. It is possible to improve creep strengths by suitable alloying or addition of elastically deforming, hard reinforcements, while maintaining a coarse grain size.

References

1. Meschter, P. J. and D. S. Schwartz. 1989. Silicide-matrix materials for high-temperature applications. *JOM* 41:52–55.
2. Keiffer, R., K. Konopicky, and F. Benesovsky. 1951. Austrian Patent 179,100.
3. Kanthal. 1953. Swedish Patent 155,836.
4. Vasudévan, A. K. and J. J. Petrovic. 1992. A comparative overview of molybdenum disilicide composites. *Mater. Sci. Eng. A* 155:1–17.

5. Maxwell, W. A. 1951. Study of molibdenum disilicide for elevated temperature application. In *Proceedings of Metallurgy and Materials Information Meet*, Oak Ridge, TN, April 16–18, 1951, Rep. TID-5061 (DEL), NSA-12.

6. Fitzer, E. 1953. Silicides of W and Mo. In *Proceedings of the 1st Plansee Seminar*, 244. Vienna, Austria: Springer.

7. Nowotny, H., E. Parthe, R. Kieffer, and P. Benesovsky. 1954. The ternary system: Molybdenum-silicon-carbon. *Monatsh. Chem.* 85:255–272.

8. Fitzer, E., O. Rubisch, J. Schlichting, and I. Sewdas. 1973. *Spec. Ceram.* 6:24.

9. Petrovic, J. J. 1997. High temperature structural silicides. *Ceram. Eng. Sci. Proc.* 18:3–17.

10. Petrovic, J. J. and A. K. Vasudevan. 1999. Key developments in high temperature structural silicides. *Mater. Sci. Eng. A* 261:1–5.

11. Berczik, D. M. 1997. US Patent No. 5,595,616, January 21, 1997.

12. Dimiduk, D. M. and J. H. Perepezko. 2003. Mo–Si–B alloys: Developing a revolutionary turbine-engine material. *MRS Bull.* 28:639–645.

13. Kumar, K. S. 1995. Silicides: Science, technology, and applications. In *Intermetallic Compounds: Principles and Practice*, eds. J. H. Westbrook and R. L. Fleischer, Vol. 2. Chichester, UK: Wiley.

14. Mitra, R. 2006. Mechanical behavior and oxidation resistance of structural silicides. *Inter. Mater. Rev.* 51:13.

15. Lemberg, J. A. and R. O. Ritchie. 2012. Mo–Si–B alloys for ultrahigh-temperature structural applications. *Adv. Mater.* 24:3445–3480.

16. Shah, D. M., D. Berczik, D. L. Anton, and R. Hecht. 1992. Appraisal of other silicides as structural materials. *Mater. Sci. Eng. A* 155:45–57.

17. Jackson, M. R., B. P. Bewlay, R. G. Rowe, D. W. Skelly, and H. A. Lipsitt. 1996. High-temperature refractory metal-intermetallic composites. *JOM* 48:39–44.

18. Zhao, J. C. and J. H. Westbrook. 2003. Ultrahigh-temperature materials for jet engines. *MRS Bull.* 28:622–630.

19. Tanaka, K., H. Onome, H. Inui, M. Yamaguchi, and M. Koiwa. 1997. Single-crystal elastic constants of $MoSi_2$ with the $C11_b$ structure. *Mater. Sci. Eng. A* 239–240:188–194.

20. Chu, F., M. Lei, S. A. Maloy, J. J. Petrovic, and T. E. Mitchell. 1996. Elastic properties of C40 transition metal disilicides. *Acta Mater.* 44:3035–3048.

21. Ito, K., H. Inui, Y. Shirai, and M. Yamaguchi. 1995. Plastic deformation of $MoSi_2$ single crystals. *Phil. Mag. A* 72:1075–1097.

22. Aikin Jr, R. M. 1992. On the ductile-to-brittle transition temperature in $MoSi_2$. *Scr. Metall. Mater.* 26:1025–1030.

23. Ito, K., T. Yano, T. Nakamoto, H. Inui, and M. Yamaguchi. 1996. Plastic deformation of $MoSi_2$ and WSi_2 single crystals and directionally solidified $MoSi_2$-based alloys. *Intermetallics* 4:S119–S131.

24. Wade, R. K. and J. J. Petrovic. 1992. Processing temperature effects on molybdenum disilicide. *J. Am. Ceram. Soc.* 75:3160–3162.

25. Gac, F. D. and J. J. Petrovic. 1985. Feasibility of a composite of SiC whiskers in an $MoSi_2$ matrix. *J. Am. Ceram. Soc.* 68:C200–C201.

26. Carter, D. H., J. J. Petrovic, R. E. Honnell, and W. S. Gibbs. 1989. SiC-$MoSi_2$ composites. *Ceram. Eng. Sci. Proc.* 10(9/10):1121–1129.

27. Suryanarayanan, R., S. M. L. Sastry, and K. L. Jerina. 1994. Mechanical properties of molybdenum disilicide based materials consolidated by hot isostatic pressing (HIP). *Acta Metall. Mater.* 42:3751–3757.

28. Mitra, R., Y. R. Mahajan, N. E. Prasad, and W. A. Chiou. 1997. Processing–microstructure–property relationships in reaction hot-pressed $MoSi_2$ and $MoSi_2/SiC_p$ composites. *Mater. Sci. Eng. A* 225:105–117.
29. Schlichting, J. 1978. Molybdenum disilicide as a component in modern high-temperature solid solutions. *High Temp. High Press.* 10:241–269.
30. Fitzer, E. 1955. Molybdandisilzid als Hochtemperatursweikstoff. In *Proceedings of 2nd Plansee Seminar*, ed. F. Benesovsky, 56. Berlin: Springer.
31. Ito, K., K. Ihara, K. Tanaka, M. Fujikura, and M. Yamaguchi. 2001. Physical and mechanical properties of single crystals of the T-2 phase in the Mo–Si–B system. *Intermetallics* 9:591–602.
32. Yoshimi, K., M. H. Yoo, A. A. Wereszczak, S. M. Borowicz, E. P. George, and R. H. Zee. 2001. Yielding and flow behavior of Mo_5Si_3 single crystals. *Scr. Mater.* 45:1321–1326.
33. Chu, F., D. J. Thoma, K. McClellan, P. Peralta, and Y. He. 1999. Synthesis and properties of Mo_5Si_3 single crystals. *Intermetallics* 7:611–620.
34. Thom, A. J., M. K. Meyer, Y. Kim, and M. Akinc. 1994. Evaluation of $A_5Si_3Z_x$ intermetallics for use as high temperature structural materials. In *Processing and Fabrication of Advanced Materials III*, eds. V. A. Ravi, T. S. Srivatsan, and J. J. Moore, 413. Warrendale, PA: TMS.
35. Meyer, M. K. and M. Akinc. 1996. Oxidation behavior of boron-modified Mo_5Si_3 at 800 degrees–1300°C. *J. Am. Ceram. Soc.* 79:938–944.
36. Choe, H., D. Chen, J. H. Schneibel, and R. O. Ritchie. 2001. Ambient to high temperature fracture toughness and fatigue-crack propagation behavior in a Mo-12Si-8.5B (at.%) intermetallic. *Intermetallics* 9:319–329.
37. Rosales, I. and J. H. Schneibel. 2000. Stoichiometry and mechanical properties of Mo_3Si. *Intermetallics* 8:885–889.
38. Ihara, K., K. Ito, K. Tanaka, and M. Yamaguchi. 2002. Mechanical properties of Mo_5SiB_2 single crystals. *Mater. Sci. Eng. A* 329–331:222–227.
39. Yoshimi, K., S. Nakatani, T. Suda, S. Hanada, and H. Habazaki. 2002. Oxidation behavior of Mo_5SiB_2-based alloy at elevated temperatures. *Intermetallics* 10:407–414.
40. Tanaka, K., K. Nawata, H. Inui, M. Yamaguchi, and M. Koiwa. 1998. Temperature dependence of single-crystal elastic constants of $Mo(Si,Al)_2$. *Intermetallics* 6:607–611.
41. Inui, H., M. Moriwaki, K. Ito, and M. Yamaguchi. 1998. Plastic deformation of single crystals of $Mo(Si,Al)_2$ with the C40 structure. *Phil. Mag. A* 77:375–394.
42. Kisley, P. S. and V. Y. Kodash. 1989. The mullite coatings on heaters made of molybdenum disilicide. *Ceram. Inter.* 15:189–191.
43. Kimura, K., M. Nakamura, and T. Hirano. 1990. High-temperature deformation-behavior of $MoSi_2$ and WSi_2 single-crystals. *J. Mater. Sci.* 25:2487–2492.
44. Hirano, T., M. Nakamura, K. Kimura, and Y. Umakoshi. 1991. Single crystal growth and mechanical properties of $MoSi_2$ and WSi_2. *Ceram. Eng. Sci. Proc.* 12:1619.
45. Inui, H., K. Ito, T. Nakamoto, K. Ishikawa, and M. Yamaguchi. 2001. Stacking faults on (001) and their influence on the deformation and fracture behavior of single crystals of $MoSi_2$–WSi_2 solid-solutions with the C11b structure. *Mater. Sci. Eng. A* 314:31–38.
46. Rosenkranz, R., G. Frommeyer, and W. Smarsly. 1992. Microstructures and properties of high melting point intermetallic Ti_5Si_3 and $TiSi_2$ compounds. *Mater. Sci. Eng. A* 152:288–294.

47. Umakoshi, Y. and T. Nakashima. 1994. High temperature deformation of Ti_5Si_3 single crystals with D88 structure. *Scr. Metall. Mater.* 30(11):1431–1436.
48. Mitra, R. 1998. Microstructure and mechanical behavior of reaction hot-pressed titanium silicide and titanium silicide-based alloys and composites. *Metall. Mater. Trans. A* 29:1629–1641.
49. Mitra, R. and V. V. R. Rao. 1998. Elevated-temperature oxidation behavior of titanium silicide and titanium silicide based alloy and composite. *Metall. Mater. Trans. A* 29:1665–1675.
50. Moriwaki, M., K. Ito, H. Inui, and M. Yamaguchi. 1997. Plastic deformation of single crystals of $NbSi_2$ with the C40 structure. *Mater. Sci. Eng. A* 239–240:69–74.
51. Pitman, S. H. 1996. Development of $NbSi_2$ base intermetallic alloys. PhD dissertation. School of Engineering, University of Surrey, Surrey, UK.
52. Zhang, L. and J. Wu. 1997. Thermal expansion and elastic moduli of the silicide based intermetallic alloys $Ti_5Si_3(X)$ and Nb_5Si_3. *Scr. Mater.* 38:307–313.
53. Nekkanti, R. K. and D. M. Dimiduk. 1990. Ductile-phase toughening in niobium-niobium silicide powder processed composite. In *Intermetallic Matrix Composites I*, eds. D. L. Anton, R. McMeeking, D. Miracle, and P. L. Martin., Materials Research Society Symposium Proceedings, Vol. 194, 175–182.
54. Mendiratta, M. G., J. J. Lewandowski, and D. M. Dimiduk. 1991. Strength and ductile-phase toughening in the 2-phase Nb/Nb_5Si_3 alloys. *Metall. Trans. A* 22:1573–1583.
55. Liu, Y., A. J. Thom, M. J. Kramer, and M. Akinc. 2003. Processing and oxidation of Nb-Si-B intermetallics. In *Processing and Fabrication of Advanced Materials for High Temperature Applications*, eds. T. S. Srivatsan and V. A. Ravi, 258–271. Materials Park, OH: ASM.
56. Inui, H., M. Moriwaki, S. Ando, and M. Yamaguchi. 1997. Plastic deformation of single crystals of $CrSi_2$ with the C40 structure. *Mater. Sci. Eng. A* 239–240:63–68.
57. Nakamura, M. 1994. Elastic constants of some transition-metal-disilicide single crystals. *Metall. Mater. Trans. A* 25:331–340.
58. Grabke, H. J. and M. Brumm. 1989. Oxidation behaviour of chromium disilicide. In *Oxidation of High Temperature Intermetallics*, eds. T. Grobstein and J. Doychak, 245. Warrendale, PA: TMS.
59. Kieffer, R. and F. Benesovsky. 1956. Symposium on powder metallurgy. Special Report No. 58, p. 292. London: The Iron and Steel Institute.
60. Fleischer, R. L. and R. J. Zabala. 1989. Mechanical properties of diverse binary high-temperature intermetallic compounds. Report No. 89CRD201. Schenectady, NY: General Electric Research and Development.
61. Chang, C. S. and D. P. Pope. 1991. High temperature deformation of single crystalline Cr_3Si (A15) intermetallic compound. In *High Temperature Ordered Intermetallic Alloys IV*, Materials Research Society Symposium Proceedings. Vol. 213, p. 745. Pittsburgh, PA.
62. Anton, D. L. and D. M. Shah. 1996. Development potential of intermettalic compounds. *Proceedings of International Conference on Intermetallic Compounds: Structure and Mechanical Properties (JIMIS-6)*, 379. Sendai, Japan: The Japan Institute of Metals.
63. Newkirk, J. W. and J. A. Sago. 1990. Ductile phase toughening of Cr_3Si with chromium. In *Intermetallic Matrix Composites*, eds. D. L. Anton, P. L. Martin, D. B. Miracle, and R. McMeeking, Materials Research Society Symposium Proceedings, Vol. 194, 183–190. Pittsburgh, PA: MRS.

64. Raj, S. V. 1995. An evaluation of the properties of Cr_3Si alloyed with Mo. *Mater. Sci. Eng. A* 201:229–241.
65. Ma, J., Y. Gu, L. Shi, L. Chen, Z. Yang, and Y. Qian. 2004. Synthesis and oxidation behavior of chromium silicide (Cr_3Si) nanorods. *J. Alloys Comp.* 375:249–252.
66. Rigney, J. D., P. M. Singh, and J. J. Lewandowski. 1992. Environmental effects on ductile-phase toughening in Nb_5Si_3-Nb composites. *JOM* 44:36–41.
67. Subramanian, P. R., M. G. Mendiratta, and D. M. Dimiduk. 1996. The development of Nb-based advanced intermetallic alloys for structural applications. *JOM* 48:33–38.
68. Subramanian, P. R., M. G. Mendiratta, D. M. Dimiduk, and M. A. Stucke. 1997. Advanced intermetallic alloys: Beyond gamma titanium aluminides. *Mater. Sci. Eng. A* 239–240:1–13.
69. Bewlay, B. P., J. J. Lewandowski, and M. R. Jackson. 1997. Refractory metal-intermetallic *in-situ* composites for aircraft engines. *JOM* 49:44–45.
70. Bewlay, B. P., M. R. Jackson, and P. R. Subramanian. 1999. Processing high-temperature refractory-metal silicide *in-situ* composites. *JOM* 51:32–36.
71. Zhao, J.-C., B. P. Bewlay, M. R. Jackson, and L. A. Peluso. 2001. Alloying and phase stability in niobium silicide *in-situ* composites. In *Proceedings of the 2001 Intermetallics Symposium on Structural Intermetallics*, eds. K. J. Hemker, D. M. Dimiduk, H. Clemens, R. Darolia, H. Inui, J. M. Larsen, V. K. Sikka, M. Thomas, and J. D. Whittenberger, 483. Warrendale, PA: TMS.
72. Balsone, S. J., B. P. Bewlay, M. R. Jackson, et al. 2001. Materials beyond superalloys: Exploiting high-temperature composites. GE Research and Development Report No. 2001CRD098.
73. Balsone, S. J., B. P. Bewlay, M. R. Jackson, et al. 2001. Materials beyond superalloys: Exploiting high-temperature composites. In *Proceedings of the 2001 Intermetallics Symposium on Structural Intermetallics*, eds. K. J. Hemker, D. M. Dimiduk, H. Clemens, R. Darolia, H. Inui, J. M. Larsen, V. K. Sikka, M. Thomas, and J. D. Whittenberger, 99. Warrendale, PA: TMS.
74. Bewlay, B. P., M. R. Jackson, J. C. Zhao, and P. R. Subramanian. 2002. A review of very high-temperature Nb-silicide based composites. GE Research and Development Report No. 2002GRC172.
75. Bewlay, B. P., M. R. Jackson, J. C. Zhao, P. R. Subramanian, M. G. Mendiratta, and J. J. Lewandowski. 2003. Ultrahigh-temperature Nb-silicide-based composites. *MRS Bull.* 28:646–653.
76. Bewlay, B. P., M. R. Jackson, J. C. Zhao, and P. R. Subramanian. 2003. A review of very-high-temperature Nb-silicide-based composites. *Metall. Mater. Trans. A* 34:2043–2052.
77. Subramanian, P. R., M. G. Mendiratta, and D. M. Dimiduk. April 21, 1998. US Patent 5,741,376.
78. Bewlay, B. P. and M. R. Jackson. November 10, 1998. US Patent 5,833,773.
79. Jackson, M. R. and B. P. Bewlay. August 3, 1999. US Patent 5,932,033.
80. Jackson, M. R., B. P. Bewlay, and J.-C. Zhao. July 16, 2002. US Patent 6,419,765.
81. Powder Diffraction File, Inorganic Phases. Swarthmore, PA: International Centre for Diffraction Data. 19073-3273, 2001, Card Nos. 17-917, 51-764, 34-371, 10-277, 9-292, 11-195, 29-1362, 8-450, 8-422, 35-781, 7-186.
82. Villars, P. and L. D. Calvert. 1991. *Pearson's Handbook of Crystallographic Data for Intermetallic Phases*, 2nd edn. Materials Park, OH: ASM International.

83. Chu, F., D. J. Thoma, K. J. McClellan, and P. Peralta. 1999. Mo_5Si_3 single crystals: Physical properties and mechanical behavior. *Mater. Sci. Eng. A.* 261:44–52.

84. Umakoshi, Y., T. Sakagami, T. Hirano, and T. Yamane. 1990. High temperature deformation of $MoSi_2$ single crystals with the $C11_b$ structure. *Acta Metall. Mater.* 38:909–915.

85. Maloy, S. A., A. H. Heuer, J. J. Lewandowski, and T. E. Mitchell. 1992. On the slip systems in $MoSi_2$. *Acta Metall. Mater.* 40:3159–3165.

86. Mason, D. P. and D. C. Van Aken. 1995. On the creep of directionally solidified $MoSi_2$–Mo_5Si3 eutectics. *Acta Metall. Mater.* 43:1201–1210.

87. Meyer, M. K., M. J. Kramer, and M. Akinca. 1996. Compressive creep behavior of Mo5Si3 with the addition of boron. *Intermetallics* 4:273–281.

88. Field, R. D., D. J. Thoma, J. C. Cooley, F. Chu, C. L. Fu, M. H. Yoo, W. L. Hults, and C. M. Cady. 2001. Dislocations in Mo_5SiB_2 T2 phase. *Intermetallics* 9:863–868.

89. Tabaru, T., K. Shobu, M. Sakamoto, H. Hirai, and S. Hanada. 2003. Oxidation behavior of $Mo(Si_{0.6},Al_{0.4})_2/HfB_2$ composites as aluminum reservoir materials for protective Al_2O_3 formation. *Scr. Mater.* 49:767–772.

90. Ito, K., T. Yano, T. Nakamoto, H. Inui, and M. Yamaguchi. 1999. Plastic deformation of single crystals of WSi_2 with the Cl1(b) structure. *Acta Mater.* 47:937–949.

91. Tanaka, K., H. Inui, M. Yamaguchi, and M. Koiwa. 1999. Directional atomic bonds in $MoSi_2$ and other transition-metal disilicides with the C11b, C40 and C54 structures. *Mater. Sci. Eng. A* 261:158–164.

92. Jeng, Y. L. and E. J. Lavernia. 1994. Processing of molybdenum disilicide. *J. Mater. Sci.* 29:2557–2571.

93. Stoloff, N. S. 1999. An overview of powder processing of silicides and their composites. *Mater. Sci. Eng. A* 261:169–180.

94. Yao, Z., J. Stiglich, and T. S. Sudarshan. 1999. Molybdenum silicide based materials and their properties. *J. Mater. Eng. Perf.* 8:291–304.

95. Trambukis, J. and Z. A. Munir. 1990. Effect of particle dispersion on the mechanism of combustion synthesis of titanium silicide. *J. Am. Ceram. Soc.* 73:1240–1245.

96. Bhattacharya, A. K. 1991. Effect of silicon-carbide reinforcement on the properties of combustion-synthesized titanium silicide. *J. Am. Ceram. Soc.* 74:2707–2710.

97. Deevi, S. C. 1991. Self-propagating high-temperature synthesis of molybdenum disilicide. *J. Mater. Sci.* 26:3343–3353.

98. Deevi, S. C. 1992. Diffusional reactions in the combustion synthesis of $MoSi_2$. *Mater. Sci. Eng. A* 149:241–251.

99. Bhaduri, S. B., R. Radhakrishnan, and Z. B. Qian. 1993. Combustion synthesis of single-phase Ti_5Si_3. *Scr. Metall. Mater.* 29:1089–1094.

100. Subrahmanyam, J. and R. Mohan Rao. 1994. Combustion synthesis of $MoSi_2WSi_2$ alloys. *Mater. Sci. Eng. A* 183:205–210.

101. Thadhani, N. N. 1994. Shock-induced and shock-assisted solid-state chemical reactions in powder mixtures. *J. Appl. Phys.* 76(4):2129–2138.

102. Deevi, S. C. and N. N. Thadhani. 1995. Reaction synthesis of high-temperature silicides. *Mater. Sci. Eng. A* 192–193:604–611.

103. Radhakrishnan, R., S. Bhaduri, and C. H. Henager. 1997. The reactive processing of silicides. *JOM* 49:41–45.

104. Schwarz, R. B., S. R. Srinivasan, J. J. Petrovic, and C. J. Maggiore. 1992. Synthesis of molybdenum disilicide by mechanical alloying. *Mater. Sci. Eng. A* 155:75–83.

105. Patankar, S. N., S. Q. Xiao, J. J. Lewandowski, and A. H. Heuer. 1993. The mechanism of mechanical alloying of MoSi₂. *J. Mater. Res.* 8:1311–1316.
106. Jayashankar, S. and M. J. Kaufman. 1993. Tailored MoSi₂/SiC composites by mechanical alloying. *J. Mater. Res.* 8:1428–1441.
107. Castro, R. G., R. W. Smith, A. D. Rollett, and P. W. Stanek. 1992. Ductile phase toughening of molybdenum disilicide by low pressure plasma spraying. *Mater. Sci. Eng. A* 155:101–107.
108. Tiwari, R., H. Herman, and S. Sampath. 1992. Vacuum plasma spraying of MoSi₂ and its composites. *Mater. Sci. Eng. A* 155:95–100.
109. Jeng, Y. L., J. Wolfenstine, E. J. Lavernia, D. E. Bailey, and A. Sickinger. 1993. Low-pressure plasma deposition of SiC-reinforced MoSi₂. *Scr. Metall. Mater.* 28:453–458.
110. Alman, D. E., K. G. Shaw, N. S. Stoloff, and K. Rajan. 1992. Fabrication, structure and properties of MoSi₂-base composites. *Mater. Sci. Eng. A* 155:85–93.
111. Bartlett, A. H., R. G. Castro, D. P. Butt, H. Kung, and J. J. Petrovic. 1996. Plasma sprayed MoSi₂/Al₂O₃ laminate composite tubes as lances in pyrometallurgical operations. *Ind. Heat.* 63(January):33–36.
112. Maloy, S. A., A. H. Heuer, J. J. Lewandowski, and J. J. Petrovic. 1991. Carbon additions to molybdenum disilicide: Improved high-temperature mechanical properties. *J. Am. Ceram. Soc.* 74:2704–2706.
113. Henager Jr, C. H., J. L. Brimhall, and J. P. Hirth. 1992. Synthesis of a MoSi₂/SiC composite *in situ* using a solid state displacement reaction. *Mater. Sci. Eng. A* 155:109–114.
114. Jayashankar, S. and M. J. Kaufman. 1992. *Insitu* reinforced MoSi₂ composites by mechanical alloying. *Scr. Metall. Mater.* 26:1245–1250.
115. Costa e Silva, A. and M. J. Kaufman. 1995. Applications of *in situ* reactions to MoSi₂-based materials. *Mater. Sci. Eng. A* 195:75–88.
116. Pan, J., M. K. Surappa, R. A. Saravanan, B. W. Liu, and D. M. Yang. 1998. Fabrication and characterization of SiC/MoSi₂ composites. *Mater. Sci. Eng. A* 244:191–198.
117. Mitra, R., R. Khanna, and V. V. Rama Rao. 2004. Microstructure, mechanical properties and oxidation behavior of a multiphase (Mo,Cr)(Si,Al)₂ intermetallic alloy–SiC composite processed by reaction hot pressing. *Mater. Sci. Eng. A* 382:150–161.
118. Hardwick, D. A., P. L. Martin, and R. J. Moores. 1992. Reaction synthesis of MoSi₂ from high purity elemental powders. *Scr. Metall. Mater.* 27:391–394.
119. Alman, D. E. and N. S. Stoloff. 1993. Preparation of MoSi₂SiC composites from elemental powders by reactive co-synthesis. *Scr. Metall. Mater.* 28:1525–1530.
120. Mitra, R., Y. R. Mahajan, N. E. Prasad, W.-A. Chiou, and C. Ganguly. 1995. Ceramic matrix composites. In *Key Engineering Materials*, eds. G. M. Newaz, H. Neber-Aeschbacher, and F. H. Wohlbier, 108–110:11. Zurich, Switzerland: Trans Tech Publications.
121. Scholl, R., A. Böhm, and B. Kieback. 1999. Fabrication of silicide materials and their composites by reaction sintering. *Mater. Sci. Eng. A* 261:204–211.
122. Ghosh, A. K., A. Basu, and H. Kung. 1992. The effect of second phase properties on the compressive creep behavior of MoSi₂ composites. In *Intermetallic Matrix Composites II*, Materials Research Society Symposium Proceedings, Vol. 273, 259–266. Pittsburgh, PA: MRS.

123. Suryanarayanan, R., S. M. L. Sastry, and K. L. Jerina. 1994. Consolidation of molybdenum disilicide based materials by hot isostatic pressing (Hip): Comparison with models. *Acta Metall. Mater.* 42:3741–3750.

124. Raman, R. V., S. V. Rele, and M. J. Paskowitz. 1993. *In-situ* synthesis and rapid consolidation of fully dense MoSi₂. *JOM* 45:54–55.

125. Weiser, M. M., S. R. Smelser, and J. J. Petrovic. 1990. In *Intermetallic Matrix Composites*, eds. D. L. Anton, R. McMeeking, D. Miracle, and P. Martin. Materials Research Society Symposium Proceedings, Vol. 194, 53. Pittsburgh, PA: MRS.

126. Dusza, J., P. Hvizdos, W. Steinkellner, and K. Kromp. 1997. Bending creep behaviour of pressureless sintered MoSi₂. *Scr. Mater.* 37:471–476.

127. Murakami, T., S. Sasaki, K. Ichikawa, and A. Kitahara. 2001. Oxidation resistance of powder compacts of the Nb–Si–Cr system and Nb₃Si₅Al₂ matrix compacts prepared by spark plasma sintering. *Intermetallics* 9:629–635.

128. Kimura, Y., H. Yamaoka, N. Sekido, and Y. Mishima. 2005. Processing, microstructure, and mechanical properties of (Nb)/Nb₅Si₃ two-phase alloys. *Metall. Mater. Trans. A* 36:483–488.

129. Summers, E., A. J. Thom, B. Cook, and M. Akinc. 2000. Extrusion and selected engineering properties of Mo–Si–B intermetallics. *Intermetallics* 8:1169–1174.

130. Jehanno, P., M. Heilmaier, H. Kestler, M. Boning, A. Venskutonis, B. Bewlay, and M. Jackson. 2005. Assessment of a powder metallurgical processing route for refractory metal silicide alloys. *Metall. Mater. Trans. A* 36:515–523.

131. Kerle, O. 2002. In *Plansee Catalogue, Refractory Metals and Alloys.*

132. Jehanno, P., M. Heilmaier, and H. Kestler. 2004. Characterization of an industrially processed Mo-based silicide alloy. *Intermetallics* 12:1005–1009.

133. Schneibel, J. H., C. T. Liu, L. Heatherly, and M. J. Kramer. 1998. Assessment of processing routes and strength of a 3-phase molybdenum boron silicide (Mo₅Si₃–Mo₅SiB₂Mo₃Si). *Scr. Mater.* 38:1169–1176.

134. Mendiratta, M. G. and D. M. Dimiduk. 1991. Phase relations and transformation kinetics in the high Nb region of the Nb–Si system. *Scr. Metall. Mater.* 25:237–242.

135. Brupbacher, J. M., L. Christodoulou, and D. C. Nagale. 1987. U.S. Patent 4,710,348.

136. Christodoulou, L., D. C. Nagale, and J. M. Brupbacher. 1988. U.S. Patent 4,774,052.

137. Nagale, D. C., L. Christodoulou, and J. M. Brupbacher. 1990. U.S. Patent 4,916,029.

138. Suzuki, M., S. R. Nutt, and R. M. Aikin Jr. 1993. Creep behavior of an SiC-reinforced XDTM MoSi₂ composite. *Mater. Sci. Eng. A* 162:73–82.

139. Bewlay, B. P., M. R. Jackson, and H. A. Lipsitt. 1996. The balance of mechanical and environmental properties of a multielement niobium-niobium silicide-based *in situ* composite. *Metall. Mater. Trans. A* 27:3801–3808.

140. Lograsso, T. A. 1992. Synthesis of MoSi₂ single crystals. *Mater. Sci. Eng. A* 155:115–119.

141. Sadananda, K., C. R. Feng, R. Mitra, and S. C. Deevi. 1999. Creep and fatigue properties of high temperature silicides and their composites. *Mater. Sci. Eng. A* 261:223–238.

142. Sadananda, K. and C. R. Feng. 1995. Effect of carbon addition on the creep of molybdenum disilicide composites. *Mater. Sci. Eng. A.* 192–193:862–867.

143. Costa e Silva, A. and M. J. Kaufman. 1993. Microstructural modification of MoSi$_2$ through aluminum additions. *Scr. Metall. Mater.* 29:1141–1145.
144. Mitra, R., V. V. R. Rao, and A. V. Rao. 1999. Effect of small aluminum additions on microstructure and mechanical properties of molybdenum di-silicide. *Intermetallics* 7:213–232.
145. Petrovic, J. J. 1995. Mechanical behavior of MoSi$_2$ and MoSi$_2$ composites. *Mater. Sci. Eng. A* 192–193:31–37.
146. Wade, R. K. and J. J. Petrovic. 1992. Fracture modes in MoSi$_2$. *J. Am. Ceram. Soc.* 75:1682–1684.
147. Petrovic, J. J. 2000. Toughening strategies for MoSi$_2$-based high temperature structural silicides. *Intermetallics* 8:1175–1182.
148. Yang, J.-M. and S. M. Jeng. 1990. Development of MoSi$_2$ based composites. In *Intermetallic Matrix Composites I*, eds. D. L. Anton, R. McMeeking, D. Miracle, and P. L. Martin, Materials Research Society Symposium Proceedings, Vol. 194, 139–146. Pittsburgh, PA: MRS.
149. Petrovic, J. J. and R. E. Honnell. 1990. Partially-stabilized ZrO$_2$ particle- MoSi$_2$ matrix composites. *J. Mater. Sci.* 25:4453–4456.
150. Petrovic, J. J., A. K. Bhattacharya, R. E. Honnell, T. E. Mitchell, R. K. Wade, and K. J. McClellan. 1992. ZrO$_2$ and ZrO$_2$–SiC particle reinforced MoSi$_2$ matrix composites. *Mater. Sci. Eng. A* 155:259–266.
151. Richardson, K. K. and D. W. Freitag. 1991. Mechanical properties of hot pressed SiC platelet-reinforced MoSi$_2$. *Ceram. Eng. Sci. Proc.* 12(9–10):1679.
152. Henager Jr, C. H., J. L. Brimhall, and J. P. Hirth. 1992. Synthesis of composites *in situ* using solid state displacement reactions. *Ceram. Eng. Sci. Proc.* 13(7–8):596.
153. Mitra, R., N. Eswara Prasad, A. Venugopal Rao, and Y. R. Mahajan. 1997. Molybdenum and titanium silicide-based composites and alloys. In *Structural Intermetallics 1997*, eds. M. V. Nathal, R. Darolia, C. T. Liu, P. L. Martin, D. B. Miracle, R. Wagner, and M. Yamaguchi, 959–968. Warrendale, PA: TMS.
154. Lee, J. I., N. L. Hecht, and T. I. Mah. 1998. *In situ* processing and properties of SiC/MoSi$_2$ nanocomposites. *J. Am. Ceram. Soc.* 81:421–424.
155. Hebsur, M. G. 1999. Development and characterization of SiC(f)/MoSi$_2$–Si$_3$N$_4$(p) hybrid composites. *Mater. Sci. Eng. A* 261:24–37.
156. Hebsur, M. G. and M. V. Nathal. 1997. Strong, tough, and pest resistant MoSi2-base hybrid composite for structural applications. In *Structural Intermetallics 1997*, eds. M. V. Nathal, R. Darolia, C. T. Liu, P. L. Martin, D. B. Miracle, R. Wagner, and M. Yamaguchi, 949–958. Warrendale, PA: TMS.
157. Schneibel, J. H. and J. A. Sekhar. 2003. Microstructure and properties of MoSi$_2$–MoB and MoSi$_2$–Mo$_5$Si$_3$ molybdenum silicides. *Mater. Sci. Eng. A* 340:204–211.
158. Mitra, R., Y. R. Mahajan, Y. S. Rao, N. Eswara Prasad, and S. Kumari. 2002. Molybdenum silicides: Processing-structure-property relationships and future trends. In *Proceedings of the Nonferrous Meet: 2002*. Nonferrous Materials Technology Development Corporation, June 28–29, 2002, Hyderabad, 170–177.
159. Bhattacharya, A. K. and J. J. Petrovic. 1991. Hardness and fracture-toughness of SiC-particle-reinforced MoSi$_2$ composites. *J. Am. Ceram. Soc.* 74:2700–2703.
160. Choe, H., T. Hsieh, Y. Wu, E. J. Lavernia, and J. Wolfenstine. 1996. The effect of powder processing on the mechanical behavior of MoSi$_2$-10 vol% Si$_3$N$_4$ composites. *J. Mater. Synth. Proc.* 4:235–243.

161. Idasetima, J. S. and J. J. Petrovic. 1997. Effects of carbon additions on $MoSi_2$ microstructure and properties. *Phil. Mag. A* 76:271–287.

162. Lu, T. C., A. G. Evans, R. J. Hecht, and R. Mehrabian. 1991. Toughening of $MoSi_2$ with a ductile (niobium) reinforcement. *Acta Metall. Mater.* 39:1853–1862.

163. Xiao, L. and R. Abbaschian. 1992. Interfacial modification of $Nb/MoSi_2$ composites and its effects on fracture toughness. *Mater. Sci. Eng. A* 155:135–145.

164. Castro, R. G., R. W. Smith, A. D. Rollett, and P. W. Stanek. 1992. Toughness of dense $MoSi_2$ and $MoSi_2$ tantalum composites produced by low pressure plasma deposition. *Scr. Metall. Mater.* 26:207–212.

165. Alman, D. E. and N. S. Stoloff. 1994. Tensile behavior of metal-intermetallic ($Ni-Ni_2Al_3$ and $Ti-Al_3Ti$) layered composites. In *Fatigue and Fracture of Ordered Intermetallic Materials I*, eds. W. O. Soboyejo, T. S. Srivatsan, and D. L. Davidson, 13. Warrendale, PA: TMS.

166. Chen, L. C., N. Bahtishi, R. Lederich, and W. Soboyejo. 1994. Ductile phase toughening of $MoSi_2$: Effect of reinforcement morphology. In *Intermetallic Matrix Composites III*, eds. R. R. Bowman, J. A. Graves, and J. J. Lewandowski, 350:183.

167. Badrinarayanan, K., A. L. McKelvey, R. O. Ritchie, and K. T. Venkateswara Rao. 1996. Fracture and fatigue-crack growth behavior in ductile-phase toughened molybdenum disilicide: Effects of niobium wire vs. particulate reinforcements. *Metall. Mater. Trans. A* 27:3781–3792.

168. Venkateswara Rao, K. T. and R. O. Ritchie. 1994. Fracture and fatigue considerations in the development of ductile-phase reinforced intermetallic-matrix composites. In *Fatigue and Fracture of Ordered Intermetallic Materials I*, eds. W. O. Soboyejo, T. S. Srivatsan, and D. L. Davidson, 3. Warrendale, PA: TMS.

169. Carter, D. H. and P. L. Martin. 1990. Ta and Nb reinforced $MoSi_2$. Materials Research Society Symposium Proceedings, Vol. 194, 131. Warrendale, PA: MRS.

170. Maloney, M. J. and R. J. Hecht. 1992. Development of continuous-fiber-reinforced $MoSi_2$-base composites. *Mater. Sci. Eng. A* 155:19–31.

171. Alman, D. E. and N. S. Stoloff. 1992. Effect of ductile phase reinforcement morphology on toughening of $MoSi_2$. In *Intermetallic Matrix Composites II*, eds. D. L. Anton, J. A. Graves, and D. B. Miracle, 247. Warrendale, PA: MRS.

172. Xiao, L., Y. S. Kim, and R. Abbaschian. 1990. Ductile phase toughening of MoSi2-chemical compatibility and fracture toughness. In *Intermetallic Matrix Composites I*, eds. D. L. Anton, R. McMeeking, D. Miracle, and P. Martin, 399. Warrendale, PA: MRS.

173. Xiao, L., Y. S. Kim, R. Abbaschian, and R. J. Hecht. 1991. Processing and mechanical properties of niobium-reinforced $MoSi_2$ composites. *Mater. Sci. Eng. A* 144:277–285.

174. Waghmare, U. V., V. Bulatov, E. Kaxiras, and M. S. Duesbery. 1999. Microalloying for ductility in molybdenum disilicide. *Mater. Sci. Eng. A* 261:147–157.

175. Peralta, P., S. A. Maloy, F. Chu, J. J. Petrovic, and T. E. Mitchell. 1997. Mechanical properties of monocrystalline C11(b) $MoSi_2$ with small aluminum additions. *Scr. Mater.* 37:1599–1604.

176. Inui, H., K. Ishikawa, and M. Yamaguchi. 2000. Effects of alloying elements on plastic deformation of single crystals of $MoSi_2$. *Intermetallics* 8:1131–1145.

177. Sharif, A. A., A. Misra, and T. E. Mitchell. 2003. Deformation mechanisms of polycrystalline $MoSi_2$ alloyed with 1 at.% Nb. *Mater. Sci. Eng. A* 358:279–287.

178. Ashby, M. F., F. J. Blunt, and M. Bannister. 1989. Flow characteristics of highly constrained metal wires. *Acta Metall.* 37:1847–1857.

179. Liu, C. T., J. H. Schneibel, and L. Heatherly. 1999. Processing, microstructure, and properties of multiphase Mo silicide alloys. In *High-Temperature Ordered Intermetallic Alloys VIII*, Materials Research Society Symposium Proceedings, Vol. 552, KK6.2.1. Warrendale, PA: MRS.

180. Schneibel, J. H., C. T. Liu, D. S. Easton, and C. A. Carmichael. 1999. Microstructure and mechanical properties of Mo–Mo$_3$Si–Mo$_5$SiB$_2$ silicides. *Mater. Sci. Eng. A* 261:78–83.

181. Schneibel, J. H., M. J. Kramer, O. Unal, and R. N. Wright. 2001. Processing and mechanical properties of a molybdenum silicide with the composition Mo–12Si–8.5B (at.%). *Intermetallics* 9:25–31.

182. Schneibel, J. H., M. J. Kramer, and D. S. Easton. 2002. A Mo–Si–B intermetallic alloy with a continuous alpha-Mo matrix. *Scr. Mater.* 46:217–221.

183. Choe, H., J. H. Schneibel, and R. O. Ritchie. 2003. On the fracture and fatigue properties of Mo–Mo$_3$Si–Mo$_5$SiB$_2$ refractory intermetallic alloys at ambient to elevated temperatures (25 degrees C to 1300 degrees C). *Metall. Mater. Trans. A* 34:225–239.

184. Kruzic, J. J., J. H. Schneibel, and R. O. Ritchie. 2004. Fracture and fatigue resistance of Mo–Si–B alloys for ultrahigh-temperature structural applications. *Scr. Mater.* 50:459–464.

185. Mitra, R., A. K. Srivastava, N. E. Prasad, and S. Kumari. 2006. Microstructure and mechanical behaviour of reaction hot pressed multiphase Mo–Si–B and Mo–Si–B–Al intermetallic alloys. *Intermetallics* 14:1461–1471.

186. Schneibel, J. H., R. O. Ritchie, J. J. Kruzic, and P. F. Tortorelli. 2005. Optimization of Mo–Si–B intermetallic alloys. *Metall. Mater. Trans. A* 36:525–531.

187. Miller, M. K., E. A. Kenik, M. S. Mousa, K. F. Russell, and A. J. Bryhan. 2002. Improvement in the ductility of molybdenum alloys due to grain boundary segregation. *Scr. Mater.* 46:299–303.

188. Schneibel, J. H., M. P. Brady, J. J. Kruzic, and R. O. Ritchie. 2005. On the improvement of the ductility of molybdenum by spinel (MgAl$_2$O$_4$) particles. *Z. Metallkd.* 96:632–637.

189. Mendiratta, M. G. and D. M. Dimiduk. 1993. Strength and toughness of a Nb/Nb$_5$Si$_3$ composite. *Metall. Trans. A* 24:501–504.

190. Rigney, J. D. and J. J. Lewandowski. 1996. Loading rate and test temperature effects on fracture of *in situ* niobium silicide–niobium composites. *Metall. Mater. Trans. A* 27:3292–3306.

191. Murayama, Y. and S. Hanada. 2002. High temperature strength, fracture toughness and oxidation resistance of Nb–Si–Al–Ti multiphase alloys. *Sci. Tech. Adv. Mater.* 3:145–156.

192. Sha, J., H. Hirai, T. Tabaru, A. Kitahara, H. Ueno, and S. Hanada. 2004. High-temperature strength and room-temperature toughness of Nb–W–Si–B alloys prepared by arc-melting. *Mater. Sci. Eng. A* 364:151–158.

193. Heian, E. M., J. C. Gibeling, and Z. A. Munir. 2004. Synthesis and characterization of Nb$_5$Si$_3$/Nb functionally graded composites. *Mater. Sci. Eng. A* 368:168–174.

194. Fujikura, M., A. Kasama, R. Tanaka, and S. Hanada. 2004. Effect of alloy chemistry on the high temperature strengths and room temperature fracture toughness of advanced Nb-based alloys. *Mater. Trans. JIM* 45:493–501.

195. Li, W., H. B. Yang, A. D. Shan, and J. S. Wu. 2004. Effect of Mo addition on microstructure and properties of Nb/Nb$_5$Si$_3$ *in-situ* composite. *Mater. Sci. Forum* 449–452:753–756.

196. Kim, W. Y., H. Tanaka, and S. Hanada. 2002. High temperature strength at 1773 K and room temperature fracture toughness of Nb-ss/Nb$_5$Si$_3$ *in situ* composites alloyed with Mo. *J. Mater. Sci.* 37:2885–2891.

197. Chattopadhyay, K., G. Balachandran, R. Mitra, and K. K. Ray. 2006. Effect of Mo on microstructure and mechanical behaviour of as-cast Nb$_{SS}$-Nb$_5$Si$_3$ *in situ* composites. *Intermetallics* 14:1452–1460.

198. Mitra, R., K. Chattopadhyay, A. K. Srivastava, K. K. Ray, and N. Eswara Prasad. 2009. Effect of ductile and brittle phases on deformation and fracture behaviour of molybdenum and niobium silicide based composites. *Key Eng. Mater.* 395:179–192.

199. Min, K. S., A. J. Ardell, S. J. Eck, and F. C. Chen. 1995. A small-specimen investigation of the fracture-toughness of Ti$_5$Si$_3$. *J. Mater. Sci.* 30:5479–5483.

200. Frommeyer, G. and R. Rosenkranz. 1991. *Proceedings of International Conference PM Aerospace Materials*, Paper 29, November 4–6, 1991, Lausanne, Switzerland.

201. Zhang, L. and J. Wu. 1998. Ti$_5$Si$_3$ and Ti$_5$Si$_3$-based alloys: Alloying behavior, microstructure and mechanical property evaluation. *Acta Mater.* 46:3535–3546.

202. Frommeyer, G., R. Rosenkranz, and C. Ludecke. 1990. Microstructure and properties of high melting point intermetallic Ti$_5$Si$_3$ and TiSi$_2$ compounds. *Z. Metallkd.* 81:307–313.

203. Shah, D. M. and D. L. Anton. 1993. Alumina fiber reinforced intermetallic matrix composites. In *First International Symposium on Structural Intermetallics (ISSI-1)*, eds. R. Darolia, J. J. Lewandowski, C. T. Liu, P. L. Martin, D. B. Martin, and M. V. Nathal. Warrendale, PA: TMS.

204. Vahldiek, F. W. and S. A. Mersol. 1968. Phase relations and substructure in single crystal MoSi$_2$. *J. Less-common Met.* 15(2):165–176.

205. Umakoshi, Y., T. Sakagami, T. Yamane, and T. Hirano. 1989. Planar faults in MoSi$_2$ single-crystals deformed at high-temperatures. *Phil. Mag. Lett.* 59:159–164.

206. Boldt, P. H., J. D. Embury, and G. C. Weatherly. 1992. Room temperature microindentation of single-crystal MoSi$_2$. *Mater. Sci. Eng. A* 155:251–258.

207. Maloy, S. A., T. E. Mitchell, J. J. Lewandowski, and A. H. Heuer. 1993. (103)(331) Slip in MoSi$_2$. *Phil. Mag. Lett.* 67:313–321.

208. Maloy, S. A., T. E. Mitchell, and A. H. Heuer. 1995. High-temperature plastic anisotropy in MoSi$_2$ single-crystals. *Acta Metall. Mater.* 43:657–668.

209. Evans, D. J., F. J. Scheltens, J. B. Woodhouse, and H. L. Fraser. 1997. Deformation mechanisms in MoSi$_2$ at temperatures above the brittle-to-ductile transition temperature. *Phil. Mag. A* 75:17–30.

210. Nakano, T., M. Azuma, and Y. Umakoshi. 2002. Tensile deformation and fracture behaviour in NbSi$_2$ and MoSi$_2$ single crystals. *Acta Mater.* 50:3731–3742.

211. Guder, S., M. Bartsch, and U. Messerschmidt. 2002. Transmission electron microscopy analysis of planar faults on (001) planes in MoSi$_2$ single crystals. *Philos. Mag.* 82(14): 2737–2754.

212. Unal, O., J. J. Petrovic, D. H. Carter, and T. E. Mitchell. 1990. Dislocations and plastic-deformation in molybdenum disilicide. *J. Am. Ceram. Soc.* 73:1752–1757.

213. Mitchell, T. E., R. G. Castro, J. J. Petrovic, S. A. Maloy, O. Unal, and M. M. Chadwick. 1992. Dislocations, twins, grain boundaries and precipitates in MoSi$_2$. *Mater. Sci. Eng. A* 155:241–249.

214. Evans, D. J., S. A. Court, P. M. Hazzledine, and H. L. Fraser. 1993. Deformation mechanisms in the intermetallic compound MoSi₂. In *High Temperature Ordered Intermetallics V*, eds. I. Baker, J. D. Whitteberger, R. Darolia, and M. H. Yoo, Materials Research Society Symposium Proceedings, Vol. 288, 567–572. Pittsburgh, PA: MRS.
215. Evans, D. J., S. A. Court, P. M. Hazzledine, and H. L. Fraser. 1993. Dislocation dissociation in the intermetallic compound MoSi₂. *Phil. Mag. Lett.* 67:331–341.
216. Campbell, J. P., H. Chang, and R. Gibala. 1995. Characterization of the high temperature dislocation substructure of MoSi₂. In *High Temperature Ordered Intermetallics VI*, eds. I. Baker, S. Hanada, J. Horton, R. D. Noebe, and D. S. Schwartz, Materials Research Society Symposium Proceedings, Vol. 364, 893–898. Pittsburgh, PA: MRS.
217. Evans, D. J., F. J. Scheltens, J. B. Woodhouse, and H. L. Fraser. 1997. Deformation mechanisms in MoSi₂ at temperatures above the brittle-to-ductile transition temperature I. Polycrystalline MoSi₂. *Phil. Mag. A* 75:1–15.
218. Mitra, R., N. E. Prasad, S. Kumari, and A. V. Rao. 2003. High-temperature deformation behavior of coarse- and fine-grained MoSi₂ with different silica contents. *Metall. Mater. Trans. A* 34:1069–1088.
219. Srinivasan, S. R., R. B. Schwarz, and J. D. Embury. 1993. Ductile-to-brittle transition in MoSi₂. In *High Temperature Ordered Intermetallics V*, eds. I. Baker, J. D. Whittenberger, R. Darolia, and M. H. Yoo, Vol. 288, 1099. Pittsburgh, PA: MRS.
220. Srinivasan, S. R. and R. B. Schwarz. 1992. Synthesis of MoSi₂-based alloys by mechanical alloying. In *Novel Powder Processing, Advances in Powder and Particulate Materials*, 7:345. Princeton, NJ: Metal Powder Industries Federation.
221. Patankar, S. N. and J. J. Lewandowski. 1993. Effect of processing on the high temperature mechanical properties of MoSi₂. In *High-Temperature Ordered Intermetallic Alloys V*, eds. I. Baker, R. Darolia, J. D. Whittenberger, and M. H. Yoo, Materials Research Society Symposium Proceedings, Vol. 288, 829. Warrendale, PA: MRS.
222. Gibala, R., H. Chang, C. M. Czarnik, K. M. Edwards, and A. Misra. 1993. Plasticity enhancement processes in NiAl and MoSi₂. In *Structural Intermetallics*, eds. R. Darolia, J. J. Lewandowski, C. T. Liu, P. L. Martin, D. B. Miracle, and M. V. Nathal, 561. Warrendale, PA: TMS.
223. Chang, H., H. Kung, and R. Gibala. 1992. Plasticity enhancement of MoSi₂ at elevated temperatures through the addition of TiC. In *Intermetallic Matrix Composites II*, eds. D. L. Anton, J. A. Graves, and D. B. Miracle, Materials Research Society Symposium Proceedings, Vol. 273, 253–258.
224. Johnston, W. G. 1962. Yield points and delay times in single crystals. *J. Appl. Phys.* 33:2716–2730.
225. Messerschmidt, U., M. Bartsch, S. Guder, D. Haussler, R. Haushalter, and M. Yamaguchi. 1998. Dynamic dislocation behaviour in the intermetallic compounds NiAl, TiAl and MoSi₂. *Intermetallics* 6:729–733.
226. Guder, S., M. Bartsch, M. Yamaguchi, and U. Messerschmidt. 1999. Dislocation processes during the deformation of MoSi₂ single crystals in a soft orientation. *Mater. Sci. Eng. A* 261:139–146.
227. Ito, K., K. Matsuda, Y. Shirai, H. Inui, and M. Yamaguchi. 1999. Brittle-ductile behavior of single crystals of MoSi₂. *Mater. Sci. Eng. A* 261:99–105.
228. Mitchell, T. E., M. I. Baskes, R. G. Hoagland, and A. Misra. 2001. Dislocation core structures and yield stress anomalies in molybdenum disilicide. *Intermetallics* 9:849–856.

229. Shan, A., J. S. Wu, H. Hashimoto, and Y. H. Park. 2004. Tensile properties of $MoSi_2$ at elevated temperatures. *Mater. Sci. Forum* 449–452:845–848.

230. Hardwick, D. A., P. L. Martin, S. N. Patankar, and J. J. Lewandowski. 1993. Processing-microstructure-property relationships in polycrystalline $MoSi_2$. In *Structural Intermetallics*, eds. R. Darolia, J. J. Lewandowski, C. T. Liu, P. L. Martin, D. B. Miracle, and M. V. Nathal, 665. Warrendale, PA: TMS.

231. Jayashankar, J. S., E. N. Ross, P. D. Eason, and M. J. Kaufman. 1997. Processing of $MoSi_2$-based intermetallics. *Mater. Sci. Eng. A* 239–240:485–492.

232. Stergiou, A. and P. Tsakiropoulos. 1995. Study of the effects of Al, Ta, W additions on the microstructure and properties of $MoSi_2$ base alloys. In *High Temperature Ordered Intermetallics VI*, eds. I. Baker, S. Hanada, J. Horton, R. D. Noebe, and D. S. Schwartz, Materials Research Society Symposium Proceedings, Vol. 364, 911–916.

233. Umakoshi, Y., T. Hirano, T. Sakagami, and T. Yamane. 1990. High temperature deformation of $MoSi_2$ single crystals with the $C11_b$ structure. In *High Temperature Aluminides and Intermetallics*, eds. S. H. Whang, C. T. Liu, D. P. Pope, and J. O. Stiegler, 111. Warrendale, PA: TMS.

234. Harada, Y., Y. Murata, and M. Morinaga. 1998. Solid solution softening and hardening in alloyed $MoSi_2$. *Intermetallics* 6:529–535.

235. Mitchell, T. E. and A. Misra. 1999. Structure and mechanical properties of (Mo, Re)Si_2 alloys. *Mater. Sci. Eng. A* 261:106–112.

236. Sharif, A. A., A. Misra, J. J. Petrovic, and T. E. Mitchell. 2001. Solid solution hardening and softening in $MoSi_2$ alloys. *Scr. Mater.* 44:879–884.

237. Aikin Jr, R. M. 1991. Structure and properties of *in-situ* reinforced $MoSi_2$. *Ceram. Eng. Sci. Proc.* 12(9–10):1643–1655.

238. Aikin Jr, R. M. 1992. Strengthening of discontinuously reinforced $MoSi_2$ composites at high temperatures. *Mater. Sci. Eng. A* 155:121–133.

239. Cotton, J. D., Y. S. Kim, and M. J. Kaufman. 1991. Intrinsic second-phase particles in powder-processed $MoSi_2$. *Mater. Sci. Eng. A* 144:287–291.

240. Inui, H., T. Nakamoto, K. Ishikawa, and M. Yamaguchi. 1999. Plastic deformation of single crystals of $(Mo1-XWX)Si_2$ with the C11b structure. *Mater. Sci. Eng. A* 261:131–138.

241. Mitra, R., K. Sadananda, and C. R. Feng. 2004. Effect of microstructural parameters and Al alloying on creep behavior, threshold stress and activation volumes of molybdenum disilicides. *Intermetallics* 12:827–836.

242. Inui, H. and M. Yamaguchi. 2001. Deformation mechanisms of transition-metal disilicides with the hexagonal C40 structure. *Intermetallics* 9:857–862.

243. Petrovic, J. J. and R. E. Honnell. 1990. SiC reinforced-$MoSi_2/WSi_2$ alloy matrix composites. *Ceram. Eng. Sci. Proc.* 11:734–744.

244. Niihara, K. and Y. Suzuki. 1999. Strong monolithic and composite $MoSi_2$ materials by nanostructure design. *Mater. Sci. Eng. A* 261:6–15.

245. Sadananda, K. and C. R. Feng. 1993. The creep of intermetallics and their composites. *JOM* 45(5):45–48.

246. Bose, S. 1992. Engineering aspect of creep deformation of molybdenum disilicide. *Mater. Sci. Eng. A* 155:217–225.

247. Maloy, S. A., T. E. Mitchell, J. J. Petrovic, A. H. Heuer, and J. J. Lewandowski. 1994. The temperature and strain rate dependence of the flow stress in $MoSi_2$ single crystals. In *High Temperature Silicides and Refractory Alloys*, eds. C. L. Briant, J. J. Petrovic, B. P. Bewlay, A. K. Vasudevan, and H. A. Lipsitt, Materials Research Society Symposium Proceedings, Vol. 322, 21. Warrendale, PA: MRS.

248. Umakoshi, Y., T. Nakashima, T. Nakano, and E. Yanagisawa. 1994. Plastic behavior and deformation structure of silicide single crystals with transition metals at high temperatures. In *High Temperature Silicides and Refractory Alloys*, eds. C. L. Briant, J. J. Petrovic, B. P. Bewlay, A. K. Vasudevan, and H. A. Lipsitt, Materials Research Society Symposium Proceedings, Vol. 322, 9. Warrendale, PA: MRS.

249. Inui, H., K. Ishikawa, and M. Yamaguchi. 2000. Creep deformation of single crystals of binary and some ternary $MoSi_2$ with the C11(b) structure. *Intermetallics* 8:1159–1168.

250. Sadananda, K., C. R. Feng, H. Jones, and J. Petrovic. 1992. Creep of molybdenum disilicide composites. *Mater. Sci. Eng. A* 155:227–239.

251. Ghosh, A. K. and A. Basu. 1993. Microstructural effects on creep strengthening of $MoSi_2$ matrix composites. In *Critical Issues in the Development of High Temperature Structural Materials*, eds. N. S. Stoloff, D. J. Duguatte, and A. F. Gramaei. Warrendale, PA: TMS.

252. French, J. D., S. M. Wiederhorn, and J. J. Petrovic. 1995. Tensile creep and creep rupture of SiC-reinforced $MoSi_2$. *Ceram. Eng. Sci. Proc.* 16:129–136.

253. Jeng, Y. L., E. J. Lavernia, J. Wolfenstine, D. E. Bailey, and A. Sickinger. 1993. Creep behavior of plasma-sprayed SiC-reinforced $MoSi_2$. *Scr. Metall. Mater.* 29:107–111.

254. Wiederhorn, S. M., R. J. Gettings, D. E. Roberts, C. Ostertag, and J. J. Petrovic. 1992. Tensile creep of silicide composites. *Mater. Sci. Eng. A* 155:209–215.

255. Sadananda, K. and C. R. Feng. 1993. Creep of high temperature composites. In *Processing and Fabrication of Advanced Materials for High Temperature Applications – II*, eds. V. A. Ravi and T. S. Srivatsan, 331. Warrendale, PA: TMS.

256. Kelly, A. and K. N. Street. 1972. Creep of discontinuous fiber composites – II. Theory for steady-state. *Proc. R. Soc. Lond. Ser. A* 328:283.

257. Sadananda, K. and C. R. Feng. 1992. Role of back stresses in the creep of molydisilicide composites. In *MID-Vol. 35, Processing, Fabrications, and Manufacturing of Composite Materials*, eds. T. S. Srivatsan and E. J. Lavernia, Book No. G00733, 231. ASME.

258. Ramamurty, U., A. S. Kim, S. Suresh, and J. J. Petrovic. 1993. Micromechanisms of creep-fatigue crack-growth in a silicide-matrix composite with SiC particles. *J. Am. Ceram. Soc.* 76:1953–1964.

259. Anton, D. L., E. Hartford, and D. M. Shah. 1991. High temperature properties of refractory intermetallics. In *High Temperature Ordered Intermetallics IV*, eds. L. A. Johnson, D. P. Pope, and J. O. Stiegler, Materials Research Society Symposium Proceedings, Vol. 213, 733–738. Pittsburgh, PA: MRS.

260. Hayashi, T., K. Ito, K. Ihara, M. Fujikura, and M. Yamaguchi. 2004. Creep of single crystalline and polycrystalline T-2 phase in the Mo–Si–B system. *Intermetallics* 12:699–704.

261. Nieh, T. G., J. G. Wang, and C. T. Liu. 2001. Deformation of a multiphase Mo–9.4Si–13.8B alloy at elevated temperatures. *Intermetallics* 9:73–79.

262. Schneibel, J. H. 2003. High temperature strength of $Mo–Mo_3Si–Mo_5SiB_2$ molybdenum silicides. *Intermetallics* 11:625–632.

263. Alur, A. P., N. Chollacoop, and K. S. Kumar. 2004. High-temperature compression behavior of Mo–Si–B alloys. *Acta Mater.* 52:5571–5587.

264. Jain, P. and K. S. Kumar. 2010. Tensile creep of Mo–Si–B alloys. *Acta Mater.* 58:2124–2142.

265. Jain, P., A. P. Alur, and K. S. Kumar. 2006. High temperature compressive flow behavior of a Mo–Si–B solid solution alloy. *Scr. Mater.* 54:13–17.

266. Jehanno, P., M. Heilmaier, H. Saage, H. Heyse, M. Boning, H. Kestler, and J. H. Schneibel. 2006. Superplasticity of a multiphase refractory Mo–Si–B alloy. *Scr. Mater.* 55:525–528.

267. Li, Z. K., J. L. Yu, X. Zheng, J. J. Zhang, H. Liu, R. Bai, H. Wang, D. H. Wang, and W. S. Wang. 2011. Superplasticity of a multiphase fine-grained Mo–Si–B alloy. *Powder Technol.* 214:54–56.

268. Jéhanno, P., M. Heilmaier, H. Saage, M. Böning, H. Kestler, J. Freudenberger, and S. Drawin. 2007. Assessment of the high temperature deformation behavior of molybdenum silicide alloys. *Mater. Sci. Eng. A* 463:216–223.

269. Yu, J. L., Z. K. Li, X. Zheng, J. J. Zhang, H. Liu, R. Bai, and H. Wang. 2012. Tensile properties of multiphase Mo–Si–B refractory alloys at elevated temperatures. *Mater. Sci. Eng. A* 532:392–395.

270. Kumar, K. S. and A. P. Alur. 2007. Deformation behavior of a two-phase Mo–Si–B alloy. *Intermetallics* 15:687–693.

271. Nakano, T., M. Kishimoto, D. Furuta, and Y. Umakoshi. 2000. Effect of substitutional elements on plastic deformation behaviour of $NbSi_2$-based silicide single crystals with C40 structure. *Acta Mater.* 48:3465–3475.

272. Subramanian, P. R., T. A. Parthasarathy, M. G. Mendiratta, and D. M. Dimiduk. 1995. Compressive creep-behavior of Nb_5Si_3. *Scr. Metall. Mater.* 32:1227–1232.

273. Subramanian, P. R., M. G. Mendiratta, and D. M. Dimiduk. 1994. Microstructures and mechanical behavior of Nb-Ti base beta + silicide alloys. In *High Temperature Silicides and Refractory Alloys*, eds. C. L. Briant, J. J. Petrovic, B. P. Bewlay, A. K. Vasudevan, and H. A. Lipsitt, Materials Research Society Symposium Proceedings, Vol. 322:491. Warrendale, PA: MRS.

274. Bewlay, B. P., P. W. Whiting, A. W. Davis, and C. L. Briant. 1999. Creep mechanisms in niobium-silicide based *in-situ* composites. Materials Research Society Symposium Proceedings, Vol. 552, KK6.11.1–KK6.11.5.

275. Henshall, G. A. and M. J. Strum. 1995. Simulations of creep in ductile-phase toughened Nb_5Si_3/Nb *in situ* composites. In *High Temperature Ordered Intermetallics VI*, eds. I. Baker, S. Hanada, J. Horton, R. D. Noebe, and D. S. Schwartz, Materials Research Society Symposium Proceedings, Vol. 364, 937–942.

276. Henshall, G. A. and M. J. Strum. 1994. Simulations of deformation in composites with 2 steady-state creeping phases. *Scr. Metall. Mater.* 30:845–850.

277. Henshall, G. A., P. R. Subramanian, M. J. Strum, and M. G. Mendiratta. 1997. Continuum predictions of deformation in composites with two creeping phases. 2. Nb_5Si_3/Nb composites. *Acta Mater.* 45:3135–3142.

278. Chan, K. S. 2002. Modeling creep behavior of niobium silicide *in-situ* composites. *Mater. Sci. Eng. A* 337:59–66.

279. Kim, J. H., T. Tabaru, H. Hirai, A. Kitahara, and S. Hanada. 2003. Tensile properties of a refractory metal base *in situ* composite consisting of an Nb solid solution and hexagonal Nb_5Si_3. *Scr. Mater.* 48:1439–1444.

7

Silicides: Oxidation Behavior

7.1 Introduction

Silicides and silicon-based ceramics derive their resistance to oxidation from a continuous and adherent film of SiO_2. However, the formation of a protective scale is not possible if simultaneous oxidation of the constituent metallic alloying elements occurs along with the formation of SiO_2 and the oxides of alloying elements remain in the scale. The fundamental mechanisms of resistance to oxidation in intermetallics have already been discussed in Chapter 4. Among the silicides, the best oxidation resistance is demonstrated by $MoSi_2$, which fully justifies its use as a heating element in oxidizing atmospheres, as well as the amount of interest shown in developing it as a structural material. A good amount of attention has also been devoted to silicides of Nb, W, Cr, and Ti, considering their potential for high-temperature applications. It is interesting to examine why $MoSi_2$ has superior oxidation resistance at elevated temperatures, and why it is not sufficiently resistant to degradation at intermediate temperatures (<700°C). The kinetics and products of oxidation of Mo, Nb, W, Cr, and Ti silicides at intermediate and elevated temperatures are discussed in this chapter with an emphasis on their operative mechanisms. A few reviews on the oxidation behavior of silicides provide an idea of the developments in this area over the last few decades.[1–3]

The oxidation behavior of silicides depends primarily on their alloy composition and the conditions of exposure. At a given temperature, the oxidation behavior depends on the characteristics of the oxide scale of the most active species. If the scale is impervious, adherent, and self-healing, it is protective against further oxidation. In contrast, a porous scale allows unabated ingress of oxygen from air to the alloy–oxide interface. Alternatively, all the alloying elements present may oxidize more or less simultaneously, which, depending on the nature of the oxides and their thermodynamic parameters, may either inhibit formation of a continuous and protective scale, or contribute to the formation of a dense and adherent scale. All these issues are discussed in this chapter.

7.2 Oxidation Behavior of Molybdenum Silicides

7.2.1 MoSi$_2$

The oxidation of MoSi$_2$ may be distributed into four major regimes, 400°C–600°C, 600°C–1000°C, 1000°C–1700°C, and above 1700°C, which is a modification of the subdivisions discussed by Berztiss et al.[3] Table 7.1 shows the results of selected studies on the oxidation behavior of MoSi$_2$ processed by different routes and having varying densities in different temperature regimes.

TABLE 7.1

Results of Selected Studies Related to the Oxidation Behavior of MoSi$_2$ in Different Temperature Regimes

Processing and Characteristics	Temperature (°C)	Mass Change	Oxidation Products	References
Powder metallurgy: Hot pressed	400–700	Gain	MoO$_3$, SiO$_2$	Wirkus and Wilder[4]
	1164–1500	Gain	Mo$_5$Si$_3$, amorphous SiO$_2$, α-cristobalite	
Powder metallurgy: Hot isostatically pressed (cyclic)	400–450	Gain	SiO$_2$ and Mo$_9$O$_{26}$	Meschter[5]
	500–600	Gain	SiO$_2$ and MoO$_3$	
	600	Loss	SiO$_2$ and MoO$_3$	
Arc melted	500–675		MoO$_3$, SiO$_2$	Chou and Nieh[6]
PM, Si rich	500	No pesting	SiO$_2$	McKamey et al.[7]
Si deficient		Pesting occurred	MoO$_3$, Mo$_5$Si$_3$, SiO$_2$	
Arc melted	500	Gain, pesting	MoO$_3$, SiO$_2$	Berztiss et al.[3]
	600–1000	No pesting Gain	Internal oxidation, SiO$_2$ formation	
	1200–1400	Gain	α-Cristobalite	
	500	Gain, no pesting	MoO$_3$, SiO$_2$, voids	
HIPped Single crystal	500	No pesting	MoO$_3$, SiO$_2$, fewer voids	
Polycrystalline (arc melted)	500	Pesting observed	MoO$_3$, SiO$_2$, Mo–Si–O film on MoSi$_2$	Chou and Nieh[8]
Single crystal		Pesting: 1000 h	Mo–Si–O followed by MoO$_3$ and SiO$_2$	
XD™ (cyclic, isothermal)	500	Gain	MoO$_3$ and SiO$_2$	Cook et al.[9]
	1200	Gain	SiO$_2$	
Reaction hot pressed	1200	Gain	Amorphous SiO$_2$	Mitra and Rama Rao[10]

7.2.1.1 Oxidation Behavior in the Range of 400°C–600°C

In the temperature range of 400°C–600°C, $MoSi_2$ undergoes accelerated oxidation ending with its disintegration into lumps of powder, which has been termed *pesting* in the literature.[5,6,8–19] Typical plots showing the kinetics of oxidation are shown in Figure 7.1.[18] It has been proposed that the oxidation of $MoSi_2$ proceeds according to the reaction:

$$2MoSi_2 + 7O_2 \rightarrow 2MoO_3 + 4SiO_2 \tag{7.1}$$

The free energy change for Reaction 7.1 is −3910.06 kJ mol^{-1} at 527°C (800 K). Transmission electron microscopy (TEM) investigation of the cross section of the oxide scale has shown an amorphous film of Mo–Si–O composition after 3 h of exposure, and crystals of MoO_3 or MoO_2 embedded in amorphous SiO_2 after a period of 51 h.[17] In addition, Mo_9O_{26} has also been observed, probably because of the insufficient diffusivity of oxygen anions at 500°C.[18] The SiO_2 scale is not continuous and protective because of competitive growth with Mo oxides, particularly MoO_3, which does not volatilize much at 500°C. Whiskers or rod- or plate-shaped crystals of MoO_3 have been found in the oxide scale after an incubation period during the exposure of $MoSi_2$ and $MoSi_2$–SiC composite at 500°C (Figure 4.7). Mitra et al. have observed peaks of Mo_5Si_3 on the surfaces of the $MoSi_2$–SiC composite oxidized at 500°C for up to 120 h, while only peaks of MoO_3 and Mo_9O_{26} could be observed after 192 h, suggesting that Reaction 7.1 may also occur in steps, with the intermediate reaction[18]:

$$5MoSi_2 + 7O_2 \rightarrow 2Mo_5Si_3 + 7SiO_2 \tag{7.2}$$

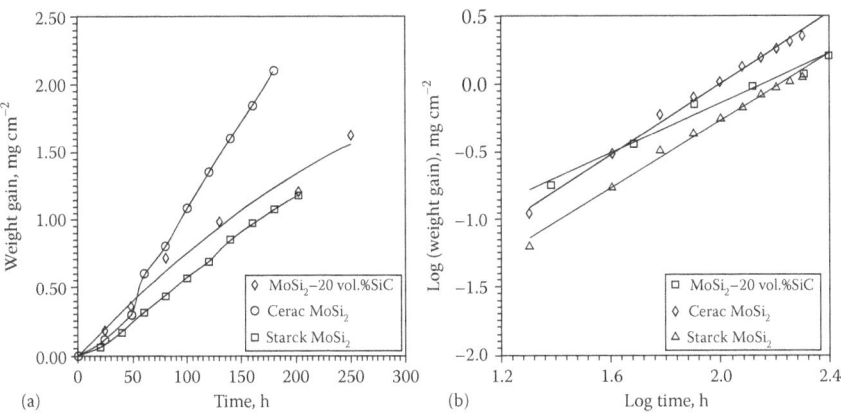

FIGURE 7.1
Plots showing the variation of (a) mass gain with time of exposure and (b) logarithm of mass gain with that of time, for $MoSi_2$ and $MoSi_2$–SiC composite at 500°C. (From Meschter, P. J., *Metall. Trans. A*, 23, 1763–1772, 1992; Mitra, R., et al., *Mater. Sci. Technol.*, 13, 415–419, 1997.)

The free energy change for Reaction 7.2 is -5084.57 kJ mol^{-1} at 527°C (800 K), suggesting that this reaction is preferable to Reaction 7.1 from the thermodynamic point of view. MoO$_3$ probably forms by the continued oxidation of Mo$_5$Si$_3$ after cracks are already initiated and contributes to the process of disintegration by further crack extension and loss in mass by its volatilization.

Pesting is known to be most severe at 500°C in the case of MoSi$_2$, and embrittlement of the grain boundaries, which provide the short-circuit paths of diffusion of oxygen, was proposed as the mechanism by Westbrook and Wood.[14] If pesting is due to grain-boundary embrittlement alone, it would occur in dense as well as porous specimens. Contrary to the above, Berkowitz-Mattuck et al. found that pesting occurred only in specimens with poorer density, and the resultant fracture was transgranular.[15] Rubisch et al. noted preferential oxidation at cracks and porosities, and found that the chances of pest disintegration increased with the increasing level of porosity. They described the wedging effect due to oxide growth as the cause of failure.[16] The activation energy for pesting has been estimated to be 27.6 kcal mol^{-1} between 375°C and 600°C from an assessment of the kinetics of volume change.[8] The much lower activation energy of pesting compared with that of 81.3 kcal mol^{-1} reported for oxidation at high temperatures by Wirkus and Wilder suggests the involvement of short-circuit diffusion of oxygen through grain boundaries, interfaces, or surfaces of cracks and open porosities.[4]

In tune with the observations of Berkowitz-Mattuck et al.[15] and Rubisch,[16] Yanagihara et al.[19] demonstrated that the initial cracks left by processing play a greater role in pest disintegration compared with the grain boundaries, because the formation of the initial oxide of Mo–Si–O formed at crack tips involves 200% volume expansion, accompanied by increased stress concentration. The calculated volume change for Reaction 7.1 is 212%. The formation of SiO$_2$ from Si and MoO$_3$ from Mo leads to an expansion in volume by 180% and 340%, respectively. Even single crystals of MoSi$_2$ have shown pest disintegration at a reduced rate, taking more than 1000 h at 500°C.[8] In the case of single crystals, the surface shows blisters constituted by the Si–Mo–O compound initially, giving way to MoO$_3$ and SiO$_2$ after a while. Thus, the slow diffusion of Si in MoSi$_2$ at low temperatures fails to maintain the concentration required for a stable film of SiO$_2$ at the alloy–oxide interface,[20] which leads to the simultaneous formation of SiO$_2$ and MoO$_3$,[19] even though the latter has a higher free energy of formation.[21] The slow diffusion of Si at lower temperatures also explains the higher parabolic rate constant[22] for the growth of MoO$_3$ when compared with that of SiO$_2$ at temperatures lower than 800°C. The diffusivity of Si increases above 800°C, which is expected due to the high concentration of Si vacancies in MoSi$_2$.[23]

7.2.1.2 Oxidation Behavior in the Temperature Range of 600°C–1000°C

Meschter has proposed that the transition from pesting to nonpesting type, that is, from nonprotective to partially protective oxidation, takes place between 500°C and 550°C.[5] According to Schlichting, $MoSi_2$ is protected from oxidation by a continuous, impervious, and amorphous film of SiO_2 between 600°C and 1700°C.[11] However, observations suggest that the oxide scales formed in the temperature range of 600°C–1000°C are nonprotective, although pesting does not occur.[3] Auger studies[17] of the oxide scale have shown evidence of the clustering of MoO_3 in the saturated SiO_2 phase in the temperature range of 800°C–1000°C. The continuous formation and volatilization of MoO_3 produce pores and do not allow a continuous film of SiO_2 to form below 1000°C.

7.2.1.3 Oxidation Behavior at Temperatures ≥1000°C

At temperatures between 1000°C and 1700°C, the oxide scale is found to contain only SiO_2 and is also protective;[4,5,10,11] therefore, the oxidation kinetics exhibits a parabolic rate law (Figure 7.2). The MoO_3 forms during heating and then volatilizes. Thereafter, an adherent, continuous, viscous, and self-healing SiO_2 scale is created as a barrier layer between the oxygen and the underlying alloy. Schlichting has proposed that the amorphous to crystalline transformation of SiO_2 is prevented at temperatures above 1000°C due to its nonstoichiometric or oxygen-deficient composition (SiO_{2-x}).[11] On the contrary, divitrification of SiO_2 into α-trydimite or cristobalite has been reported for studies involving extended exposure in the temperature range of 1100°C–1400°C.[24] Above 1700°C, the viscosity of SiO_2 decreases sharply with an increased concentration of defects, as a result of which the oxidation resistance decreases.

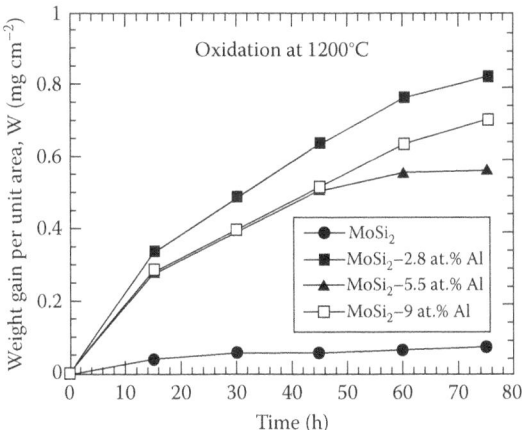

FIGURE 7.2
Plots showing the variation of mass gain with time of exposure for $MoSi_2$ and $MoSi_2$–Al alloys at 1200°C. (From Mitra, R. and V. V. Rama Rao, *Mater. Sci. Eng. A*, 260, 146–160, 1999. Reprinted with permission from Elsevier.)

7.2.1.4 Effect of Alloying Elements on Oxidation Behavior

7.2.1.4.1 *Temperature Range of 400°C–600°C*

The alloying of $MoSi_2$ with elements such as Al, Ti, Zr and Y that have a stronger affinity for oxidation than that of Si suppresses pesting, because these oxidize selectively, arresting the internal oxidation of Si.[25] A typical oxidation reaction would be

$$280Mo\left(Si_{0.85}Al_{0.15}\right)_2 + 63O_2 \rightarrow 12Mo_5Si_3 + 220MoSi_2 + 42Al_2O_3 \quad (7.3)$$

Though Yanagihara et al.[25] initially proposed that the cause for arrest of pesting on alloying with Al is the much reduced volume expansion from the formation of Al_2O_3 compared with that of SiO_2, the explanation was withdrawn in a subsequent publication,[19] because SiO_2 and Al_2O_3 were found to form simultaneously. Hence, the lower rate of pesting in the $MoSi_2$–Al alloy has been attributed to the increased plasticity of the amorphous Mo–Si–Al–O scale and a reduced initial crack density.

Alloying with Ta or W has not improved the pest resistance at all,[25,26] because SiO_2 is formed in preference to oxides of Ta, which have a higher free energy of formation.[21] Alloying with Ge has been found to assist in improving the oxidation behavior under cyclic conditions by increasing the coefficient of thermal expansion of the amorphous SiO_2 scale and decreasing the mismatch with the underlying alloy.[27–29] Furthermore, the scale of SiO_2–GeO_2 alloy has higher fluidity at lower temperatures,[30] so that the cracks can be healed easily.

7.2.1.4.2 *Temperature Range ≥1000°C*

Yanagihara et al. have studied the high-temperature oxidation of $MoSi_2$–X (X = Al, Ti, Ta, Zr, and Y) alloys under isothermal or cyclic conditions up to 1685°C.[31] The free energy of formation of the oxide of Ta is higher than that of SiO_2, while those of the oxides of Zr, Y, Al, and Ti are lower.[21] While $(Mo,Ta)Si_2$ has shown oxidation behavior similar to $MoSi_2$ with the formation of SiO_2 as the protective scale, $(Mo,Zr)Si_2$ and $(Mo,Y)Si_2$ have been found to be susceptible to severe internal oxidation. $(Mo,Ti)Si_2$ forms a duplex scale composed of SiO_2 and TiO_2 below the eutectic temperature of the SiO_2–TiO_2 system. Above the eutectic temperature, TiO_2 dissolves in SiO_2, causing hypereutectic composition, which crystallizes to form cristobalite, resulting in severe oxidation and spallation. $Mo(Si,Al)_2$ also forms a fluid scale of SiO_2, with Al_2O_3 dissolved in it above the eutectic temperature of 1595°C in the SiO_2–mullite system. The SiO_2–Al_2O_3 scale on the $Mo(Si,Al)_2$ has been found to be hypereutectic and strongly adherent. Yanagihara et al. have concluded that the scale of SiO_2 alloyed with the oxide of a third element should be hypereutectic in character to prevent devitrification.[31]

In oxidation studies at temperatures up to 1775°C, Yanagihara et al. have reported a higher rate of oxidation in the $Mo(Si_{1-x}Al_x)_2$ intermetallic alloy compared with $MoSi_2$.[31] However, in the case of cyclic oxidation up to 1750°C, the spallation of $Mo(Si_{1-x},Al_x)_2$ has been found to be reduced in comparison with that of $MoSi_2$, because amorphous to β-cristobalite phase transformation, accompanied by volume expansion and internal stresses, is inhibited in the case of the former alloy due to kinetic considerations. As the solubility of Al in β-cristobalite is very low, the formation of β-cristobalite requires the precipitation of Al-rich mullite, which is a slow process. The suppression of the crystallization of SiO_2 in $Mo(Si_{1-x},Al_x)_2$ explains the better adherence of the scale at high temperatures.

For minor alloying with 2.8, 5.5, and 9 at.% Al,[10] the isothermal oxidation rates at 1200°C in air have been found to be higher than that of $MoSi_2$. The oxide scale in the $MoSi_2$–Al alloys consisted of a mixture of α-Al_2O_3 and SiO_2 in the case of $MoSi_2$–2.8 at.% Al alloy, and primarily α-Al_2O_3 in the case of $MoSi_2$–5.5 and 9 at.% Al alloys (Figure 7.3a–d). The presence of Al_2O_3 in place of SiO_2 may be justified by the fact that the free energy of formation of the former is lower than that of the latter, and there is a reasonably higher activity of Al in the $MoSi_2$–5.5 and 9 at.% Al alloys. The planar character of the

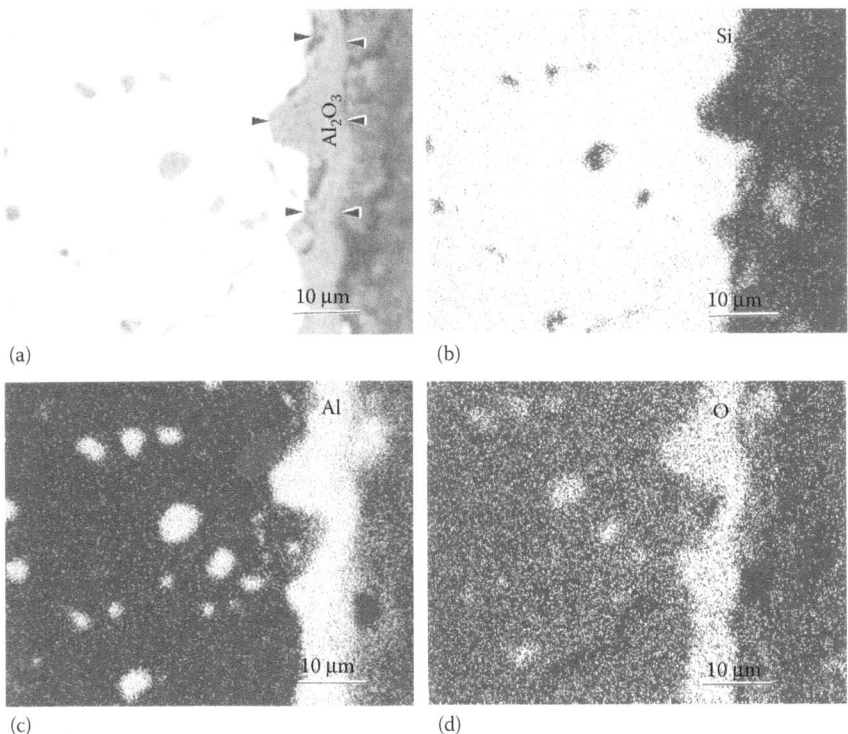

(a)   (b)   (c)   (d)

FIGURE 7.3
Oxide scale of the $MoSi_2$–5.5 at.% Al alloy after exposure in air at 1200°C: (a) SEM (back-scatter electron [BSE]) image, and WDS x-ray maps of (b) Si, (c) Al, and (d) O. (From Mitra, R. and V. V. Rama Rao, *Mater. Sci. Eng. A*, 260, 146–160, 1999. Reprinted with permission from Elsevier.)

metal–oxide scale interface observed in the case of $MoSi_2$ (Figure 4.4a–b) and a wavy interface in the case of $MoSi_2$–Al alloys (Figure 7.3a–d) is consistent with parabolic kinetics for the growth of a stable and protective scale in the case of the former and an unstable one in the case of the latter (Figure 7.3a–d). In another study, Stergiou and Tsakiropoulos have shown that $Mo(Si_{1-x}Al_x)_2$ with 29.3 at.% Al possessed higher oxidation resistance compared with $MoSi_2$ at 1200°C, while the rate of oxidation in the alloy with 21.8 at.% was faster.[26] The higher oxidation rate of the alloy bearing 21.8 at.% Al has been attributed to the depletion in the Al concentration in the $Mo_5(Si_{1-x}Al_x)_3$ layer adjacent to the oxide scale, which does not happen in the case of the alloy with a higher concentration of Al. Therefore, alloys with a minimum supply of Al^{3+} ions to the oxide scale being maintained continuously show protection from further oxidation.

Recently, the high-temperature oxidation behavior of an *in situ* processed, SiC-reinforced $(Mo,Cr)(Si,Al)_2$ intermetallic composite,[32] comprising a $(Mo_{0.23}Cr_{0.10})(Si_{0.67-x}Al_x)$ phase as the matrix and having a dispersion of $(Mo_{0.16}Cr_{0.17})(Si_{0.67-x}Al_x)$ (where $x=0.0002–0.0008$), β-SiC, and α-Al_2O_3 phases, has been studied to understand the effect of the competitive growth of different oxide scales: Cr_2O_3, Al_2O_3, and SiO_2. X-ray diffraction (XRD) analyses of an oxidized surface after 8 h at 1600°C have shown evidence of α-Al_2O_3, Cr_2O_3, SiO_2 (α-cristobalite), Cr_2MoO_6, Mo_5Si_3, and $C11_b$ $MoSi_2$; the presence of all of these has been analyzed using thermodynamic considerations. Analyses of thermodynamic data[21] lead to the conclusion that Cr and Mo have less affinity for oxygen than Si, whereas the tendency of Al to oxidize is higher (Figure 4.1). Hence, α-Al_2O_3 has been found in the oxide scale, despite the low activity of Al. The oxidation of Cr takes place preferentially, due to the higher diffusivity of Cr^{3+} ions as well as a higher parabolic rate constant of Cr_2O_3, which is subsequently lost by its further oxidation to form CrO_3 (g) and its escape. As a result, the alloy matrix is depleted in Cr concentration. Furthermore, the internal oxidation of SiC causes the formation of pores and cracks in the oxide scale with the escape of CO gas.

7.2.1.4.3 *Effect of Reinforcement*

The results of selected recent publications on the oxidation of $MoSi_2$-based composites are shown in Table 7.2. The isothermal and cyclic oxidation behavior of $MoSi_2$-based composites reinforced with 30 vol.% TiB_2, ZrB_2, HfB_2, or SiC have been studied by Cook et al.[9] Oxidation resistance is excellent in the temperature range of 800°C–1500°C due to the formation of SiO_2 film. It has been proposed that B_2O_3 forms in the scale of an $MoSi_2$–TiB_2 composite and protects against pesting in the temperature range of 400°C–600°C. Here, too, SiO_2 has lower free energy of formation per mole of oxygen (Figure 4.1), but B_2O_3 forms due to kinetic considerations. However, with increasing temperature, B_2O_3 reacts with SiO_2 to form a fluid borosilicate glass, reducing the viscosity of the SiO_2-rich scale, as a result of which $MoSi_2$ reinforced with borides has slightly poorer oxidation resistance than $MoSi_2$. At higher temperatures, B_2O_3 evaporates and the

TABLE 7.2

Oxidation Behavior of Some of the Selected $MoSi_2$-Based Composites Shown with the Temperature Range of Study and Oxidation Products

Material	Temperature (°C)	Mass Change	Oxidation Products	References
$MoSi_2$–30 vol.% Pressureless sintered ZrO_2	1500	10 times that of $MoSi_2$	ZrO_2, $ZrSiO_4$, SiO_2 (glassy)	Petrovic and Honnel[33]
$MoSi_2$ and composites with 30 vol.% TiB_2, ZrB_2, HfB_2, and SiC	500–1500			Cook et al.[9]
	500: Static			
	$MoSi_2$–TiB_2	Gain	B_2O_3, TiO_2, MoO_3	
	$MoSi_2$–SiC	Gain	SiO_2, MoO_3	
	1200 (C)		B_2O_3, TiO_2, SiO_2	
	$MoSi_2$–TiB_2	Gain	SiO_2	
	$MoSi_2$–SiC	Microcracks		
$MoSi_2$–SiC composite	500	Gain	<120 h: Mo_5Si_3, Mo–Si–O	Mitra et al.[18]
			After 120 h, Mo_9O_{26}, MoO_3, SiO_2	
$MoSi_2$–$SiC_{(p)}$– SiC_f,$MoSi_2$–Al_2O_3	500	Gain	MoO_3, SiO_2	Maloney and Hecht[13]
$MoSi_2$–Si_3N_4	500–800	Gain followed by loss	MoO_3, SiO_2	Natesan and Deevi[34]
	1100–1400	Gain	SiO_2	
$MoSi_2$	400–600 (C)	Gain, pesting	SiO_2, MoO_3	Hebsur[35] and Hebsur et al.[36]
$MoSi_2$–SiC_f		Gain, pesting	SiO_2, MoO_3	Kowalik and Hebsur[37]
$MoSi_2$–$Si_3N_{4(p)}$	1250	No pesting	Si_2ON_2	
$MoSi_2$–$Si_3N_{4(p)}$–SiC_f		Gain	α-Cristobalite	
$MoSi_2$	1400	Gain	SiO_2	Schneibel and Sekhar[38]
$MoSi_2$–Mo_5Si_3	Cyclic	Loss, pesting	SiO_2, MoO_3	

Note: Cyclic oxidation is identified as (C).

viscosity of the oxide scale increases, making it more impermeable to oxygen. Cook et al. have analyzed data using Relation 4.3. The kinetics of oxidation of the $MoSi_2$–TiB_2 composites at 500°C has been found to follow between linear ($n=1$) and parabolic ($n=2$) relationships. In another investigation,[18] the isothermal oxidation of $MoSi_2$–SiC composites at 500°C has also shown a value of n exceeding 1.0, whereas the values of n calculated for the oxidation of Cerac and Starck $MoSi_2$ have been found to be between 0.7 and 0.8, implying that the oxidation rate of the composite is less severe.[18] An examination of the surfaces of the $MoSi_2$–20 vol.% SiC composite oxidized at 500°C by wavelength-dispersive spectroscopy (WDS) microanalyses has shown clear evidence of preferential oxidation near the grain boundaries or interfaces with the SiC particles. However, cyclic oxidation has led to microcracking of the oxide scale due to the residual strains imposed by the coefficient of thermal expansion (CTE) mismatch between the substrate and the oxide scale.[9]

$MoSi_2$–Si_3N_4 composites have been found by Natesan and Deevi to be resistant to oxidation at temperatures between 800°C and 1400°C, but are susceptible to loss of mass at 500°C.[34] The loss of mass is more significant in the case of composites with 60 and 80 vol.% Si_3N_4, and is not as severe in the case of composites with lower volume fractions of Si_3N_4. In yet another but significant study, Hebsur and coworkers[35–37] have reported excellent resistance to isothermal or cyclic oxidation at 500°C in the case of the $MoSi_2$–30 vol.% Si_3N_4 composite in stressed or unstressed conditions. While $MoSi_2$–SiC_f (SCS-6) composites disintegrated to powder when subjected to 24 cycles between ambient temperature and 500°C, the $MoSi_2$–30 vol.% Si_3N_4–SiC_f (SCS-6) hybrid composite maintained its integrity after 200 cycles. The effect of reinforcement with Si_3N_4 may be appreciated from the fact that $MoSi_2$–40 vol.% SiC_p–SCS-6 and $MoSi_2$–Al_2O_3 composites have shown pest disintegration.[13] $MoSi_2$–30 and 50 vol.% Si_3N_4 composites and the $MoSi_2$–50 vol.% Si_3N_4–Hi Nicalon SiC fiber hybrid composite maintained their integrity against cyclic oxidation at 900°C and showed negligible weight gain after 500 h.[37] However, $MoSi_2$–50 vol.% Si_3N_4–SCS-6 fiber composites disintegrated on similar thermal cycling. The carbon core of SCS-6 fibers oxidized and the fiber surface converted to SiO_2, causing delamination and longitudinal cracking due to the mismatch in CTE.[35] Protection against oxidation in the case of $MoSi_2$–Si_3N_4 composites is due to the formation of a passive scale of $SiON_2$.

7.2.2 Oxidation Behavior of Mo_5Si_3

Mo_5Si_3 also undergoes severe pest disintegration at 800°C,[39] and its oxidation resistance is much poorer than that of $MoSi_2$ because of the inability of a continuous scale of SiO_2 to form due to the lower activity of Si. The possible oxidation reactions are

$$2Mo_5Si_3 + 21O_2 \rightarrow 10MoO_3 + 6SiO_2 \qquad (7.4)$$

$$3Mo_5Si_3 + 4O_2 \rightarrow 5Mo_3Si + 4SiO_2 \qquad (7.5)$$

$$Mo_5Si_3 + 3O_2 \rightarrow 5Mo + 3SiO_2 \qquad (7.6)$$

The free energy changes per mole of oxygen involved in Reactions 7.4 through 7.6 at 1027°C are −212, −566, and −343 kJ mol⁻¹, respectively.[21]

On analyses of thermodynamic data, it is found that the changes in free energy for Reactions 7.5 and 7.6 are more negative than that for Reaction 7.4. If oxidation takes place by Reactions 7.5 and 7.6, the transport of oxygen to the alloy–oxide interface will be rate controlling and a net mass gain is expected. However, oxidation involving mass loss, and the fact that MoO_3 is observed as being trapped throughout the oxide scale, suggests that Reaction 7.4 occurs and the oxidation rate is controlled by the outward transport of MoO_3 away from the oxidation interface through the scale. Anton and Shah have detected α-cristobalite as the residue left on subjecting arc-cast Mo_5Si_3 to cyclic oxidation of 20 1 h cycles between 1149°C and ambient temperature.[40]

7.2.3 Oxidation Behavior of Mo–Si–B Alloys

Considerable research has been devoted to understanding the oxidation mechanisms and the improvement in the resistance of Mo_5Si_3 through boron additions and Mo–Si–B alloys. The results of selected investigations on the oxidation behavior of Mo_5Si_3 and B-doped Mo_5Si_3 are presented in Table 7.3. It has been observed[39] that the alloying of Mo_5Si_3 with about 2 wt.% B leads to a decrease in the rate of oxidation by five orders of magnitude at 1200°C and also arrests catastrophic pest disintegration at 800°C. Normally, the entire time span of oxidation is classified into a period of rapid mass loss, followed by steady state or a reduced rate of oxidation. It has been proposed that the formation of B_2O_3 and its fluidity at 800°C enable viscous sintering of the oxide scale to close the pores that form during the transient oxidation period due to volatilization of MoO_3.[39] However, the mechanism of oxidation is different in various temperature regimes. The oxidation behavior of Mo–Si–B alloys with phase composition of $Mo_5Si_3B_x$, $MoSi_2$, and MoB has been studied by Thom et al. in the temperature range of 600°C–1600°C.[43] Rapid mass loss took place between 600°C and 760°C, such that sample degradation was at its worst between 662°C and 708°C. The formation of a glassy oxide scale was initiated at 756°C and completed at 804°C, during which the transition from mass loss to near-zero change in mass occurred. High-temperature oxidation studies between 1150°C and 1600°C have shown mass gain accompanied by the formation of a 10–20 μm thick scale of B-doped SiO_2. The top surface of this oxide scale primarily contains SiO_2, as it is fully depleted in boron. Nomura et al. have reported that a coating of Mo–25Si–8.4B alloy (containing a phase mixture of Mo_3Si, Mo_5Si_3, and Mo_5SiB_2) on an Mo–ZrC composite shows excellent resistance to oxidation at 1400°C due to the formation of the protective scale of SiO_2.[44]

The oxidation behavior of Mo–Si–B alloys containing α-Mo, Mo_3Si, and Mo_5SiB_2 as constituent phases (Table 7.4) is quite different from that of alloys having a phase mixture of Mo_3Si, Mo_5Si_3, and Mo_5SiB_2. It is intuitive that the Si/B ratio plays a very critical role in the oxidation behavior of Mo–Si–B alloys. Supatarawanich et al. have recently studied the oxidation behavior of near-eutectic Mo–Si–B alloys with Si/B ratios of 0.5, 0.71, and 1.0, and have reported that the oxidation resistance at 800°C is better in an alloy with an Si/B ratio of 0.5 and a high volume fraction of the Mo_5SiB_2 phase.[48] On the other hand, the alloy with an Si/B ratio of 1.0, having a higher volume fraction of Mo_3Si, has shown the best oxidation resistance at 1300°C. While the

TABLE 7.3

Results of Oxidation Experiments on Mo_5Si_3 and Mo_5Si_3–B Alloys, Shown with Temperature of Oxidation, Nature of Mass Change, and Oxidation Products

Material	Temperature (°C)	Mass Change	Oxidation Products	References and Salient Features
Mo_5Si_3	800	Pest in 35 h	MoO_3, SiO_2	Meyer and Akinc[39]
	900–1200	Loss	Same	Diffusion of O in borosilicate layer
Mo_5Si_3–B (1.24 wt.% B)	900–1000	Loss		
	1050–1300	Gain		
Mo_5Si_3	600–1000	Gain at 600°C	MoO_3	Natesan and Deevi[34]
		Loss at >700°C	MoO_3, SiO_2	
$Mo_5Si_3B_y$	500	Gain	MoO_3, B–SiO_2	Maximum loss at 800°C
	600–700	Loss	MoO_3, B–SiO_2	
Mo–Si–B	800–1000	Transient loss and steady state	MoO_2, B–SiO_2	
	1000–1200		Mo, B–SiO_2	
Mo_5SiB_2	800	Transient loss and steady state	MoO_2, B–SiO_2	
Mo_5Si_3–B	1450			Meyer and Akinc[41]
1.6 wt.% B		Gain	SiO_2	
0.14 wt.% B		Loss	SiO_2, MoO_3	Rate constants
Mo–Si–B B/Si = 0.02–0.24	600–1000	Loss increases with increasing B/Si ratio	MoO_3 (<800°C) SiO_2 (>800°C)	Meyer et al.[42] Rate constants
$70Mo_5Si_3B_x$– $20MoSi_2$– 10MoB (by vol.%)	617	Gain followed by loss	MoO_3, SiO_2	Thom et al.[43]
	660–760	Pesting, worst: 730°C	MoO_3, SiO_2	
	756–804	Transition to zero loss	Borosilicate	
	1150–1600	Loss	B content low	

oxidation of Mo_5SiB_2 contributes to the formation of B_2O_3 or B-doped SiO_2, Mo_3Si contributes to the formation of SiO_2. At lower temperatures, the protective scale of SiO_2 cannot form and doping with B increases fluidity and allows the self-healing of cracks and pores in the scale. At higher temperatures, the viscous scale of SiO_2 is responsible for protection, and higher B content is effective in reducing the viscosity of the oxide scale, thereby increasing the diffusivity of oxygen anions and making the scale nonprotective. Also, the escape of B_2O_3 (g) leaves open channels inside the oxide scale cross sections,

TABLE 7.4

Results on Oxidation Experiments of Mo–Si–B Alloys, Shown with Temperature of Oxidation, Nature of Mass Change, and Oxidation Products

Material	Temperature (°C)	Mass Change	Oxidation Products	References
Mo14.2Si9.6B αMo,Mo$_3$Si,T2	1000, 1200		Top: SiO$_2$, B$_2$O$_3$ Inner: MoO$_2$	Park et al.[45]
Mo$_5$SiB$_2$	700–800	Gain followed by loss	MoO$_3$, B$_2$O$_3$, SiO$_2$	Yoshimi et al.[46]
	900–1400	Continuous loss of MoO$_3$, B$_2$O$_3$	Outer: SiO$_2$ Inner: Mo	
Mo–11Si–11B	600–650	Transient gain	MoO$_3$	Mendiratta et al.[47] No oxidation protection
	650–750	Loss	MoO$_3$, B–SiO$_2$	Porous, MoO$_3$ escapes
	800–1000	Loss	B–SiO$_2$	B-depleted scale is protective
	1000–1300	Loss	Outer: B–SiO$_2$ Inner: Mo	Less protective: Viscosity is low
	1300–1400 Cyclic	Loss	B–SiO$_2$	
Mo–Si–B alloys	600	Gain	MoO$_3$	Supatarwanich et al.[48]
	800–1000	Loss	SiO$_2$, B$_2$O$_3$	
	1300	Loss	SiO$_2$	
Near eutectic Mo–Si–B	800	Loss	MoO$_3$, SiO$_2$, B$_2$O$_3$	Supatarwanich et al.[49]
	Cyclic			800°C: high vol.% of T2 is better
Si/B = 0.5, 0.71, 1.0	1300	Loss	SiO$_2$	Protection by SiO$_2$ at 1300°C (Si/B = 1 best)
	Cyclic			
T1–MoSi$_2$–MoB T1–T2–Mo$_3$Si	1000–1100 Dry air	Loss Loss	B–SiO$_2$, subscale: Mo, MoO$_2$	Mandal et al.[50] Amorphous scale is protective
αMo–T2–Mo$_3$Si	Wet air	Loss		Moisture helps B loss
Mo–Si–B–O: Mo, Mo$_3$Si, Mo$_5$Si$_3$, Mo$_5$SiB$_2$, SiO$_2$	600 1000–1400	Gain Loss	MoO$_3$, MoO$_2$, Mo–Si–B–O, SiO$_2$, B$_2$O$_3$	Yoshimi et al.[51] Volatile: MoO$_3$ & B$_2$O$_3$ B–SiO$_2$ scale >727°C

which creates the formation of pores and delays the occurrence of steady-state oxidation. In general, the oxidation behavior of the alloys containing α-Mo, Mo_3Si, and Mo_5SiB_2 as constituent phases[41,42,45,47-54] varies significantly in different temperature regimes. In the temperature range of 600°C–750°C, increasing volatilization of MoO_3 and little protection against oxidation are observed. Between 800°C and 900°C, a transient stage of mass loss due to the volatilization of MoO_3, followed by a steady-state stage associated with the porous fluid of the borosilicate layer allowing the escape of gaseous MoO_3 (g), is observed. On the other hand, at temperatures between 1000°C and 1300°C a short transient period showing mass loss followed by a steady-state regime of negligible mass change accompanied by the formation of a protective, continuous, and well-adherent SiO_2 scale is observed. This oxide scale is found to be depleted in B on the upper surface and has an inner reaction zone of Mo and SiO_2. The oxidation of the Mo–14.2Si–9.6B alloy at 1000°C and 1200°C has led to the formation of borosilicate and MoO_2 as oxidation products.[44] Surfaces coated using a plasma spray of SiO_2 did not even show the MoO_2 phase on oxidation. The presence of Mo or MoO_2 in preference to MoO_3 in the oxide scale proves that the borosilicate or SiO_2 layer acts as an effective barrier limiting oxygen flux to the oxide–alloy interface.

Studies on the oxidation behavior of Mo–14Si–10B alloys have shown a small mass gain in the range of 400°C–600°C and unabated mass loss at 700°C due to the formation and subsequent vaporization of MoO_3 (Figure 4.8).[52] On the other hand, on exposure in the range of 800°C–1300°C, mass loss has been observed in the initial stages, which is followed by a stable regime of no change in mass (Figure 7.4).[53] Exposure at 1150°C leads to the formation of a more or less stable and protective scale of B_2O_3–SiO_2 after noticeable mass loss during the initial transient stage, as shown in Figure 7.5.[53-55] The formation of this protective scale on prior exposure at 1150°C has been found to resist the damage due to thermal shock caused by air cooling, and also significantly reduces the oxidation rate at the intermediate critical temperature range of 700°C–900°C. A recent study[54] has shown the transient stage at 1150°C to last for 8 min, when mass loss occurs due to the preferential oxidation of the Mo_{ss} phase to form volatile MoO_3. The pores formed in the oxide scale can be attributed to the escape of MoO_3 (g). Interestingly, the rate of mass loss by the vaporization of MoO_3 appears to resemble the kinetics of the formation of cone-shaped cavities on the oxide scale.[55] Whereas the pores permit the penetration of atmospheric oxygen into the alloy, the oxidation of B and Si leads to the formation of a glassy borosilicate scale. This scale covers the top surface after about 8 min of exposure by viscoplastic flow, and thereby provides protection against further oxidation (Figures 7.6 through 7.8).[55] The growth of this outer protective scale is found to follow the parabolic rate law (Figure 7.9).[55]

Surfaces coated using a plasma spray of SiO_2 did not even show the MoO_2 phase on oxidation. The presence of Mo or MoO_2 in preference to MoO_3 in the oxide scale proves that the borosilicate or SiO_2 layer acts as an effective

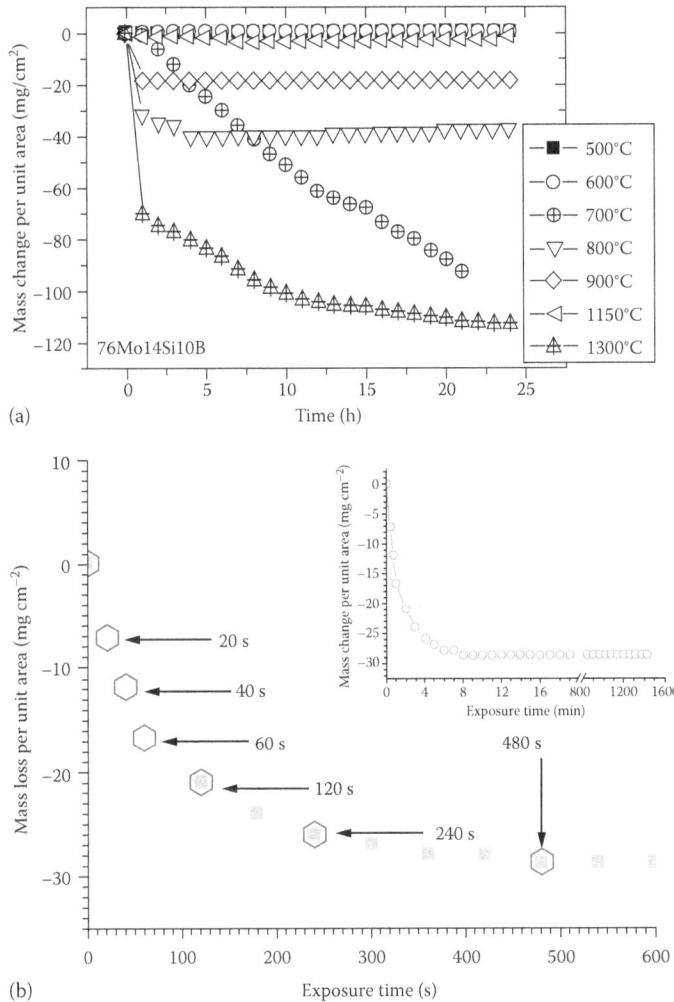

FIGURE 7.4
Plots showing (a) the change in mass of the 76Mo–14Si–10B alloy isothermally exposed for 24 h in the temperature range of 500°C–1300°C and (b) mass change in the initial stages of isothermal exposure at 1150°C. (From Roy, B., et al., *Corros. Sci.*, 68, 231–237, 2013. Reprinted with permission from Elsevier.)

barrier limiting oxygen flux to the oxide–alloy interface. Yoshimi et al. have shown that the incorporation of SiO_2 in the microstructure leads to a significant improvement in oxidation resistance.[51] The oxidation resistance of Mo–Si–B alloys in the temperature regime of 600°C–900°C may be enhanced by the choice of a proper composition or the preoxidation of the alloys at high temperatures above 1000°C, so as to form an adherent scale of SiO_2 to protect the alloy from pest disintegration.

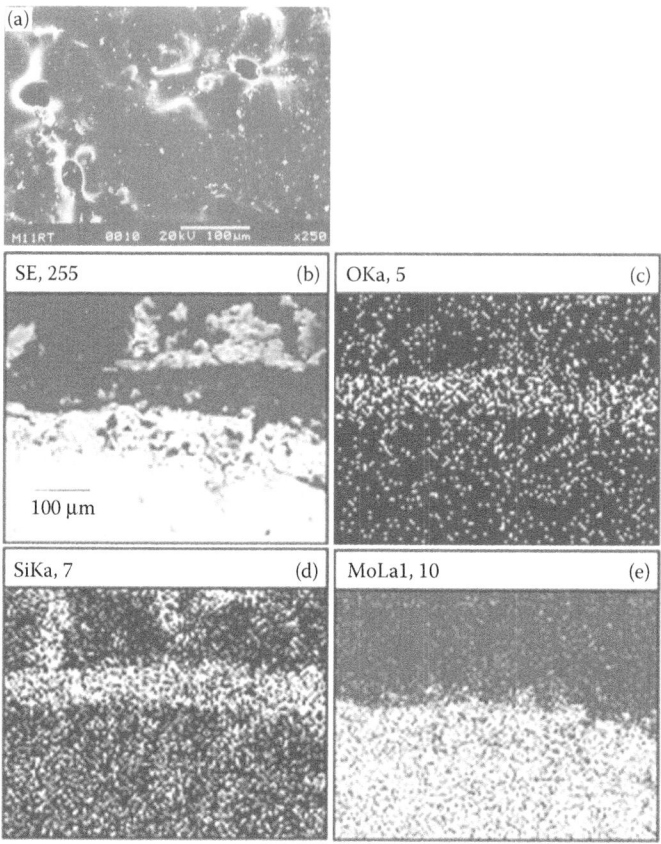

FIGURE 7.5
Oxide scale of the 76Mo–14Si–10B alloy subjected to cyclic exposure at 1150°C with air cooling at 1 h intervals. Here, a protective scale of B_2O_3–SiO_2 is observed: SEM (SE) images of (a) top surface and (b) alloy–oxide cross section, as well as energy-dispersive spectroscopy (EDS) x-ray maps of (c) O, (d) Si, and (e) Mo. (From Paswan, S., et al., *Metall. Mater. Trans. A*, 40, 2644–2658, 2009. Reprinted with permission from Springer Science + Business Media.)

7.3 Oxidation Behavior of Niobium Silicides

$NbSi_2$ shows pest disintegration between 500°C and 850°C.[56,57] Arc-melted $NbSi_2$ has been found to disintegrate on exposure in air at 500°C and 750°C for 160 h and 3 h, respectively.[58] On the other hand, defect-free single crystals of $NbSi_2$ have not shown any evidence of pesting, indicating that grain boundaries and porosities have a significant role in promoting accelerated oxidation. Pesting of $NbSi_2$ has shown three regimes, which are associated with the formation of various Nb oxide forms: NbO_2, α-Nb_2O_5, and β-Nb_2O_5.[56] It is interesting to note

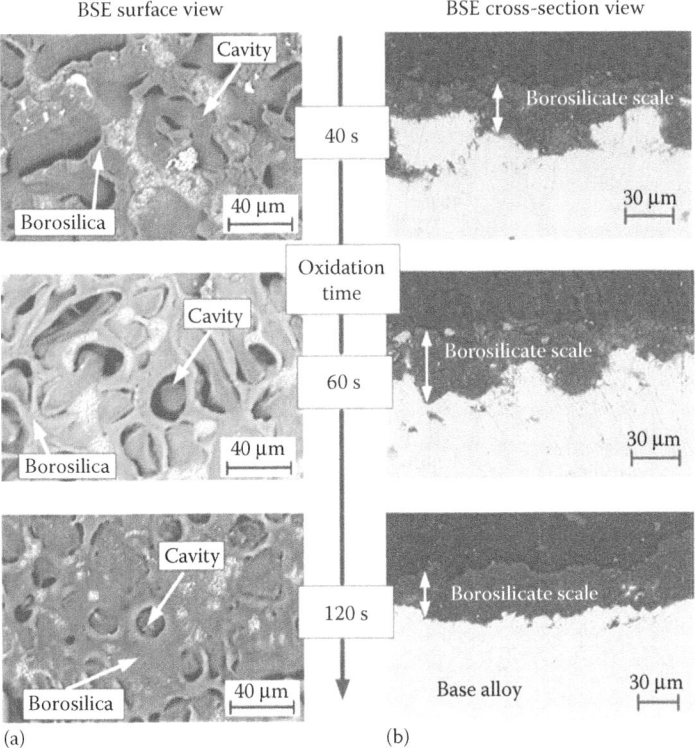

FIGURE 7.6
SEM (BSE) images depicting the morphology of the top surface of the oxide scale formed on the 76Mo–14Si–10B alloy after 40 s, 60 s, 120 s of exposure (a) and BSE images of the oxide-alloy cross-section (b). (From Roy, B., et al., *Corros. Sci.* 68:231–237, 2013. Reprinted with permission from Elsevier.)

that, although the free energies of formation of all the Nb oxides are higher than that of SiO_2, the Nb oxides are formed at lower temperatures. This trend may be attributed to the lower growth rate of SiO_2, as expected from its lower parabolic rate constant. Unlike MoO_3, the oxides of Nb are not volatile. However, due to the nonprotective character of the oxide scale, ingress of oxygen is possible through cracks and by short-circuit diffusion through grain boundaries. This causes embrittlement of the material, followed by failure.

Pest oxidation is significantly reduced in the Nb–60Si–20Cr alloy.[58–60] Duplex oxide layers of Cr_2O_3 and SiO_2 form, which protect the alloy from further oxidation and prevent the formation of Nb oxides. However, alloying with Cr is ineffective for protection against oxidation at temperatures exceeding 1000°C due to the formation and volatilization of CrO_3. Alloying $NbSi_2$ with Al to obtain the composition of Nb–47Si–20Al leads to excellent oxidation resistance between 1100°C and 1300°C because of protection by the scale of Al_2O_3.[60] However, the Nb–Si–Al alloy has shown poor resistance to oxidation in the temperature range of 800°C–1000°C. Alloying the Nb–Si–Al alloy with Cr, as

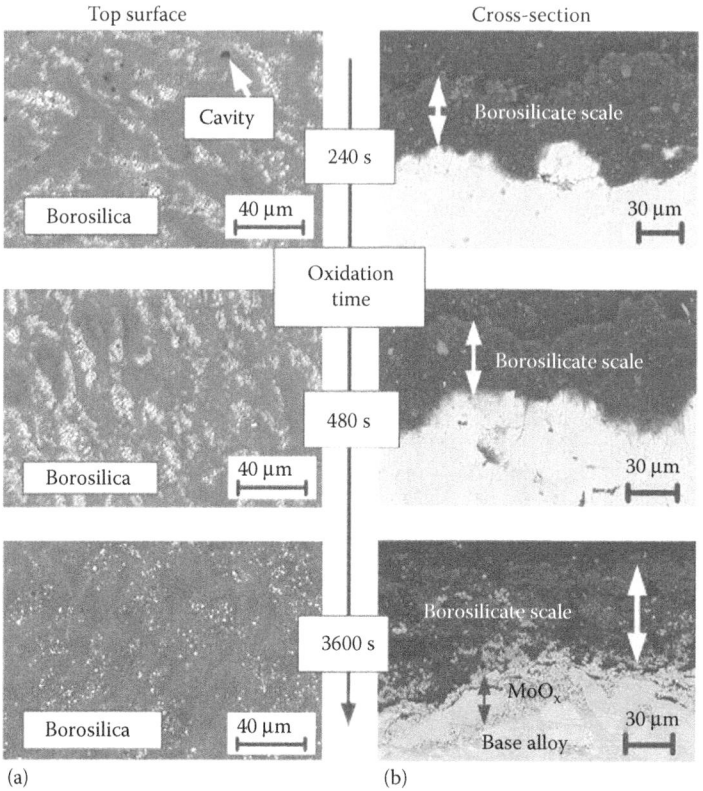

FIGURE 7.7
SEM (BSE) images depicting the morphology of the top surface of the oxide scale formed on the 76Mo–14Si–10B alloy after 240 s, 480 s, 3600 s of exposure (a) and BSE images of the oxide-alloy cross-section (b). (From Roy, B., et al., *Corros. Sci.* 68:231–237, 2013. Reprinted with permission from Elsevier.)

in the Nb–56Si–11Al–3Cr alloy, leads to excellent oxidation resistance at temperatures between 500°C and 1400°C. On exposure at 750°C, the scale that is formed contains the oxides of Nb, Si, Al, and Cr, while only a scale of Al_2O_3 forms at 1000°C. As Al_2O_3 is more stable than the oxides of Nb, Si, and Cr, its formation can be easily justified. The formation of other oxides in the scale at 750°C could be due to an increase in the activities of other alloying elements with the depletion of Al to form Al_2O_3. Unlike $MoSi_2$, $NbSi_2$ fails to form a continuous and protective oxide scale of SiO_2 at temperatures above 1000°C, because the oxides of Nb are less volatile. Instead, this oxide scale contains mixtures of SiO_2 and α-Nb_2O_5 as well as cracks during growth leading to continued oxidation, as has been reported by Pitman and Tsakiropoulos.[58]

Nb_5Si_3 undergoes accelerated pest oxidation in the temperature regime of 750°C–1000°C, forming Nb_2O_5.[60,61] Complete disintegration of α-Nb_5Si_3 has been observed on exposure at 1000°C for 1–3 h. The susceptibility to pest disintegration of Nb-silicides is due to the mismatch in the CTEs of the

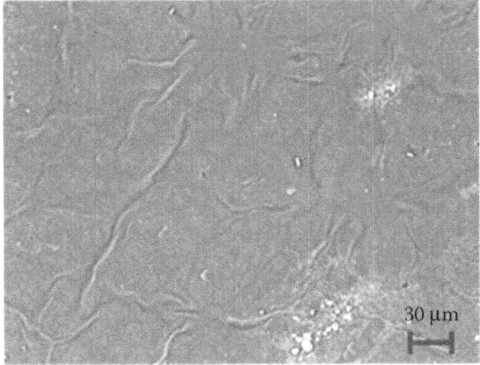

FIGURE 7.8
SEM (SE) micrograph of the oxide scale formed over the 76Mo–14Si–10B alloy after 24 h of exposure at 1150°C. It shows the presence of ripple-like features on the surface, which suggest the viscous flow of borosilicate glass. (From Roy, B., et al., *Corros. Sci.* 68, 231–237, 2013. Reprinted with permission from Elsevier.)

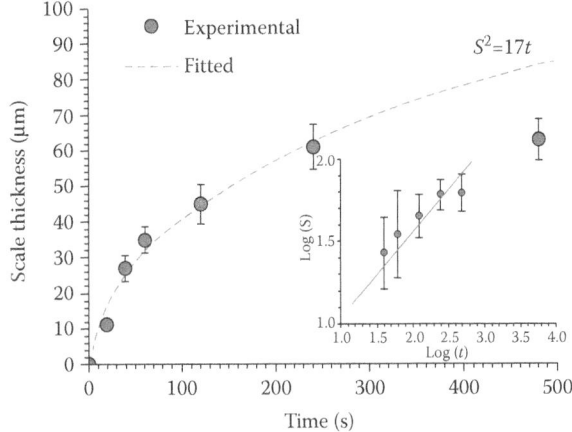

FIGURE 7.9
Plot of scale thickness (S) against duration of exposure (t) in the course of the transient-stage oxidation of the Mo76Si14B10 alloy (*Inset*: Plots showing log S vs. log t with the best-fit line used to calculate the exponent). (From Roy, B., et al., *Corros. Sci.* 68, 231–237, 2013. Reprinted with permission from Elsevier.)

oxidation products. It may be noted that the CTE of Nb_2O_5 is just one-quarter of that of Nb_5Si_3.[62] The lower CTE of the oxide scale, coupled with the volume expansion accompanying its formation, leads to compressive residual stresses inside the scale and its spallation during cooling. Again, the diffusion coefficient of oxygen in Nb_2O_5 is greater than that in SiO_2 at 1000°C by nearly three orders of magnitude. Inspired by success in the case of B-doped Mo_5Si_3, doping of Nb_5Si_3 with B has also been attempted, but has not been very successful.[60,63] Nb_5SiB_2 does not form a continuous and protective scale of borosilicate, because Nb_2O_5 is not volatile and remains as a prominent constituent of the scale. Compared with Nb_5Si_3, Nb_5SiB_2 is reported to be more resistant to oxidation, while it is inferior compared with $NbSi_2$.

In the case of Nb_{ss}–Nb_5Si_3 composites, some of the alloying elements,[65] such as Mo, W, V, Zr, and Ta, have shown a positive influence on creep and an adverse effect on oxidation resistance. Alloys with V, Zr, and Mo have shown the worst oxidation characteristics, though it should be noted that Mo within 1 at.%, and Ta or W within 2 at.%, did not have much impact on the oxidation rate. The recently developed Nb-silicide alloys have shown an impressive gain in oxidation resistance, by about an order of magnitude, when compared with their counterparts developed earlier than 1998, indicating the remarkable pace of development. Alloy development efforts involving the use of Ti, Al, and Hf as alloying elements, the oxides of which have a lower free energy of formation compared with SiO_2, have shown significant success.[66–68] Although alloying with Al and Hf has been found to restrict damage due to pesting in Nb–Si alloys, the use of Sn has been most effective in eliminating damage due to pesting in the temperature regime of 750°C–950°C, although it has no visible effect on oxidation kinetics beyond 1200°C.[69] The beneficial effect of alloying with Sn has been established using cyclic oxidation studies involving 100 h of exposure at temperatures of 760°C, 870°C, and 980°C. Pest disintegration of niobium silicides not containing Sn could be observed on exposure at 870°C, while alloys having 1.5 at.% Sn were found to be pest resistant. The pest resistance decreases on increasing the volume fraction of the Nb_{ss} phase.

Research on the high-temperature oxidation behavior of Nb–Si–Al–Ti alloys[68] containing three phases (Nb_{ss}, Nb_5Si_3, and Nb_3Al) has shown a distinct reduction in mass gain on reducing the volume fraction of Nb_{ss}. The isothermal oxidation of Nb–Ti–Si–X… alloys at temperatures of 900°C or less leads to the formation of cracks just below the oxide scale, originating from the presence of residual stresses contributed by the mismatch between the CTEs of the constituents.[67,68] The oxidation products in multicomponent Nb-silicides with Ti and Cr as alloying elements have been found to be $TiNb_2O_7$, $CrNbO_4$, SiO_2 (cristobalite), $3Nb_2O_5.TiO_2$, and Nb_2O_5.[67] Chan has investigated the cyclic oxidation behavior of multiphase niobium silicides alloyed with Ti, Hf, Cr, and Ge between ambient temperature and peak temperatures ranging between 900°C and 1400°C, and has reported spallation with weight loss.[70] Ti, Cr, and Ge-containing alloys formed scales containing $CrNbO_4$, Nb_2O_5, and $Nb_2O_5.TiO_2$, with smaller quantities of SiO_2 or GeO_2. Oxidation resistance improves with an increasing volume fraction of $CrNb_2O_5$ in the oxide scale at the expense of Nb_2O_5 and TiO_2, and a decreasing volume fraction of Nb_{ss}.

The alloying of Nb–Si-based alloys with Mo has been found to contribute to the high-temperature oxidation resistance of these alloys, which is obvious from the isothermal oxidation kinetics of alloys with varying Mo concentrations, as depicted in Figure 7.10.[71] First of all, it reduces the activity

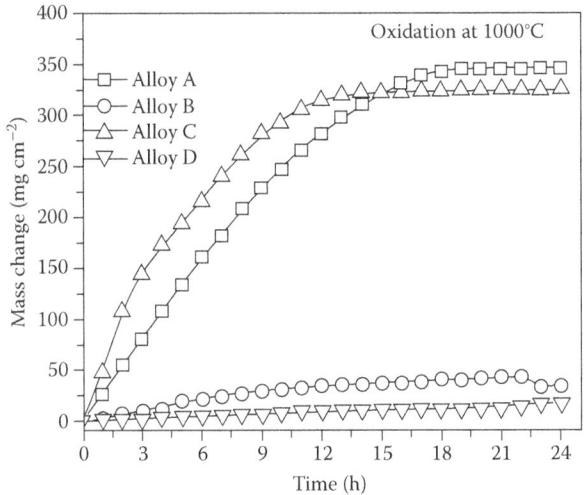

FIGURE 7.10

Plots depicting the kinetics of isothermal oxidation during exposure for 24 h at 1000°C for alloys with compositions: Nb–19Si–5Mo (Alloy A), Nb–18Si–26Mo (Alloy B), Nb–13Si–4Mo (Alloy C), and Nb–12Si–15Mo (Alloy D). (From Chattopadhyay, K., et al., *Metall. Mater. Trans. A,* 39, 577–592, 2008. Reprinted with permission from Springer Science + Business Media.)

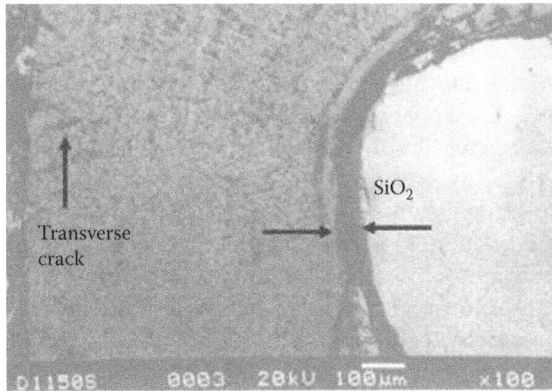

FIGURE 7.11

SEM (BSE) image of the cross section of the oxide scale formed on the Nb–12Si–15Mo alloy due to exposure at 1150°C for 24 h. The SiO_2 layer at the metal–oxide interface is shown with a pair of arrows. Transverse cracks (single arrow) are visible in the oxide scale. (From Shibata, T., et al., *Appl. Phys. Lett.,* 40, 77–80, 1982. Reprinted with permission from Springer Science + Business Media.)

of Nb, thereby causing the formation of a continuous and impervious layer of SiO_2 at the alloy–oxide interface (Figure 7.11). Moreover, its dissolution in Nb_2O_5 is expected to lower the diffusivity of oxygen anions. The free energy of formation of MoO_2 is the closest among the oxides of Mo to that of Nb_2O_5.[21] Moreover, the radii of Mo^{4+} and Nb^{5+} ions are 0.70 and 0.69 Å,

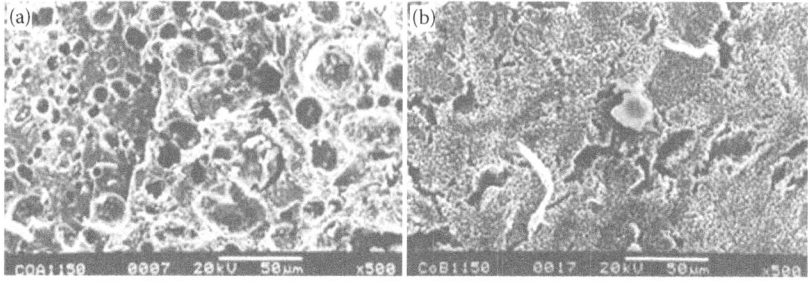

FIGURE 7.12
SEM (SE) images of the top surfaces of oxide scales formed on (a) Nb–19Si–5Mo (Alloy A) and (b) Nb–18Si–26Mo (Alloy B). (From Chattopadhyay, K., et al., *Metall. Mater. Trans. A*, 39, 577–592, 2008. Reprinted with permission from Springer Science + Business Media.)

respectively, whereas the Mo^{6+} ion has a radius of 0.62 Å. Hence, it is intuitive that the Mo atom would dissolve in the Nb_2O_5 in the form of Mo^{4+}, thereby substituting the Nb^{5+} sites. It should also be noted that, in contrast to the n-type character of Nb_2O_5 and SiO_2, MoO_2 is an amphoteric conductor. Because Mo^{4+} cations have a lower valence than Nb^{5+}, substitution of the former type of cation in Nb_2O_5 is expected to lower the concentration of anion vacancies and reduce the rate of oxygen anion diffusion through the oxide scale. Furthermore, the oxides of Mo have lower melting points than Nb_2O_5, and therefore the sinterability of the oxide scale is improved significantly. The reduction of void surface area through sintering of the oxide scale is expected to reduce the rate of oxygen penetration through surface diffusion significantly. The contribution of Mo addition to oxide scale densification has been confirmed by the observation of higher smoothness and less porosity on the top surfaces of the oxide scale formed on Nb–Si–Mo alloys with higher Mo content (Figure 7.12).

The combined effects of the alloying elements, including Ti, Hf, Si, Cr, Fe, and Al, on high-temperature oxidation behavior have been investigated.[65,69,70,72–74] Optimum oxidation resistance, coupled with high creep strength, has been achieved for compositions with Nb:(Ti + Hf) ratios of 1.8–2.1 and Si levels of 17%–19%. Alloying with Cr and Al has been found to be the most beneficial for oxidation resistance.[73,74] Increasing the concentration of Cr to 5 at.% stabilizes the Cr-rich Laves phases. Ge additions show the best properties at 6 at.% concentration, while B additions are beneficial in the range of 2–6 at.%, beyond which they are detrimental. The addition of Fe as a substitute for Cr has also been found to be beneficial.

The short-term target[74] of loss in mass by less than 200 μm in 10 h of exposure at 1370°C is attainable, while the long-term goal of loss in mass by less than 25 μm in 100 h at 1315°C is yet to be achieved.[66] The long-term goal has been set in analogy with the second-generation Ni-base superalloys, which show similar loss in mass at 1150°C. Presently, the Nb-silicides show a loss of less than 25 μm in 100 h at 1200°C.

7.4 Oxidation Behavior of WSi_2

The oxidation behavior of WSi_2 is similar to that of $MoSi_2$,[3] displaying accelerated oxidation at the intermediate temperature range of 400°C–600°C[8] and developing a semiprotective scale of SiO_2 at temperatures $\geq$900°C. Dry and wet oxidation experiments[75,76] on WSi_2 thin films have shown that kinetics involving gain in mass with time of exposure follows a linear rate to begin with and then a parabolic rate law. However, dry oxidation at 1000°C and 1200°C has shown evidence of the simultaneous formation of WO_3 and SiO_2, whereas WO_3 has not been found on oxidation using steam. WSi_2 has shown loss in mass during oxidation at 1200°C and 1500°C in dry air, probably by volatilization of WO_3, in contrast to the minor gain in mass observed in the case of $MoSi_2$ under similar conditions of oxidation.[77] The formation of W_5Si_3 by the reaction of WSi_2 with oxygen has also been confirmed.[75,76] The oxidation resistance of W_5Si_3 has been found to be relatively poorer, and characterized by a large mass loss at 1200°C.[4]

7.5 Oxidation Behavior of Ti_5Si_3

The oxidation behavior of Ti_5Si_3 has been found by Rosenkranz et al. to follow a parabolic rate law.[78] The oxide scale contains mixtures of TiO_2 and SiO_2. At a given temperature, the free energy of formation of TiO_2 is slightly less than that of SiO_2, as shown in Figure 4.1. A study by Thom et al. on the effect of the grain size of Ti_5Si_3 on oxidation rates in the temperature range of 700°C–1000°C has shown that oxidation rates increase on decreasing grain size, because of the enhanced role of grain-boundary diffusion.[79] The oxide film contains predominantly TiO_2 (rutile) with a relatively lower amount of SiO_2, and the thickness grows with increasing temperature.

Oxidation behavior in the temperature range of 1200°C–1400°C has been examined by Mitra and Rao.[80] Although the oxidation kinetics follows a rate slower than that predicted by parabolic rate law at 1200°C, it is significantly faster at 1300°C (Figure 4.2). The values of n (in $(\Delta W)^n = kt$) have been found to be 3.819 at 1200°C and 0.732 at 1300°C. The oxidation rates are much higher at 1400°C. The oxide scale formed at 1200°C has shown an outer layer of TiO_2 and an inner layer of SiO_2 (Figure 7.13a–d). Considering the fact that TiO_2 has a slightly lower free energy of formation and a higher parabolic rate constant than SiO_2, and the partial pressure of oxygen required for Ti/TiO_2 thermodynamic equilibrium is less than that for Si/SiO_2, it is appropriate to infer that the rate of TiO_2 formation supersedes that of SiO_2. Thom et al. have attributed the higher growth rate of TiO_2 to the faster diffusion of oxygen anions in TiO_2.[79] However, the presence of TiO_2 as the outer layer (Figure 7.13a–d) suggests that

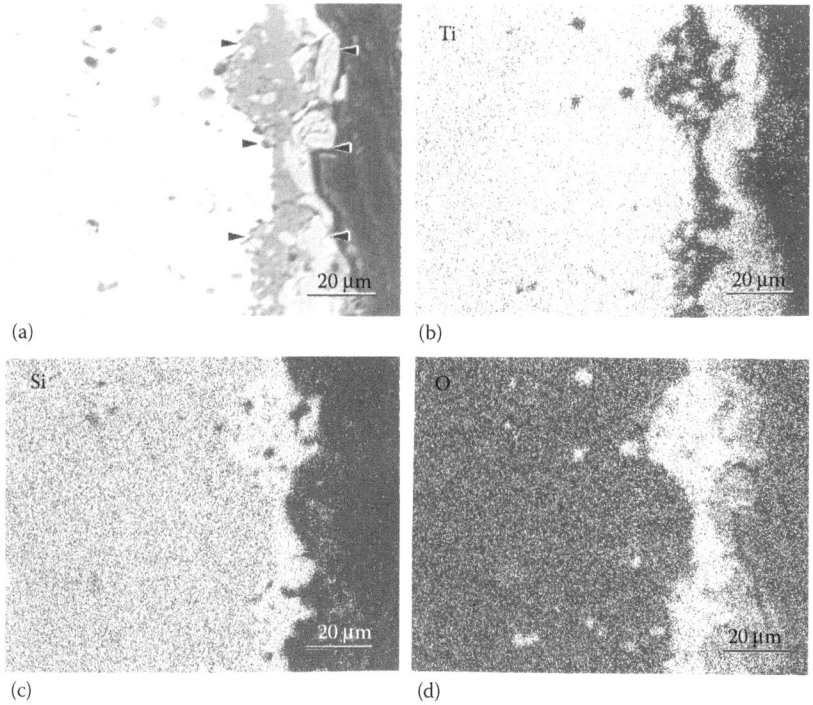

(a)   (b)

(c)   (d)

FIGURE 7.13
Oxide scale formed on Ti_5Si_3 on exposure at 1200°C for 80 h: (a) SEM (BSE) image and EDS x-ray maps of (b) Ti, (c) Si, and (d) O. (From Mitra, R. and V. V. R. Rao, *Metall. Mater. Trans. A*, 29, 1665–1675, 1998. Reprinted with permission from Springer Science + Business Media.)

the rate of oxidation is controlled by the outward diffusion of Ti^{4+} through the oxide scale from the oxide–alloy interface to the oxide–air interface. Even the Ti^{4+} ions at the oxide–metal interface have to diffuse through the intermediate SiO_2 layer, resulting in a growth rate slower than that predicted by the parabolic rate law at 1200°C. The activation energy for the oxidation of Ti_5Si_3 in the temperature range of 1000°C–1300°C has been found to be 46.09 kcal mol^{-1},[80] which is only 0.57 times that measured for $MoSi_2$ in a similar temperature regime.[3] Figure 7.14 shows the plots comparing the oxidation rates of $MoSi_2$ and Ti_5Si_3 at 1200°C. From the results in these plots, it is obvious that $MoSi_2$ is much more oxidation resistant than Ti_5Si_3.

The scale of the Ti_5Si_3–8 wt.% Al alloy subjected to oxidation at 1200°C in air contained α-Al_2O_3 and TiO_2, and no SiO_2.[80] The higher rate of oxidation of the Ti_5Si_3–8 at.% Al alloy compared with Ti_5Si_3 is probably due its inability to form an impervious Al_2O_3 scale. A study of the oxidation of the Ti_5Si_3–20 vol.% TiC composite at 1200°C has shown its oxidation resistance to be much poorer than that of Ti_5Si_3. SiO_2 is not observed at all; the entire scale showed porous TiO_2, which can be explained by the much higher activity of

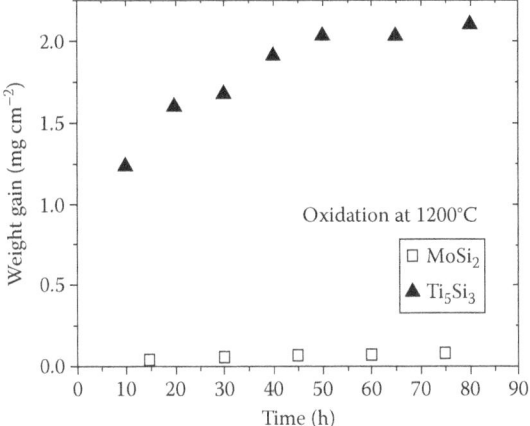

FIGURE 7.14

Plots of mass gain against temperature for $MoSi_2$ and Ti_5Si_3 subjected to cyclic exposure at 1200°C. (From Mitra, R. and V. V. Rama Rao, *Mater. Sci. Eng. A*, 260, 146–60, 1999; Mitra, R. and V. V. R. Rao, *Metall. Mater. Trans. A*, 29, 1665–1675, 1998.)

Ti in the composite. The oxidation of TiC particles leads to the evolution of CO gas through the scale, which would lead to the formation of porosities and cracks in the oxide scale, rendering it nonprotective and leading to an enhanced rate of oxidation.

7.6 Oxidation Behavior of Chromium Silicides

The oxidation behavior of $CrSi_2$ films grown by electron beam evaporation on Si (100) substrates has been studied by Frampton et al.[81] It has been observed that in comparison to films of $MoSi_2$ and WSi_2, $CrSi_2$ oxidizes much more rapidly during exposure at temperatures between 700°C and 900°C. The oxidation is controlled by the diffusion of oxygen through the SiO_2 scale, as a result of which a parabolic rate law is followed. The density of the SiO_2 scale is also less in the $CrSi_2$ film when compared with $MoSi_2$ and WSi_2, probably due to the competitive growth of Cr_2O_3. The preference for the formation of SiO_2 rather than Cr_2O_3 on the $CrSi_2$ film is probably because of the higher activity of Si.

Oxidation experiments[82] on Cr_3Si have shown negligible gain in weight on heating from room temperature to 650°C, above which gradual oxidation has been observed. The rate of oxidation is faster between 700°C and 900°C, and is modest close to 1000°C, due to the formation of a protective scale of Cr_2O_3. Even if the SiO_2 is more stable, the formation of Cr_2O_3 is preferred because of the higher activity of Cr in Cr_3Si, the higher parabolic rate constant, and the lower diffusivity of Si. Studies by Raj on the oxidation of $Cr_{40}Mo_{30}Si_{30}$[83] have shown the formation of protective scales of Cr_2O_3 at low temperatures and α-cristobalite (SiO_2) at high temperatures. Pesting at 500°C is prevented completely due to the formation of the protective scale of Cr_2O_3, even in the samples with pores and cracks. The oxidation involves a small gain in mass at 1100°C, above which loss takes place due to the volatilization of CrO_3 formed from further oxidation of Cr_2O_3 at higher temperatures. Oxidation experiments in isothermal conditions at 1200°C or in cyclic conditions have shown steady weight loss, while those in the burner rig have shown an initial rapid weight loss, followed by weight gain according to the parabolic rate law. The high-velocity combustion flame in the burner rig drives the rapid volatilization of CrO_3, as a result of which the protective scale of SiO_2 can form to prevent further oxidation. Oxidation studies on this alloy reinforced with 15 vol.% of Pt–6%Rh have revealed Cr_2O_3 scales at 900°C and 1000°C similar to the alloy matrix, but a continuous inner layer of SiO_2 doped with Cr, in addition to the outer layer of Cr_2O_3, at 1100°C. The oxidation is characterized by distinct mass gain at 1000°C under isothermal conditions and gain followed by loss under cyclic conditions. At 1100°C, the oxidation involves primarily mass loss after an initial gain in both isothermal and cyclic conditions, as expected from the volatilization of CrO_3. Studies on the C/SiC–Cr_3Si composites[84] processed by the melt infiltration process have shown reasonable oxidation resistance, as well as retention of strength, when subjected to the burner rig for 20 h at 1300°C.

Oxidation studies by Tomasi et al.[85] of an alloy of composition 50.55Mo36.05Cr13.4Si (at.%) and containing Cr_3Si and Cr_5Si_3 phases have shown oxide scales primarily constituted by Cr_2O_3 at 900°C and 1000°C, while the presence of cristobalite could be detected in the oxide scales corresponding to 1100°C, similar to the report by Raj.[83] The oxide scale was fully protective in nature up to 900°C, while loss of mass occurred at higher temperatures, as expected. The spallation of the oxide scale takes place during oxidation under cyclic conditions, as expected.

7.7 Summary

The oxidation behavior of all silicides is strongly dependent on their composition and the temperature of oxidation. The constitution of the oxide scale is a function of not only the composition, but also the thermodynamic and kinetic parameters. For pest disintegration under isothermal or cyclic oxidation, it is necessary that samples are not fully dense and the oxide scale is nonprotective, allowing internal oxidation. Residual stresses build up with internal oxidation or due to mismatch in the CTEs, leading to crack initiation and catastrophic propagation. For high-temperature applications, it is preferable to have a continuous scale of SiO_2 or Al_2O_3, the formation of which often competes with the oxides of other alloying elements. Hence, careful control of composition is required.

References

1. Mitra, R. 2006. Mechanical behaviour and oxidation resistance of structural silicides. *Inter. Mater. Rev.* 51:13–64.
2. Vasudevan, A. K. and J. J. Petrovic. 1992. A comparative overview of molybdenum disilicide composites. *Mater. Sci. Eng. A* 155:1–17.
3. Berztiss, D. A., R. R. Cerchiara, E. A. Gulbransen, F. S. Pettit, and G. H. Meier. 1992. Oxidation of $MoSi_2$ and comparison with other silicide materials. *Mater. Sci. Eng. A* 155:165–181.
4. Wirkus, C. D. and D. R. Wilder. 1966. High-temperature oxidation of molybdenum disilicide. *J. Am. Ceram. Soc.* 49:173–177.
5. Meschter, P. J. 1992. Low-temperature oxidation of molybdenum disilicide. *Metall. Trans. A* 23:1763–1772.
6. Chou, T. C. and T. G. Nieh. 1992. New observations of $MoSi_2$ pest at 500°C. *Scr. Metall. Mater.* 26:1637–1642.
7. McKamey, C. G., P. F. Tortorelli, J. H. Devan, and C. A. Carmichael. 1992. A study of pest oxidation in polycrystalline $MoSi_2$. *J. Mater. Res.* 7:2747–2755.
8. Chou, T. C. and T. G. Nieh. 1993. Pesting of the high-temperature intermetallic $MoSi_2$. *JOM* 45:15–21.
9. Cook, J., A. Khan, E. Lee, and R. Mahapatra. 1992. Oxidation of $MoSi_2$-based composites. *Mater. Sci. Eng. A* 155:183–198.
10. Mitra, R. and V. V. Rama Rao. 1999. Effect of minor alloying with Al on oxidation behaviour of $MoSi_2$ at 1200°C. *Mater. Sci. Eng. A* 260:146–160.
11. Schlichting, J. 1978. Molybdenum disilicide as a component in modern high-temperature solid solutions. *High Temp. High Press.* 10:241–269.
12. Fitzer, E. and J. Schwab. 1955. The chemical stability of molybdenum disilicides as a raw material. *Metallurgy* 9:1062.
13. Maloney, M. J. and R. J. Hecht. 1992. Development of continuous-fiber-reinforced $MoSi_2$-base composites. *Mater. Sci. Eng. A* 155:19–31.
14. Westbrook, J. H. and D. L. Wood. 1964. "PEST" degradation in beryllides, silicides, aluminides, and related compounds. *J. Nucl. Mater.* 12:208–215.
15. Berkowitz-Mattuck, J. B., P. E. Blackburn, and E. J. Felton. 1965. The intermediate-temperature oxidation behavior of molybdenum disilicide. *Trans. AIME* 233:1093.
16. Fitzer, E. 1955. Molybdenum disilicide as high-temperature material. In *Proceedings of the 2nd Plansee Seminar*, ed. F. Benesovsky, 56–79. Vienna: Springer-Verlag.
17. Bartlett, R. W., J. W. Mccamont, and P. R. Gage. 1965. Structure and chemistry of oxide films thermally grown on molybdenum silicides. *J. Am. Ceram. Soc.* 48:551–558.
18. Mitra, R., V. V. R. Rao, and Y. R. Mahajan. 1997. Oxidation behaviour of reaction hot pressed $MoSi_2$-SiC composites at 500°C. *Mater. Sci. Technol.* 13:415–419.

19. Yanagihara, K., K. Przybylski, and T. Maruyama. 1997. The role of microstructure on pesting during oxidation of MoSi$_2$ and Mo(Si,Al)$_2$ at 773 K. *Oxid. Metall.* 47(3–4):277–293.
20. Liu, Y. Q., G. Shao, and P. Tsakiropoulos. 2001. On the oxidation behaviour of MoSi$_2$. *Intermetallics* 9:125–136.
21. Barin, I. 1989. *Thermochemical Data of Pure Substances*, Vols. 1 and 2. Weinheim, Germany: VCH.
22. Meier, G. H. 1996. Research on oxidation and embrittlement of Intermetallic Compounds in the U.S. *Mater. Corros.* 47:595–618.
23. Guder, S., M. Bartsch, M. Yamaguchi, and U. Messerschmidt. 1999. Dislocation processes during the deformation of MoSi$_2$ single crystals in a soft orientation. *Mater. Sci. Eng. A* 261:139–146.
24. Yanagihara, K., T. Maruyama, and K. Nagata. 1995. High-temperature oxidation of Mo–Si–X intermetallics (X=Al, Ti, Ta, Zr and Y). *Intermetallics* 3:243–251.
25. Yanagihara, K., T. Maruyama, and K. Nagata. 1996. Effect of third elements on the pesting suppression of Mo–Si–X intermetallics (X=Al, Ta, Ti, Zr and Y). *Intermetallics* 4:S133–S139.
26. Stergiou, A. and P. Tsakiropoulos. 1997. The intermediate and high-temperature oxidation behaviour of (Mo,X)Si$_2$ (X=W,Ta) intermetallic alloys. *Intermetallics* 5:117–126.
27. Mueller, A., G. Wang, R. A. Rapp, E. L. Courtright, and T. A. Kircher. 1992. Oxidation behavior of tungsten and germanium-alloyed molybdenum disilicide coatings. *Mater. Sci. Eng. A* 155:199–207.
28. Cockeram, B. V. 1995. Growth and oxidation resistance of boron-modified and germanium-doped silicide diffusion coatings formed by the halide-activated pack cementation method. *Surf. Coat. Technol.* 76:20–27.
29. Glass, D. E., R. N. Shenoy, Z. Wang, and M. C. Halbig. 2001. Effectiveness of diffusion barrier coatings for Mo–Re embedded in C/SiC and C/C composites. NASA/TM-2001-211264.
30. Kingery, W. D., H. D. Bowen, and D. R. Uhlmann. 1975. *Introduction to Ceramics*, 92, 257. New York: Wiley.
31. Yanagihara, K., T. Maruyama, and K. Nagata. 1993. Isothermal and cyclic oxidation of Mo(Si$_{1-x}$, Al$_x$)$_2$ up to 2048 K. *Mater. Trans. JIM* 34:1200–1206.
32. Mitra, R., R. Khanna, and V. V. Rama Rao. 2004. Microstructure, mechanical properties and oxidation behavior of a multiphase (Mo,Cr)(Si,Al)$_2$ intermetallic alloy–SiC composite processed by reaction hot pressing. *Mater. Sci. Eng. A* 382:150–161.
33. Petrovic, J. and R. E. Honnell. 1990. MoSi$_2$ particle reinforced-SiC and Si$_3$N$_4$ matrix composites. *J. Mater. Sci.* 25:4453–4456.
34. Natesan, K. and S. C. Deevi. 2000. Oxidation behavior of molybdenum silicides and their composites. *Intermetallics* 8:1147–1158.
35. Hebsur, M. G. 1999. Development and characterization of SiC$_{(f)}$/MoSi$_2$–Si$_3$N$_4$(p) hybrid composites. *Mater. Sci. Eng. A* 261:24–37.
36. Hebsur, M. G. and M. V. Nathal. 1997. Strong, tough, and pest resistant MoSi$_2$-base hybrid composite for structural applications. In *Second International Symposium on Structural Intermetallics*. Seven Springs Mountain Resort, Champion, PA, September 1997, eds. M. V. Nathal, R. Darolia, C. T. Liu, P. L. Martin, D. B. Miracle, R. Wagner and M. Yamaguchi, p. 949. Warrendale, PA: TMS.
37. Kowalik, R. W. and M. G. Hebsur. 1999. Cyclic oxidation study of MoSi$_2$–Si$_3$N$_4$ base composites. *Mater. Sci. Eng. A* 261:300–303.

38. Schneibel, J. H. and J. A. Sekhar. 2003. Microstructure and properties of $MoSi_2$–MoB and $MoSi_2$–Mo_5Si_3 molybdenum silicides. *Mater. Sci. Eng. A* 340:204–211.

39. Meyer, M. K. and M. Akinc. 1996. Oxidation behavior of boron-modified Mo_5Si_3 at 800 degrees-1300°C. *J. Am. Ceram. Soc.* 79:938–944.

40. Anton, D. L. and D. M. Shah. 1991. High temperature properties of refractory intermetallics. *Mater. Res. Soc. Symp. Proc.* 213:733–739.

41. Meyer, M. K. and M. Akinc. 1996. Isothermal oxidation behavior of Mo-Si-B intermetallics at 1450°C. *J. Am. Ceram. Soc.* 79:2763–2766.

42. Meyer, M. K., A. J. Thom, and M. Akinc. 1999. Oxide scale formation and isothermal oxidation behavior of Mo–Si–B intermetallics at 600–1000°C. *Intermetallics* 7:153–162.

43. Thom, A. J., E. Summers, and M. Akinc. 2002. Oxidation behavior of extruded Mo_5Si_3Bx–$MoSi_2$–MoB intermetallics from 600 degrees–1600°C. *Intermetallics* 10:555–570.

44. Nomura, N., T. Suzuki, K. Yoshimi, and S. Hanada. 2003. Microstructure and oxidation resistance of a plasma sprayed Mo–Si–B multiphase alloy coating. *Intermetallics* 11:735–742.

45. Park, J. S., R. Sakidja, and J. H. Perepezko. 2002. Coating designs for oxidation control of Mo-Si-B alloys. *Scr. Mater.* 46:765–770.

46. Yoshimi, K., S. Nakatani, T. Suda, S. Hanada, and H. Habazaki. 2002. Oxidation behavior of Mo_5SiB_2-based alloy at elevated temperatures. *Intermetallics* 10:407–414.

47. Mendiratta, M. G., T. A. Parthasarathy, and D. M. Dimiduk. 2002. Oxidation behavior of alpha Mo–Mo_3Si–Mo_5SiB_2 (T2) three phase system. *Intermetallics* 10:225–232.

48. Supatarawanich, V., D. R. Johnson, and C. T. Liu. 2004. Oxidation behavior of multiphase Mo–Si–B alloys. *Intermetallics* 12:721–725.

49. Supatarawanich, V., D. R. Johnson, and C. T. Liu. 2003. Effects of microstructure on the oxidation behavior of multiphase Mo–Si–B alloys. *Mater. Sci. Eng. A* 344:328–339.

50. Mandal, P., A. J. Thom, M. J. Kramer, V. Behrani, and M. Akinc. 2004. Oxidation behavior of Mo–Si–B alloys in wet air. *Mater. Sci. Eng. A* 371:335–342.

51. Yoshimi, K., S. Nakatani, S. Hanada, S.-H. Ko, and Y.-H. Park. 2002. Synthesis and high temperature oxidation of Mo–Si–B–O pseudo *in situ* composites. *Sci. Technol. Adv. Mater.* 3:181–192.

52. Paswan, S., R. Mitra, and S. K. Roy. 2006. Isothermal oxidation behaviour of Mo–Si–B and Mo–Si–B–Al alloys in the temperature range of 400–800°C. *Mater. Sci. Eng. A* 424:251–265.

53. Paswan, S., R. Mitra, and S. K. Roy. 2007. Oxidation behaviour of the Mo–Si–B and Mo–Si–B–Al alloys in the temperature range of 700–1300°C. *Intermetallics* 15:1217–1227.

54. Paswan, S., R. Mitra, and S. K. Roy. 2009. Nonisothermal and cyclic oxidation behavior of Mo–Si–B and Mo–Si–B–Al alloys. *Metall. Mater. Trans. A* 40:2644–2658.

55. Roy, B., J. Das, and R. Mitra. 2013. Transient stage oxidation behavior of $Mo_{76}Si_{14}B_{10}$ alloy at 1150°C. *Corros. Sci.* 68:231–237.

56. Pitman, S. H. 1996. Development of NbSi2 base intermetallic alloys. PhD dissertation. Surrey, UK: School of Engineering, University of Surrey.

57. Zhang, F., L. T. Zhang, J. X. Yu, and J. S. Wu. 2005. Growth of single crystals and low temperature oxidation behaviors of $MoSi_6$ and $NbSi_2$. *Mater. Sci. Forum* 475–479:729–732.

58. Pitman, S. H. and P. Tsakiropoulos. 1995. Study of the microstructure and oxidation of $NbSi_2$ base alloys. *Mater. Res. Soc. Symp. Proc.* 364:1321–1326.

59. Murakami, T., S. Sasaki, K. Ichikawa, and A. Kitahara. 2001. Microstructure, mechanical properties and oxidation behavior of Nb–Si–Al and Nb–Si–N powder compacts prepared by spark plasma sintering. *Intermetallics* 9:621–627.

60. Murakami, T., S. Sasaki, and K. Ito. 2003. Oxidation behavior and thermal stability of Cr-doped $Nb(Si,Al)_2$ and $Nb_3Si_5Al_2$ matrix compacts prepared by spark plasma sintering. *Intermetallics* 11:269–278.

61. Liu, Y., A. J. Thom, M. J. Kramer, and M. Akinc. 2003. Processing and oxidation of Nb-Si-B intermetallics. In *Processing and Fabrication of Advanced Materials for High Temperature Applications*, eds. T. S. Srivatsan and V. A. Ravi. Warrendale, PA: TMS.

62. Jackson, M. R., R. G. Rowe, and D. W. Skelly. 1995. High-temperature ordered intermetallic alloys VI. *Mater. Res. Soc. Symp. Proc.* 364:1339–1344.

63. Douglass, D. L. 1963. The thermal expansion of niobium pentoxide and its effect on the spalling of niobium oxidation films. *J. Less-Common Met.* 5(2):151–157.

64. Murakami, T., C. N. Xu, A. Kitahara, M. Kawahara, Y. Takahashi, H. Inui, and M. Yamaguchi. 1999. Microstructure, mechanical properties and oxidation behavior of powder compacts of the Nb–Si–B system prepared by spark plasma sintering. *Intermetallics* 7:1043–1048.

65. Balsone, S. J., B. P. Bewlay, M. R. Jackson, P. R. Subramanian, J.-C. Zhao, A. Chatterjee, and T. M. Heffernan. 2001. Materials beyond superalloys: Exploiting high-temperature composites, July, Report No. 2001CRD098. Niskayuna, NY: GE Research and Development.

66. Zhao, J.-C., B. P. Bewlay, M. R. Jackson, and L. A. Peluso. 2001. Alloying and phase stability in niobium silicide in-situ composites. In *Proceedings of the International Symposium on Structural Intermetallics*, eds. K. J. Hemker, D. M. Dimiduk, H. Clemens, R. Darolia, H. Inui, J. M. Larsen, V. K. Sikka, M. Thomas, and J. D. Whittenberger, 483. Warrendale, PA: TMS.

67. Menon, S. K. and M. G. Mendiratta. 2005. High temperature oxidation in multicomponent Nb alloys: E. *Mater. Sci. Forum* 475–476:717–720.

68. Menon, E. S. K., M. G. Mendiratta, and D. M. Dimiduk. 2001. High temperature oxidation mechanisms in Nb-silicide bearing multicomponent alloys. In *Structural Intermetallics 2001*, eds. K. J. Hemker, D. M. Dimiduk, H. Clemens, R. Darolia, H. Inui, J. M. Larsen, V. K. Sikka, M. Thomas, and J. D. Whittenberger, 591. Warrendale, PA: TMS.

69. Bewlay, B. P., M. R. Jackson, J. C. Zhao, and P. R. Subramanian. 2003. A review of very-high-temperature Nb-silicide-based composites. *Metall. Mater. Trans. A* 34:2043–2052.

70. Chan, K. S. 2004. Cyclic oxidation response of multiphase niobium-based alloys. *Metall. Mater. Trans. A* 35:589–597.

71. Chattopadhyay, K., R. Mitra, and K. K. Ray. 2008. Non-isothermal and isothermal oxidation behaviour of the Nb–Si–Mo alloys. *Metall. Mater. Trans. A* 39:577–592.

72. Balsone, S. J., B. P. Bewlay, M. R. Jackson, P. R. Subramanian, J.-C. Zhao, A. Chatterjee, and T. M. Heffernan. 2001. Materials beyond superalloys-exploiting high-temperature composites. *Proceedings of the 2001 Intermetallics Symposium*

on Structural Intermetallics, eds. K. J. Hemker, D. M. Dimiduk, H. Clemens, R. Darolia, H. Inui, J. M. Larsen, V. K. Sikka, M. Thomas, and J. D. Whittenberger, 99. Warrendale, PA: TMS.

73. Bewlay, B. P., M. R. Jackson, J. C. Zhao, and P. R. Subramanian. 2002. A review of very high-temperature Nb-silicide based composites, Report No. 2002GRC172, September. Niskayuna, NY: GE Research and Development.

74. Bewlay, B. P., M. R. Jackson, J. C. Zhao, P. R. Subramanian, M. G. Mendiratta, and J. J. Lewandowski. 2003. Ultrahigh-temperature Nb-silicide-based composites. *MRS Bull.* 28:646–653.

75. Shibata, T., A. Wakita, T. W. Sigmon, J. F. Gibbons, and T. R. Cass. 1982. Investigation of the oxidation properties of Cw laser formed WSi_2. *Appl. Phys. Lett.* 40:77–80.

76. Mohammadi, F., K. C. Saraswat, and J. D. Meindl. 1979. Kinetics of the thermal-oxidation of WSi_2. *Appl. Phys. Lett.* 35:529–531.

77. Vasudévan, A. K. and J. J. Petrovic. 1992. A comparative overview of molybdenum disilicide composites. *Mater. Sci. Eng. A* 155:1–17.

78. Rosenkranz, R., G. Frommeyer, and W. Smarsly. 1992. Microstructures and properties of high melting point intermetallic Ti_5Si_3 and $TiSi_2$ compounds. *Mater. Sci. Eng. A* 152:288–294.

79. Thom, A. J., Y. Kim, and M. Akinc. 1993. Effect of processing on oxidation of Ti_5Si_3. In *High-Temperature Ordered Intermetallic Alloys V*, MRS Proceedings, Vol. 288, pp. 1037–1042.

80. Mitra, R. and V. V. R. Rao. 1998. Elevated-temperature oxidation behavior of titanium silicide and titanium silicide based alloy and composite. *Metall. Mater. Trans. A* 29:1665–1675.

81. Frampton, R. D., E. A. Irene, and F. M. Dheurle. 1987. A study of the oxidation of selected metal silicides. *J. Appl. Phys.* 62:2972–2980.

82. Ma, J., Y. Gu, L. Shi, L. Chen, Z. Yang, and Y. Qian. 2004. Synthesis and oxidation behavior of chromium silicide (Cr_3Si) nanorods. *J. Alloys Comp.* 375:249–252.

83. Raj, S. V. 1995. An evaluation of the properties of Cr_3Si alloyed with Mo. *Mater. Sci. Eng. A* 201:229–241.

84. Yin, X., L. Cheng, L. Zhang, Y. Xu, and C. You. 2000. Microstructure and oxidation resistance of carbon/silicon carbide composites infiltrated with chromium silicide. *Mater. Sci. Eng. A* 290:89–94.

85. Tomasi, A., R. Ceccato, M. Nazmy, and S. Gialanella. 1997. Microstructure and oxidation behaviour of chromium–molybdenum silicides. *Mater. Sci. Eng. A* 239–240:877–881.

8

Aluminides: Processing and Mechanical Behavior

8.1 Introduction

The aluminides of nickel, titanium, and iron have attracted significant attention from researchers for several decades. These aluminides are attractive due to various combinations of desirable properties, such as high melting temperatures, high specific strength and stiffness, strength retention at elevated temperatures, and reasonably low brittle-to-ductile transition temperatures (BDTTs), as well as the possibility of fabricating near-net-shaped components by casting, powder metallurgy processing, or specific metal-forming operations.[1] Further improvement in mechanical properties has become possible by addition of different alloying elements or reinforcing these aluminides with various types of reinforcements, both brittle and ductile. One of the major challenges for high-temperature application of some of these materials is the requirement for high-temperature oxidation resistance. The presence of Al in aluminide-based intermetallic alloys leads to the formation of passive scale of Al_2O_3, which protects against further oxidation.[2,3] The primary motivation for development of these materials is weight reduction and increase in the operating temperature of various aero-engine components in order to increase the efficiency of jet engines by increasing the thrust-to-weight ratio. The mechanical properties and BDTTs of the aluminides are strongly dependent on the stoichiometry, as well as the strength and directionality of interatomic bonding. The aluminides of interest for various high-temperature structural applications that are discussed in this chapter are Ni_3Al and NiAl among the nickel aluminides; Al_3Ti, γ-TiAl, and Ti_3Al among the titanium aluminides; and FeAl and Fe_3Al among the iron aluminides. The physical and room-temperature mechanical properties are shown in Table 8.1,[4-6] while the positive attributes along with BDTT and possible maximum temperatures of use are summarized in Table 8.2.[4-12] It may be noted that there is a gap between the maximum temperatures for strength retention and environmental degradation, with the latter being higher for all the intermetallics except Ti_3Al. This observation points to the necessity

TABLE 8.1

Crystal Structure, Melting Point, and Physical and Mechanical Properties of
Selected Structural Aluminides of Ni, Ti, and Fe.

Intermetallic	Crystal Structure	Melting Temperature (°C)	Density (g cm^{-3})	Young's Modulus (GPa)	CTE ($\times 10^{-6}$ °C^{-1})	RT Tensile Yield Strength (MPa) and Elongation
Ni$_3$Al	fcc (L1$_2$, cP4)	1397	7.5	179	16	200–900 1%–54%
NiAl	bcc (B2, cP2)	1638	5.9	294	14–15	175–300 0%–2%
Al$_3$Ti	Tetragonal (D0$_{22}$, tI8)	1350	3.4	215	12–15	120–425
	fcc (L1$_2$ stabilized cP4)					0.1%–0.5%
TiAl	fcc (L1$_0$, tP4)	1480	3.8	176	11	400–775 1–4%
Ti$_3$Al	hcp (D0$_{19}$, hP8)	1680	4.1	145	12	700–990 2%
Fe$_3$Al	Tetragonal (D0$_3$, cF16)	1380	6.7	141		350–800
	bcc (B2, cP2)					1%–8%
FeAl	bcc (B2, cP2)	1255	5.5	261		400–800 1%–8%

Source: Deevi, S. C., et al., *Prog. Mater. Sci.*, 42, 177–192, 1997; WardClose, C. M., et al., *Intermetallics*, 4, 217–229, 1996; Cinca, N., et al., *J. Mater. Res. Technol.*, 2, 75–86, 2013.

for use of suitable alloying additions and microstructural modifications, not only to bridge the gap between these temperature limits but also to enhance their operating temperatures.

8.2 Processing

Alloys and composites based on aluminides of Ni, Ti, and Fe can be processed by both ingot and powder metallurgy routes. Of course, the ingot metallurgy route involving melting and casting is relatively inexpensive and is preferred for fabricating near-net-shaped and large components, where machining can be avoided. For example, Ni$_3$Al-based alloys can be prepared by induction melting in air, arc-melting in inert gas, and electron beam melting in vacuum.[13] As a high concentration of Al is present in the melt, the

TABLE 8.2

Positive Attributes, BDTT, and Maximum Temperatures for Use of Selected Structural Aluminides of Ni, Ti, and Fe

			Maximum Temperature for Use	
Intermetallic	Positive Attributes	BDTT (°C)	Strength Limit (°C)	Corrosion Limit (°C)
Ni$_3$Al	Oxidation, carburization, and nitridation resistance; high-temperature strength	RT	1100	1150
NiAl	High melting point; high thermal conductivity; oxidation, carburization, and nitridation resistance	400	1200	1400
Al$_3$Ti	High specific strength and stiffness, outstanding oxidation resistance			
TiAl	Low density; good specific-strength properties and wear resistance	600	1000	800
Ti$_3$Al	Low density; high specific strength	700–800	760	650
Fe$_3$Al	Oxidation and sulfidation resistance	RT	700	1200
FeAl	Resistance to oxidation, sulfidation, molten salt, and carburization	300–400	800	1200

Source: Deevi, S. C., et al., *Prog. Mater. Sci.*, 42, 177–192, 1997; WardClose, C. M., et al., *Intermetallics*, 4, 217–229, 1996; Cinca, N., et al., *J. Mater Res. Technol.*, 2, 75–86, 2013; Liu, C. T., *Scr. Metall. Mater.*, 27, 25–28, 1992; Miracle, D. B., *Acta Metall. Mater.*, 41, 649–684, 1993; Risanti, D., et al., *Intermetallics*, 13, 1337–1342, 2005; Yoshimi, K., *Mater. Trans. JIM*, 35, 51–57, 1994; Lipsitt, H. A., et al., *Metall. Trans. A*, 11, 1369–1375, 1980; Imayev, V. M., et al., *Intermetallics*, 8, 1–6, 2000.

molten alloys form a surface layer of Al$_2$O$_3$, which lowers the amount of oxygen contamination to <50 wppm in the air-melted alloys.

For melting the aforementioned aluminides, the following issues need to be considered:[4] (1) the melting points of aluminum and transition metals such as nickel, iron, and titanium; and (2) the exothermic nature of formation of the intermetallic compound, as shown in Figure 1.2. As a sufficiently large amount of Al is present in the intermetallic alloys, melting of aluminum leads to the onset of aluminide formation. The exothermic nature of the aluminide-forming reactions aids in the process of melting and thereby drives it to completion. The idea of combustion synthesis has been used to modify the melting process. It has been observed that addition of the Al melt stock to the molten Ni bath leads to instantaneous occurrence of a vigorous exothermic reaction with the peak temperature of the crucible reaching

2300°C.[4] This type of melting is carried out in crucibles made of alumina, zirconia, or mullite. For commercial utilization of this exothermic reaction-aided process, the Exo-Melt™ method involving division of the melt stock into several parts has been developed.[4] In this technique, the Al melt stock is arranged vertically inside the crucible as the top layer, followed by layers of Ni and other alloying elements, and then a bottom layer of Ni. The power required to melt a batch of metal using the Exo-Melt™ method is 47% less than in the case of a conventional melting process.

Directional solidification and single-crystal growth are preferred processing routes to fabricate turbine blades and vanes, where grain boundaries are not desirable. Single crystals have been grown by the float zone process, Czochralski crystal growth, and modified Bridgman method. Single crystals of Al_3Ti have been grown using the floating zone method of melting.[14] It is possible to grow a shaped single crystal of NiAl by drawing liquid metal through a shaped die.[15] Near-net-shaped NiAl single-crystal airfoils have been produced by a modified Bridgman technique. The challenge for processing single crystals of NiAl is due to its melting temperature being 300°C higher than that of Ni-based superalloys. The available ceramic core and mold materials for investment casting are silica (1630°C), mullite (1800°C), and alumina (1900°C).[15]

It is not possible to fabricate single-phase γ-TiAl by conventional solidification due to the occurrence of two peritectic reactions near the equiatomic (Ti/Al = 1) composition, as is evident from the phase diagram shown in Figure 1.10. The microstructures obtained by peritectic solidification show a significant amount of Al segregation. The phase obtained first through solidification of equiatomic TiAl is α, whereas the γ forms as a secondary phase at the grain boundaries peritectically (L + α → γ) due to Al segregation.[16,17] Therefore, the nonequilibrium microstructure containing α + γ requires homogenization treatment to obtain single-phase γ. Wrought processing of γ-TiAl-based alloys is carried out with the twin objectives of breaking up the dendritic structure of ingots and obtaining near-net-shaped products. In binary TiAl-based alloys, the hot workability is enhanced with decrease in $α_2$ content, which is achievable with reduction in the Al concentration.[18] The addition of Cr is reported to enhance the hot workability by forming the B2 phase, which promotes grain-boundary sliding at the temperature of forging.[19] Efforts have also been made to process γ-TiAl alloys by sheet casting,[20] superplastic forming,[19,21] and directional solidification.[22–24]

Melting of Fe-aluminides is generally carried out in a controlled atmosphere. Conventional casting of Fe-aluminides has been found to be made difficult by extensive segregation of Al.[25] The Al-enriched locations are usually more brittle, and are further susceptible to environmental embrittlement if exposed to moisture. Therefore, microcracks are formed during cooling from the processing temperature and subsequent deformation processing. It is possible to fabricate Fe_3Al-based alloys with impressive room-temperature ductility by adding Cr, Zr, Nb, and C.[26,27] These alloys can be processed by arc-melting using a nonconsumable electrode, induction melting in either argon atmosphere or vacuum, or electroslag refining. It is possible to prepare laboratory-scale (500 g) to commercial-scale (230 kg) quantities of Fe-aluminides by using the aforementioned methods. Carbon-containing Fe-aluminide alloys have been processed by a two-step process involving argon induction melting followed by electroslag remelting,[28,29] or even by air induction melting under slag cover.[30,31] Use of slag cover leads to lower sulfur content in the alloy. Electroslag-remelted alloys also have lower inclusion content, as well as finer precipitate and inclusion size. Fe-aluminides have been also processed by pressure-assisted reactive infiltration of molten Al into a preform of steel fibers.[32,33] Fiber preforms of volume fractions of 40% and 80% (corresponding closely to FeAl and Fe_3Al stoichiometry) were prepared from steel wool. After homogenization for 5 h at 1000°C, the intermediate reaction products are eliminated. The major limitation of this process is that a nonuniform residual porosity distribution remains, which has been attributed to a nonuniform distribution of steel fibers used.

Aluminides can be fabricated by a wide variety of casting techniques, which include sand casting, investment casting, centrifugal casting, and directional solidification. Subsequent processing of cast aluminides is carried out using different metal-forming methods such as hot extrusion, hot forging, hot flat and bar rolling, hot swaging, cold flat and bar rolling, and cold drawing into tube, rod, and wire, which lead to microstructural refinement and enhancement of mechanical properties. For example, Ni_3Al alloys containing <0.3 at.% Zr are hot-forged in the temperature range of 1050°C–1150°C.[34]

Powder metallurgy processing involves pressureless sintering, uniaxial or multiaxial hot pressing, liquid phase-assisted or reactive sintering with or without application of pressure, or spark plasma sintering. Aluminide powders such as Ni_3Al are produced by atomization in an inert atmosphere. Subsequently, these powders are canned and hot extruded using a reduction ratio of ≥8:1 at 1100°C–1200°C.[34,35] The products obtained by such powder metallurgy processing usually have fine grain size formed through dynamic recrystallization, and are therefore amenable to superplastic forming to obtain near-net shapes.

Reaction sintering has been developed as a novel process to obtain dense near-net-shaped products, starting from elemental metallic powders.[36–38] The formation of aluminides occurs through a self-sustaining exothermic reaction. If the reaction is rapid and occurs under adiabatic conditions, it is referred to as self-propagating high-temperature synthesis or combustion synthesis. This method can be considered as equivalent to the Exo-Melt™ process in the ingot metallurgy route in terms of energy efficiency. Cold-pressed powder compacts are first preheated in vacuum in the temperature range of 300°C–600°C in a controlled atmosphere (vacuum) for degassing, followed by reaction initiation. The heat evolved due to exothermic reaction leads to a localized increase in temperature, which further triggers the process of combustion synthesis. This process has been described in detail in Chapter 2. Aluminide-based intermetallic alloy products with up to 97% of theoretical density have been obtained by optimizing the reaction synthesis parameters, including particle size, heating rate, and atmosphere.[36]

Powders of γ-TiAl obtained by rapid solidification processing contain relatively smaller length scales of segregation, and it is possible to obtain chemically homogeneous products after consolidation such as hot isostatic pressing and extrusion.[39,40] Here, the main challenge is incorporation of oxygen as a contaminant in the powders used as raw materials. It has been observed that powders processed using a plasma rotating electrode are cleaner than inert gas-atomized powders, with oxygen content <1000 wppm.

Fe_3Al-(2–5 at.%)Cr-based alloy powders obtained by inert gas (N_2 or Ar) atomization have been consolidated by hot extrusion with a 9:1 ratio at 1000°C. It has also been possible to obtain thin sheets (with <1 mm thickness) of Fe_3Al by hot forging at 1000°C or by hot rolling first at 850°C and then warm rolling at 650°C.[41,42] For FeAl, hot extrusion of the canned powders has been carried out successfully at 900°C with extrusion ratios between 8:1 and 12:1 to obtain fine-grained and dense products.[43] Vacuum hot pressing (uniaxial) and hot isostatic pressing have also been used to successfully consolidate FeAl powders.

For aluminide-based composites, innovating processing techniques have been attempted to obtain the desired distribution of the reinforcement phase, which could be in the form of particles, whiskers, or short fibers. Particulate-reinforced composites have been processed by powder mixing and consolidation, liquid/particle cospraying,[44,45] and *in situ* precipitation,[46,47] as well as reaction synthesis techniques involving the *in situ* formation of the reinforcement phase. A composite of Fe–28(at.%)Al–5Cr dispersed with 2.5 vol.% Al_2O_3 was prepared by reaction synthesis from elemental Fe, Al, and Cr powders, where the alumina was formed during synthesis by scavenging the oxygen content of the powders.[48] Oxide dispersion-strengthened (ODS) FeAl has been produced by mechanical alloying of elemental Fe and Al powders.[49] A stable dispersion of $FeAl_2O_4$ (15 nm) has been formed *in situ* in an FeAl matrix with fine grain size using a low temperature (1050°C) for sintering to prevent coarsening of microstructure. A large (~15 vol.%) volume fraction of oxide was obtained by using starting materials with high oxygen content. The techniques for long-fiber-reinforced intermetallic matrix composites include melt infiltration of fiber bundles[50,51] and solid-state consolidation of fibers with metal in the form of powder,[52,53] foil,[54] or prefabricated tapes.

8.3 Nickel Aluminides

8.3.1 Ni₃Al

Interest in Ni_3Al is due to its role in precipitation hardening of nickel-based superalloys as well as the potential to be used as a structural material for high-temperature applications. Ni_3Al with ordered face-centered cubic (fcc) ($L1_2$, cP4) structure usually exhibits $[111]\langle110\rangle$ slip, and has sufficient number of slip systems for extensive plastic deformation.[55] One of the advantages of Ni_3Al, both as a precipitate in nickel-based superalloys and as a structural material, is the increase in yield strength with increase in temperature, and this behavior can be well explained on the basis of dislocation-based mechanisms in $L1_2$ alloys.[7,55-61] Other interesting observations regarding the mechanical behavior of Ni_3Al-based alloys is the sharp difference in the ductility of single and polycrystalline material, as well as the strong influence of the moisture content of atmosphere on ductility.[13] These issues will be discussed in Sections 8.3.1.1 and 8.3.1.2.

8.3.1.1 Dislocation-Based Mechanism and Anomalous Increase in Yield Strength

Superdislocations with $b = \langle110\rangle$, as expected in $L1_2$-structured Ni_3Al, are expected to dissociate involving the formation of a superlattice intrinsic stacking fault (SISF), APB, and/or CSF, as discussed in Chapter 3. The typical dislocation dissociation reactions in the case of Ni_3Al are discussed here.[13]

1. The dislocation dissociation reaction with formation of APB and CSF (Figure 3.3) is[62]

$$[\bar{1}01] = \tfrac{1}{6}[\bar{1}\bar{1}2] + \tfrac{1}{6}[\bar{2}11] + \tfrac{1}{6}[\bar{1}\bar{1}2] + \tfrac{1}{6}[\bar{2}11] \tag{8.1}$$

2. The dissociation reaction accompanied by formation of APB, SISF, and CSF on the (111) plane is (Figure 3.3b–d)[63]

$$[\bar{1}01] = \tfrac{1}{6}[\bar{1}\bar{1}2] + \tfrac{1}{6}[\bar{2}11] + \tfrac{1}{6}[\bar{1}2\bar{1}] + \tfrac{1}{6}[\bar{1}2\bar{1}] + \tfrac{1}{6}[\bar{1}\bar{1}2] + \tfrac{1}{6}[\bar{2}11] \tag{8.2}$$

3. The dislocation dissociation with APB on the (111) plane is shown as (Figure 3.3b)

$$[\bar{1}01] = \tfrac{1}{2}[\bar{1}01] + \tfrac{1}{2}[\bar{1}01] \tag{8.3}$$

4. The dislocation dissociation with SISF on the (111) plane is shown as (Figures 3.3c and 8.1)

$$[\bar{1}01] = \tfrac{1}{3}[\bar{2}11] + \tfrac{1}{3}[\bar{1}\bar{1}2] \tag{8.4}$$

5. The dislocation dissociation with APB on the (010) plane is shown as[7]

$$[\bar{1}01] = \tfrac{1}{2}[\bar{1}01] + \tfrac{1}{2}[\bar{1}01] \tag{8.5}$$

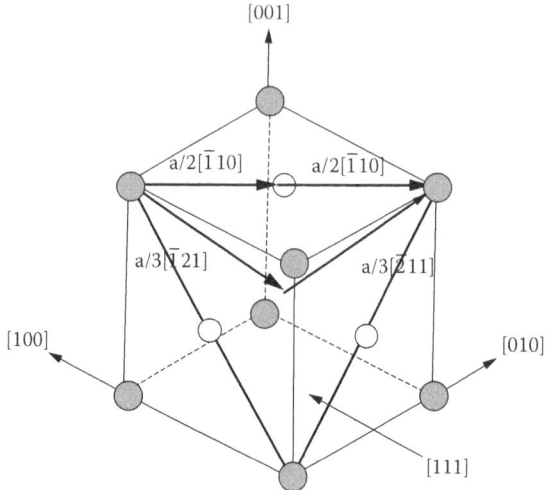

FIGURE 8.1
Schematic illustration of splitting of ⟨110⟩ superdislocation into superpartials separated by SISF in L1$_2$-structured Ni$_3$Al.

The dislocation dissociation reaction (8.1) is favored if the APB energy on the {111} planes is lower than that in other planes. However, if the APB energy on the {111} planes is high, Reaction 8.4 is preferred, and this type of dislocation splitting is schematically shown in Figure 8.1. The APB energies of the {111} and {010} planes have been found to be 111 ± 15 mJ m^{-2} and 90 ± 5 mJ m^{-2}, respectively, by Douin et al.[64] On the other hand, a study by Dimiduk[65] on Ni–22.9Al has shown the APB energies to be 169 ± 19 and 104 ± 8 mJ m^{-2} for the {111} and {010} planes, respectively. Observation of $\gamma_{\{111\}}/\gamma_{\{010\}} > 1$ indicates that dissociation of $[\bar{1}01]$ dislocations would be preferred on the {010} planes. However, Reaction 8.1 is most commonly observed, as {111} planes are the slip planes and the energy for this reaction is low.

Probably the most interesting characteristic in the deformation behavior of Ni$_3$Al-based alloys is anomalous increase in yield strength with temperature up to about 600°C–800°C, depending on the alloy composition and orientation of the single-crystal sample being tested. This type of behavior has also been reported in polycrystalline samples containing different types of alloying elements, as shown in Figure 8.2.[66–68] The anomalous increase in yield strength of Ni$_3$Al with increase in temperature was first reported on the basis of hardness measurements by Westbrook.[55] This observation was confirmed by separate investigations on flow behavior of polycrystalline[56,57] and single-crystalline[58] samples of Ni$_3$Al. The following characteristics have been observed regarding the anomalous increase in yield strength:

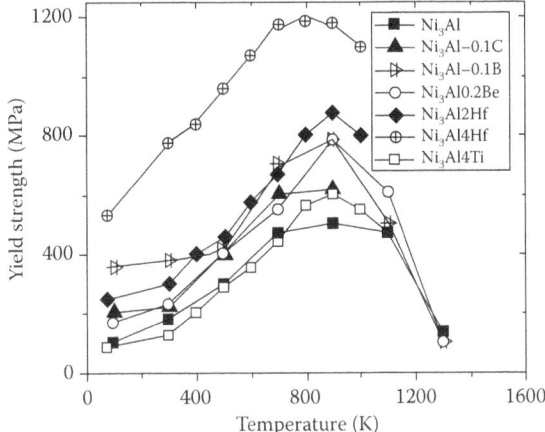

FIGURE 8.2

Plots of yield strength of Ni$_3$Al-based alloys against temperature depicting anomalous behavior using published data. (From Bonneville, J., et al., *Mater. Sci. Eng. A*, 234–236, 770–773, 1997; Masahashi, N., et al., *Acta Metall.*, 36, 7, 1823–1836, 1988; Mishima, Y., et al., *Trans. Jpn Inst. Mater.*, 27, 41–50, 1986.)

1. The flow stress increases with increasing temperature to a peak (termed *anomalous behavior* as this is the opposite of what is observed normally), and then it decreases.
2. In the temperature regime of increasing flow stress, slip occurs on the {111}⟨10$\bar{1}$] systems.[59]
3. At temperatures above the peak, slip takes place on the {100}⟨011⟩ slip systems, except for single-crystal samples oriented near ⟨001⟩, which continue to exhibit {110}⟨$\bar{1}$01] slip.[60]
4. The dislocation structure formed in the samples deformed in the temperature regime of anomalous behavior is found to comprise long straight ⟨$\bar{1}$01] screw dislocations on {111} planes.[69]

The anomalous increase in yield strength has been explained on the basis of the "cross-slip model" first proposed by Kear and Wilsdorf to explain the work-hardening behavior of L1$_2$-structured materials.[61] This model was first used by Thornton et al. to explain the anomalous increase in yield strength of Ni$_3$Al.[70] The model was subsequently refined for explaining orientation dependence of {110}⟨$\bar{1}$01] slip in Ni$_3$Al single crystals.[71] Further modifications were later introduced by Paider et al.[72] and Yoo[73] by including the effects of elastic anisotropy in addition to the difference in APB energies. The model suggests that there is a net driving force for cross-slip of the screw segments of {111}⟨$\bar{1}$01] dislocations into {010} planes, because of the net differences in APB and elastic energies of dislocations. Cross-slip, being a thermally activated process, becomes easier with increasing temperature. Therefore, there is an increase in the number of cross-slipped segments with increasing temperature. These cross-slipped segments of superdislocations find it energetically favorable to dissociate further, forming APBs, according to Reaction 8.5. As {010} is not a close-packed plane, the cross-slipped and dissociated dislocation segments on these planes tend to be sessile in the regime of anomalous behavior, and therefore effectively pin the original {111}⟨$\bar{1}$01] dislocations. Therefore, for single crystals, the critical resolved shear stress (CRSS) for {111} ⟨$\bar{1}$01] slip increases until {001}⟨$\bar{1}$10] slip begins and the flow stress decreases.

8.3.1.2 Grain-Boundary Embrittlement

It is interesting to note that single crystals of Ni_3Al are ductile, whereas pure polycrystalline Ni_3Al is brittle at room temperature because of intergranular fracture. Brittle intergranular fracture in conventional materials is usually accompanied by segregation of impurity elements such as S, P, and O, causing embrittlement at grain boundaries. However, no evidence of such segregation has been observed in sufficiently pure polycrystalline Ni_3Al, which suggests that the grain boundaries are intrinsically brittle. It has been shown that the brittleness of grain boundaries is associated with poor grain-boundary cohesion as well as environmental embrittlement.[74] The absence of grain-boundary cohesion is related to ordering energy, electronegativity difference, valency difference, and atomic size difference between the atomic constituents of the intermetallics.

Tensile experiments carried out by Liu and Oliver on polycrystalline Ni_3Al showed fracture strain of $\approx$2.5%–2.6% in air and $\approx$7.2%–8.2% in dry oxygen environments.[75] The results of this study confirmed the susceptibility of Ni_3Al to environmental embrittlement, which in turn is responsible for poor ductility. It has been shown that Al present in Ni_3Al could react with moisture and generate atomic hydrogen by the following chemical reaction[75–78]:

$$2Al + 3H_2O \rightarrow Al_2O_3 + 6H \tag{8.6}$$

The presence of atomic hydrogen at grain boundaries is expected to cause embrittlement. On the other hand, the following reaction is expected in dry oxygen:

$$4Al + 3O_2 \rightarrow 2Al_2O_3 \tag{8.7}$$

Here, formation of Al_2O_3 is preferred to generation of atomic hydrogen. However, formation of Al_2O_3 at the grain boundaries also promotes intergranular failure. Alloying with 0.26 at.% Zr (Ni–22.65Al–0.26Zr) has shown tensile ductility of 6%–9% in water, 11%–13% in air, and 48%–51% in oxygen, but fracture is reported to be intergranular even in the most ductile tensile sample.[79] It is now well accepted that grain-boundary embrittlement is caused by atomic hydrogen formed by interaction of Ni_3Al with moisture.

TABLE 8.3

Tensile Properties of Ni₃Al and Ni₃Al–B Alloys under Different Environments

Alloy Composition	Strain Rate (s⁻¹) and Environment	Yield Strength (MPa)	Ultimate Tensile Strength (MPa)	Elongation to Failure (%)
Tests at Room Temperature				
Ni–24 Al	3.3×10^{-3}, air	280	333	2.6
Ni–24 Al	3.3×10^{-3}, oxygen	279	439	7.2
Ni–24 Al–500 ppm B	3.3×10^{-3}, air	290	1261	41.2
Ni–24 Al–500 ppm B	3.3×10^{-3}, oxygen	289	1316	39.4
Ni–24.8 Al–500 ppm B	3.3×10^{-3}, air	290	671	18.1
Ni–24.8 Al–500 ppm B	3.3×10^{-3}, oxygen	306	801	25.4
Ni–25.2 Al– 500 ppm B	3.3×10^{-3}, air	221	300	8.4
Ni–25.2 Al– 1000 ppm B	3.3×10^{-3}, air	344	552	10.2
Tests at −196°C (77 K)				
Ni–23.4 Al		254–269	672–762	31.3–31.8

Source: Nobuki, M. and T. Tsujimoto, *ISIJ Intl.*, 31, 931–937, 1991; Tiberghien, D., et al., *Mater. Sci. Eng. A*, 323, 427–435, 2002.

Alloying with B has been found to be effective in significantly improving the tensile ductility of polycrystalline Ni₃Al containing <25 at.% Al tested in air at room temperature, such that the mode of fracture changes from intergranular to transgranular.[79–82] Tensile fracture strain of >50% in air was observed in Ni₃Al microalloyed with 0.1 wt.% B.[79] Selected results from tensile tests carried out on Ni₃Al with and without B in different environments are shown in Table 8.3.[79,81–83] The results in this table show that the ductilizing effect of B is reduced significantly with even a minor increase in the Al content, and a greater amount of B needs to be added to Ni₃Al with higher Al content to obtain similar values for elongation to failure. Studies using Auger electron spectroscopy have shown evidence for segregation of B at the Ni₃Al grain boundaries. As the preferred sites of B and H are similar, segregation of the former species at grain boundaries inhibits diffusion of atomic hydrogen. In this manner, embrittlement of grain boundaries is prevented in the Ni₃Al–B alloy. Two other mechanisms proposed to explain the ductility enhancement through B addition are: (i) increase in the grain-boundary cohesive strength through B addition[84–87]; and (ii) slip transfer across the grain boundary.[88–90] It has also been reported that columnar grain structure in Ni₃Al can effectively reduce its environmental embrittlement at elevated temperatures.[75,76]

Interestingly, high ductility has been observed on tensile testing of Ni₃Al at liquid nitrogen temperatures (−196°C or 77 K), which has been attributed to limited moisture content of air as well as reduced diffusivity of atomic hydrogen at very low temperatures.[83,91,92] While Chiba et al.[91] have reported >45% tensile elongation with transgranular fracture mechanism for strain-annealed Ni–23(at.%)Al (no boron added and grain size = 676 µm), the results obtained by Lee and White[92] have shown elongations of 6.9% and 19.7% in the case of Ni–24(at.%)Al (grain size = 129 µm) and Ni–24Al–700(wppm)B alloys (grain size = 62 µm), respectively. The difference in these results can be attributed to the large volume fraction of coincidence site lattice (CSL) boundaries in the samples subjected to strain annealing. The net fraction of low-angle and Σ3 boundaries in strain-annealed Ni_3Al has been found to be nearly 30%. The results of the study by Chiba et al. have confirmed the significant influence of the grain-boundary structure of polycrystalline Ni_3Al on its ductility.[91] Another study has shown ≈32% elongation to fracture in the case of fine-grained (15 µm grain size) Ni_3Al tested in tension at −196°C.[83]

8.3.1.3 Effect of Alloying on Strength and Ductility

The mechanical properties of Ni_3Al are strongly influenced by deviation from stoichiometry (expressed as Al/Ni ratio) and ternary alloying additions.[13] Alloying of Ni_3Al has been attempted for both solid-solution hardening and increase in ductility. It may be noted that Ni_3Al has significant solid solubility for a number of elements, which undergo dissolution without forming a second phase.[93] The alloying elements can be classified on the basis of their preference for site substitution. For example, Si, Ge, Ti, V, Hf, Zr, Nb, Zn, V, Ta, Ti, Mo, and W are known to substitute in the Al sublattice sites, whereas Cu, Co, and Pt substitute in the Ni sublattice sites. Furthermore, Fe, Mn, and Cr substitute in both Ni and Al sublattice sites. The preference of a given alloying element for a particular type of sublattice site depends on how its atomic size and electronic structure match with that of the parent lattice.[94] The extent of solid solubility of the element X in Ni_3Al has been correlated with the atomic size misfit and difference between the heats of formation of Ni_3Al and Ni_3X.[39]

Solid-solution hardening of Ni_3Al is dependent on factors such as sublattice occupation of solute atoms, atomic size misfit, and deviation from the base alloy stoichiometry. The following observations have been reported regarding the effect of stoichiometry on solid-solution hardening: (i) Al-rich deviations from stoichiometry cause more significant hardening and greater temperature dependence of strength than Ni-rich deviations at low and intermediate temperatures[95]; and (ii) the defect hardening in off-stoichiometric alloys with either Ni-rich or Al-rich compositions is dependent on the symmetry of antisite defects.[96] In other words, higher solid-solution hardening is caused by substitution of alloying elements in the Al sublattice sites. Among the above mentioned substitutional solute elements for the Al sublattice, Hf and Zr lead to maximum strengthening, whereas Ti is the least effective. It should be noted that Hf possesses the highest atomic size misfit with Al, and therefore alloying of Ni_3Al with the former element leads to maximum strengthening at ambient temperature. Interaction of the hydrostatic compressive strain field around the oversize solute atom with edge dislocations contributes to the strengthening. On the other hand, interstitial alloying of Ni_3Al with B also leads to significant strengthening.[97]

The mechanism governing solid-solution hardening is expected to be different in the temperature regimes of normal and anomalous behavior regarding the dependence of yield strength on temperature. Interestingly, Ni_3Al has shown anomalous behavior at 77 K or −196°C. The compressive yield strength of Ni_3Al alloyed with different transition metals has been found to vary with solute concentration (c) following a linear relationship,[68] which contradicts the well-known dependence of solid-solution hardening on $\sqrt{c}$ or $c^{1/3}$.[98] The relationship of yield strength with solute concentration for undersize transition metals has been found to be similar to that of other metals. However, this relationship is found to deviate for oversize transition metals, as the yield strength is increased beyond a limit. This observation indicates that oversize solute atoms not only interact with the strain field of edge dislocations, but also affect the superdislocation core structure and cross-slip processes.[97]

Similarly to binary Ni_3Al, ternary $Ni_3Al–X$ alloys also show anomalous increase in yield strength. In Ni_3Al alloyed with either Hf (most effective solute for hardening) or Ti (least effective solute for hardening), the solutes are more effective in stress increments during anomalous behavior at elevated temperatures than at the ambient temperature.[98] It is possible that anomalous behavior of solid-solution hardening is enhanced by promotion of cross-slip by addition of the ternary solute. It has been proposed that solute addition affects the activation barrier for core transformation of superdislocations, thereby promoting cross-slip and resulting in solute-hardening anomaly.

8.3.1.4 Creep Behavior

Single and polycrystalline Ni_3Al shows typical "inverse creep" behavior in the intermediate temperature range, as reported in some studies.[99–101] Here, the creep curves contain a brief primary stage up to <1% strain characterized by decrease in strain rate with increasing strain, which is followed by an "inverse" tertiary region exhibiting accelerated creep leading toward failure. The steady-state creep stage is not found in these creep curves. In single-crystal samples, creep failure is preceded by necking rather than formation of voids.[101] However, a subsequent study of creep behavior by Rong et al. has shown the occurrence of inverse creep to be dependent on stress and temperature of the creep test.[102] Furthermore, creep tests carried out on single crystals of Ni_3Al (1 at.% Ta) have shown the presence of a steady-state creep stage for all orientations.[103] Interestingly, the steady-state creep rate of differently oriented single-crystal specimens has been found to scale with the resolved shear stress for the cube cross-slip planes. This is as expected, as transmission electron microscopy (TEM) studies have provided evidence for slip on octahedral planes during the primary stage of creep, and on cube cross-slip planes during the secondary creep regime.

It has been suggested by Hemker et al.,[101] on the basis of a creep experiment in the range of 257°C–700°C, that slip is operative on primary and secondary octahedral planes during the primary stage, and hardening occurs by cross-slip of screw segments of dislocations into cube planes, resulting in formation of Kear–Wilsdorf locks.[61] Thereafter, glide on octahedral planes is arrested due to formation of these locks, which subsequently act as sources of generation of dislocations in the cube cross-slip planes (on the [010] plane). These dislocations are able to bow out of the locks, glide along the cube cross-slip planes, and lead to accelerated creep, termed *inverse creep*. These dislocations face a Peierls-type barrier during glide on non-close-packed cube planes, and therefore their movement requires thermal activation. Hence, the inverse creep rate is found to alter sharply with temperature. The creep rates are also strongly dependent on orientation of single-crystal samples, as the dislocation mobility on cube cross-slip planes varies with the resolved shear stress. Therefore, if slip on the (010) plane is considered during inverse creep, the crystal with loading axis along the [001] direction is expected to be the most resistant to creep.

Results contrary to the anomalous increase in yield strength have been reported in some studies,[101,104] where the creep strain rate has been found to decrease with increasing temperature. However, a study by Rong et al. has shown anomalous increase in creep strength with increasing temperature in the range of 380°C–580°C[105]. Study of the postcreep microstructures has shown evidence for ease of cross-slip of screw segments as well as movement of superdislocations ($b = \langle 110 \rangle$) on the cube cross-slip planes at relatively lower temperatures.[102,105] Therefore, operation of cube cross-slip appears to be a necessary condition for inverse creep, but need not be the sufficient condition.

Attempts have been made to enhance the creep strength by use of ODS alloys. The creep strength of Ni_3Al–B–Cr alloy (76Ni18.9Al5Cr0.1B) dispersed with 2 vol.% Y_2O_3 has exhibited dependence on matrix grain size.[106] However, the creep strength of this material has been found to be inferior to that of Ni-based superalloys in use. Another approach is use of multiphase alloys based on Ni_3Al. Directionally solidified (DS) Ni–20.2Al–8.2Cr–2.44Fe (at.%) alloy containing columnar γ' grains (oriented along the load axis) along with a small amount of γ phase in the interlamellar region as well as uniformly distributed α-Cr precipitates in the microstructure has been creep tested at temperatures between 777°C and 877°C under a stress range of 150–350 MPa.[107] The steady-state creep of this alloy has been found to be controlled by dislocation climb, as suggested by an apparent stress exponent of 4.1 and activation energy of 316 ± 9 kJ mol^{-1}. The true activation energy ($\approx 301 \pm 9$ kJ mol^{-1}) for creep has been found to match the value required for diffusion of Ni in Ni_3Al. The tertiary stage is characterized by inhomogeneous deformation, evidenced by formation of slip bands followed by formation of voids within the γ' lamellae and cavities at grain boundaries.

Initial efforts to develop creep-resistant alloys based on Ni_3Al have involved the use of the composition as that of γ' phase (Ni–14.4Al–5.19Ta–2.73Co–2.52Ti–0.89W–1.94Cr) of the high-strength, single-crystal Ni-based superalloy (PWA1480).[108] The creep properties of this alloy evaluated as a function of Ni/Al ratio and concentration of various refractory elements have not been found to be superior to those of conventional Ni-based superalloy. In contrast, the DS alloy with the composition Ni–16.3Al–8.2Mo–0.2B (in at.%) has shown higher yield and creep strengths in the temperature range of 700°C–1050°C compared with those of the commercial Ni-based superalloy PWA 1422 (Ni–11.23Al–11.89Cr–4.93Co–3.95Ta–1.76Ti–1.29W [in at.%]).[109] For example, the rupture life under the applied stress of 88.3 MPa at 1100°C is reported to be 254 h for the aforementioned Ni_3Al-based alloy in comparison with 43 h for the PWA 1422 alloy. In this Ni_3Al-based alloy, solid-solution strengthening by Mo contributes most to its high-temperature strength superiority. Furthermore, the DS IC6 alloy having the composition (in at.%) (73.51–75.26) Ni–(16.34–17.58)Al–(8.14–8.25)Mo–(0.16–0.78)B containing γ, γ', and boride particles in the microstructure has exhibited high strength and ductility up to 1200°C, as well as excellent creep resistance in the range of 700°C–1100°C.[110] The rupture life under the applied stress of 88.3 MPa at 1100°C has been found to be >250 h, and this is comparable to the advanced Ni-based superalloys CMSX-2 (Ni-8Cr–4.6Cr–7.9W–0.6Mo–5.6Al–0.9Ti–5.8Ta–0.1Hf) and NASIR100 (Ni–9Cr–10.5W–1.0Mo–5.8Al–1.2Ti–3.3Ta). This alloy has also exhibited excellent resistance to thermal cycle fatigue in the range of 700°C–1100°C, and has been successfully tested for use in turbine vanes of jet engines.

8.3.2 NiAl

Research on NiAl-based intermetallics has been driven by the motivation to develop suitable materials for gas turbine applications, primarily blades and vanes in both high- and low-pressure turbine sections. The interest in NiAl is rooted in several attractive properties, such as high melting temperature (1638°C), moderate density (5.86 g cm^{-3}), resistance to environmental degradation, high thermal conductivity, attractive stiffness, modest BDTT, and low cost of raw materials.[1,111] It can be processed relatively easily by ingot or powder metallurgy techniques, and then near-net shaped by hot working. The major shortcomings of this material are poor fracture toughness and limited damage tolerance at ambient temperature, as well as inadequate strength and creep resistance at elevated temperatures. Significant efforts have been made by researchers to overcome these shortcomings by alloying or addition of suitable reinforcements. Whereas it is possible to obtain a 30% reduction in the weight of the turbine rotor stage (turbine blade and disk) due to density reduction, high thermal conductivity provides excellent cooling efficiency, as well as a reduction in airfoil temperatures and thermal gradients. The physical and mechanical properties of NiAl are discussed in the following subsections.

8.3.2.1 Physical Properties

The density of binary, stoichiometric NiAl is nearly two-thirds that of the Ni-based superalloy. On the other hand, the density of nonstoichiometric NiAl ranges from 5.35 g cm^{-3} at the Al-rich boundary of the NiAl phase field to 6.50 g cm^{-3} at the Ni-rich boundary. The weight reduction provides advantages of lower self-induced stresses in rotating turbine airfoils, which may also justify reducing the size of turbine disks. Although the high congruent melting point of 1638°C may be considered to be encouraging, the increase in bulk diffusivity above $0.5T_m$ ($\approx$680°C) leads to poor creep and stress rupture properties. Higher diffusivity of NiAl may be attributed to its ordered body-centered cubic (bcc) structure, which is not close packed in nature. The Young's modulus of polycrystalline stoichiometric NiAl is 188 GPa at room temperature, which decreases less with increasing temperature than is observed for Ni-based superalloys.[111,112] Moreover, NiAl is elastically anisotropic, with $E_{\langle 111 \rangle}/E_{\langle 100 \rangle} = 2.9$ at ambient temperature, and this value is higher than that of the Ni-based superalloy. The thermal conductivity of NiAl has been found to be in the range of 70–80 Wm^{-1} K^{-1} within the temperature range of 20°C–1100°C, which is three to eight times greater than that of Ni-based superalloys.[113] High thermal conductivity lowers the temperature gradient in a component, and thereby contributes to thermal-shock resistance by lowering thermal stresses. Furthermore, the coefficient of thermal expansion of NiAl is comparable to that of Ni-based superalloys (15.1×10^{-6}°C^{-1} from 820°C to 1560°C).[111]

8.3.2.2 Mechanical Behavior

The nature of bonding and defect structure has a strong influence on mechanical behavior and anisotropy of NiAl. Nearly 2% of the lattice sites are found to be vacant in NiAl, as vacancy formation is energetically favorable. Besides thermal vacancies (equilibrium concentration of point defects at a given temperature), a large concentration of constitutional point defects is possible in NiAl. Deviations from stoichiometry are accommodated by the formation of antisite defects, where Ni atoms occupy Al sites in Ni-rich compositions, whereas there are Ni vacancies in the Al-rich NiAl. It has been suggested that these defects form clusters with short-range order, which aid in heterogeneous nucleation of phases adjacent to NiAl.

Based on the requirement for the dislocation self-energy ($\approx$0.5 Gb^2) to be the lowest, dislocations with the shortest possible Burgers vectors are usually preferred. This logic justifies the operation of $\langle 100 \rangle \{010\}$ and $\langle 100 \rangle \{011\}$ slip systems in NiAl, as is obvious from the schematic illustration of the slip systems depicted in Figure 8.3. The screw dislocation with $b = [100]$ is stabilized under stress, and undergoes cross-slip. The CRSS for slip of [100] dislocations on {011} planes is $\approx$70 MPa in stoichiometric NiAl at ambient temperature.[114,115] However, the resolved shear stress on these dislocations for crystals having

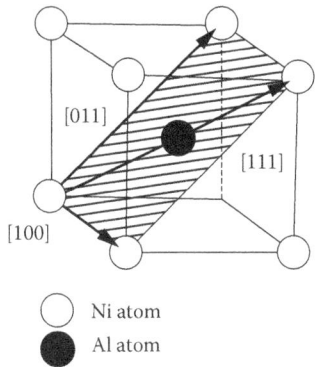

FIGURE 8.3
Schematic illustration of slip vectors in B2-structured NiAl.

loading axis along ⟨100⟩ is zero, and therefore these cube orientations are usu-ally referred to as "hard" orientations. During compression of single crystals with cube orientations (i.e., hard orientations), ⟨111⟩ dislocations are opera-tive on {$\bar{1}$10} or {11$\bar{2}$} planes at temperatures <300°C.[116–119] Interestingly, this phenomenon has not been observed for tensile deformation of samples. The CRSS for ⟨111⟩ dislocations is 600 MPa and 575 MPa at ambient temperature and 300°C, respectively.[119] However, the ⟨111⟩ dislocations are operative above 300°C, as the CRSS decreases sharply beyond this temperature and is reported as ≈100 MPa above 700°C. Such a drop in CRSS has been attributed to the involvement of diffusional mechanisms in deformation.[8]

Slip of ⟨100⟩ dislocation requires lower stress than that of ⟨111⟩ superdislo-cation, which also has a larger Burgers vector compared with that of the for-mer type of dislocation. It make sense to infer that reduction of APB energy for splitting ⟨111⟩ superdislocation into pairs of ½⟨111⟩ dislocations would promote the activation of these dislocations on the {$\bar{1}$10} planes. The APB energy of unalloyed NiAl obtained by various computational methods such as embedded atom approach has been found to be in the range of ≈240–460 mJ m^{-2},[120,121] whereas the value obtained through first-principle calcu-lation is ≈900 mJ m^{-2}.[122,123] The tensile ductility of polycrystalline NiAl at ambient temperature has been found to be in the range of 0%–2%,[111] depend-ing upon stoichiometry,[124] texture,[124,125] grain size,[126] and impurity content and substructure. The activation of ⟨100⟩ dislocations provides only three independent slip systems, which can explain the limited ductility of NiAl. Besides the absence of five independent slip systems, poor grain-boundary cohesivity and slip localization are considered as the causes of poor ductility of this material. Polycrystalline NiAl is found to undergo brittle-to-ductile transition at 400°C, and elongation of 60% is observed at 600°C.[124]

The tensile ductility of NiAl single crystals has been found to be aniso-tropic. In tension, single crystals with hard orientation (⟨100⟩) have shown only elastic strain, whereas those with soft orientations (⟨110⟩ and ⟨111⟩)

have exhibited failure strains up to 2%. The compressive deformation of the single crystals has been found to occur by kinking, involving shear instability through glide of $\langle 100 \rangle$ dislocations. The BDTTs of the NiAl crystals with "soft" and "hard" orientations are 200°C and 400°C, respectively.[127] At 400°C, NiAl single crystals have exhibited elongation exceeding 20%.

It is obvious that activation of $\langle 111 \rangle$ dislocations at ambient temperature by reduction of APB energy of the corresponding slip planes would improve the ductility by increasing the number of slip systems. Using *ab initio* calculations, it has been shown that the APB energy is reduced significantly by ternary additions of Cr and Mn as well as deviations from stoichiometry.[118] Of course, the role of Cr addition in favoring the activation of $\langle 111 \rangle$ dislocations has been confirmed.[128,129] Moreover, microalloying with Fe, Ga, or Mo has been found to improve ductility of single crystals loaded along the $\langle 110 \rangle$ direction.[130] Addition of 0.25 at.% Ga or Fe has been found to increase the plastic strain to failure of the NiAl single crystals to 4.5% or 6%, respectively, from a low value of 1% observed in the case of unalloyed NiAl.[127] Although Fe addition has not been found to improve the ductility of $\langle 100 \rangle$-oriented single crystals, it has been found to lower the BDTT.

Besides poor room-temperature ductility, sharp decrease in strength with increase in temperature is a major shortcoming of NiAl, and this behavior is observed even in single crystals with hard orientations.[113,131] The strength of NiAl with hard orientation has been found to be higher than that of the cast Ni-based superalloy René 80 at the ambient temperature, but lower at temperatures >400°C.[131] Interestingly, it has been observed that deviations from stoichiometry, as in Ni–40(at.%)Al, can lead to increase in compressive strength up to 700°C.[111] However, the difference between the strengths of stoichiometric and nonstoichiometric NiAl is reported to be significantly reduced at 1200°C.[132] Solid-solution strengthening by ternary additions of Ti, Hf, Zr, V, and Ta has been applied to improve the elevated temperature yield strength.[111] Addition of these elements (X) to beyond the solid-solubility limits leads to the formation of hard ternary intermetallic precipitates having compositions of Heusler phase (β'), N_2AlX (L2_1, cF16), or Laves phase, NiAlX (hP12). Formation of β' precipitates with coherent or semicoherent interfaces leads to high strength as in Ni-based superalloys, but ductility is worsened.[133–135] Limited low-temperature solubility (<1%) also exists for elements such as Cr, Mo, and Re. When added beyond the solubility limit, disordered bcc (β) phase is precipitated, which leads to significant strengthening. Moreover, the eutectic reactions in Ni–Al–R (R being a refractory metal like Mo) can be used to tailor directionally solidified microstructures, which contribute to a significant rise in high-temperature strength.[111] Further efforts for strengthening have been made by adding incoherent dispersoids such as Al_2O_3 and TiB_2, but the strength of these composites is found to be less than that of the alloys with β' precipitates.[136,137]

Studies on creep behavior of NiAl-based materials in the range of 800°C–1000°C have shown evidence of dislocation climb being the rate-controlling mechanism, which can be inferred from the value stress exponent (≈ 5) and apparent activation energy (250–300 kJ mol^{-1}).[138] This value of apparent activation energy for creep is in close agreement with that for diffusion of Ni in stoichiometric NiAl.[139] The creep of NiAl single crystals in the range of 850°C–1200°C has shown strong orientation dependence, as confirmed by the steady-state creep rates of crystals with hard orientations ($\langle 100 \rangle$) being half an order of magnitude lower than those of soft-oriented (such as $\langle 110 \rangle$, $\langle 111 \rangle$) crystals. Crystals with soft orientations have shown creep with activation energies much lower than that for lattice diffusion of Ni, which is controlled by glide of $\langle 100 \rangle$ dislocations. However, the creep rate of crystals with hard orientations has been reported to be controlled by dislocation climb. As expected, the creep resistance of single crystals has been found to be superior to that of polycrystalline NiAl. Furthermore, alloying NiAl with 1 at.% Hf leads to a significant increase in creep resistance, similar to that observed in the case of a single-crystalline Ni-based superalloy.

8.4 Titanium Aluminides

8.4.1 Al$_3$Ti

8.4.1.1 Physical Properties

Al$_3$Ti is a line compound with little off-stoichiometry. Interest in Al$_3$Ti has originated from the drive to develop lightweight structural materials for high-temperature applications. These intermetallic alloys are particularly attractive for their low density (3.4 g cm^{-3}) and high oxidation resistance. The directional bonds and high melting points of Al$_3$Ti contribute to a significantly large Young's modulus. Al$_3$Ti is reported to have a Young's modulus 3.5 times that of Al and 2.5 times that of Ti, whereas its density is only 23% higher than that of pure Al. Therefore, Al$_3$Ti is considered attractive for its high specific strength and stiffness.[140]

8.4.1.2 Deformation Mechanisms

The major shortcoming of this alloy is its inherent brittleness, which can be attributed to its ordered character, complex crystal structure (DO$_{22}$, tI8), and directional bonding. The DO$_{22}$ structure is derived from L1$_2$ (cP4) structure (Figure 1.9) by introducing APB with a displacement vector, $\frac{1}{2}\langle 110 \rangle$ on every (001) plane. The periodic introduction of APB into L1$_2$ structure transforms the triangular arrangement of the B atoms on {111} planes in L1$_2$ structure into a rectangular arrangement. The relatively short lattice translation vectors in the DO$_{22}$ structure occur along the directions of $\langle 110]$ and $\langle 112]$ on {111}, and along $\langle 110]$ and $\langle 100]$ on (001).[140] Studies using TEM have shown the major deformation mode in Al$_3$Ti to be ordered twinning of {111}$\langle 112 \rangle$ type. At elevated temperatures, four $\langle 112 \rangle$-type ordered twinning systems are augmented by slip along the $\langle 110 \rangle$ and $\langle 100 \rangle$ directions. The superdislocations with $b=[110]$ on (001) are dissociated into a pair of superlattice partial dislocations with $b=\frac{1}{2}[110]$ separated by an antiphase domain. Similar splitting of superdislocation is also observed on the {111} planes, but separation between the superlattice partial dislocations is smaller than that in (001) due to much higher APB energy. On the basis of separation between superlattice partial dislocations with $b=\frac{1}{2}[110]$ on the (001) and {111} planes, the APB energies have been found to be 25–32 mJ m^{-2} and 190–200 mJ m^{-2}, respectively.[141,142] In a similar manner, the energy of the APB between the superlattice partial dislocations with $b=\frac{1}{2}\langle 110 \rangle$ has been found to be lowest in the case of the (001) plane. Therefore, the operating slip system at elevated temperatures in both L1$_2$ and DO$_{22}$ (L1$_2$ derivative) is (001)$\langle 110 \rangle$. Screw dislocations, such as $\frac{1}{3}\langle 112 \rangle$-type superpartial dislocation formed by splitting of $\frac{1}{2}\langle 112 \rangle$ dislocations by a dissociation reaction, $\frac{1}{2}\langle 112 \rangle \longrightarrow 1/6\langle 112 \rangle + SISF + \frac{1}{3}\langle 112 \rangle$ on the non-close-packed plane (001), are sessile, and therefore thermal activation is required for their movement.[141] The abundance of twins observed in the postdeformation microstructures is in tune with the difficulty of dislocation movement and ease of twinning. Bands of dislocations with $b=\frac{1}{2}\langle 110 \rangle$ with APB have been found on the (001) plane at the points of twin intersections. Formation of these dislocations appears to relax the stress concentration.[141,143,144]

Single crystals of Al$_3$Ti grown using the floating zone method of melting can be plastically deformed at temperatures exceeding 400°C.[140] As expected, the stress–strain behavior of the Al$_3$Ti single crystals depends strongly on crystal orientation. For certain crystal orientations, compressive strain is caused by extensive twinning, and the corresponding stress–strain curve shows serrations. On the other hand, limited or no twinning has been observed in crystals with certain other orientations. The samples undergoing limited twinning exhibit partially serrated and partially smooth stress–strain curves. The onset of twinning is also reported to be dependent on the local stress environment within the Al$_3$Ti single crystal. The presence of heterogeneities appears to be responsible for twinning for certain orientations. In other words, it is possible to define a critical twinning stress that needs to be reached for the onset of twinning in Al$_3$Ti single crystals. Twinning is preferred to slip, as the stress required for onset of twinning is less than the CRSS for slip on the (001) $\langle 110 \rangle$ system. As the temperature is increased, the CRSS for slip decreases very sharply, whereas the critical stress for twinning is more or less independent of temperature. The strong dependence of CRSS for (001) $\langle 110 \rangle$ slip on temperature can be attributed to low mobility of dislocations due to a high Peierls barrier caused by their sessile core structure.[145,146]

8.4.1.3 Strategies for Improving Ductility

The major strategy for improvement of ductility of Al$_3$Ti with D0$_{22}$ structure involves microalloying in order to enhance the ordered twinning on {111} $\langle 112 \rangle$ or slip on (001)$\langle 110 \rangle$ systems.[144,147,148] It has been postulated that reduction of the SISF energy on {111} or APB energy on (001) would promote twinning or slip, respectively. It has been suggested that addition of alloying elements stabilizing either DO_{23} or $L1_2$ phase would decrease the APB energy for $\frac{1}{2}\langle 110]$ on the (001) plane. Significant research has been devoted to ternary alloying of Al_3Ti for stabilization of $L1_2$ structure. These alloying elements include Fe,[149,150] Mn,[151] Nb,[152] Pd,[153] and W.[152] Quaternary and quinary alloys based on $L1_2$ Al_3Ti have been also prepared, as Cr-, Mn-, and Fe- alloyed trialuminides have exhibited continuous solubility in each other.[154] The maximum solid solubility of substitutional alloying elements in the $L1_2$ phase is reported to be in the range of 4–12 at.%. For forming a substitutional solid solution, the difference between the shortest distance between the atoms in the ternary alloying element X and that in the $L1_2$ phase has been found to have the best correlation with solubility of the former element.[152]

As Al_3Ti forms by peritectic reaction, polycrystalline specimens prepared by the ingot metallurgy route contain a dispersed secondary phase comprising Al. This dispersion of Al makes the ingot plastically deformable. The Al_3Ti/Al composite is reported to be amenable to near-net shaping by hot-working operations such as extrusion at 600°C.[143] This composite with fine-grained microstructure has also exhibited excellent compressive ductility even at the ambient temperature.[146] Plastic deformation of the ductile, soft Al surrounding the Al_3Ti grains is able to relax the internal stresses, which is responsible for enhanced formability of the Al_3Ti/Al composite.[155]

Conversion of $D0_{22}$ to $L1_2$ structure of Al_3Ti by ternary alloying addition of Cr or Mn does not affect the density, but leads to lower hardness as well as higher compressive and tensile ductility.[140] In the $L1_2$-structured Al_3Ti alloyed with Cr or Mn, noticeable tensile plastic strain is observed. While the melting point remains high ($\approx$1350°C) in spite of alloying, the creep properties have been found to be comparable to those of TiAl.[156,157] Studies on compressive behavior of Al_3Ti matrix composites reinforced with TiB_2 particles have shown their strength-to-density ratio to be four times greater than that of Ni-based superalloys up to 600°C. However, this advantage is not maintained above 700°C. Nonetheless, the creep strength of Al_3Ti–Fe composite at 730°C and 830°C has been found to be comparable to that of Ni-based superalloys. Al_3Ti-based alloys are also of interest for their oxidation resistance, which is ensured by the formation of passive scale of Al_2O_3.

8.4.2 TiAl

8.4.2.1 Physical Properties

γ-TiAl remains ordered until melting starts at $\approx$1440°C. The presence of a strong Ti–Al bond raises the Young's modulus as well as the activation energy for diffusion, which in turn contributes to enhancement of high-temperature stiffness, strength, and creep resistance. The Young's modulus of γ-TiAl is $\approx$170 GPa at ambient temperature, and it decreases by a relatively small amount ($\approx$17%) on increase in the operating temperature to 900°C.[158] The resulting high stiffness over a wide temperature range makes static load-bearing applications possible, where elastic deflection needs to be restrained. Moreover, high thermal conductivity promotes faster cooling of hot-end components made of γ-TiAl.

8.4.2.2 Microstructure

On the basis of composition and type of heat treatment, it is possible to tailor wrought γ-TiAl-based alloys with microstructures containing single-phase γ (52–54 at.% Al), a combination of γ and lamellar morphology (45–51(at.%) Al) termed *duplex structure*, or fully lamellar morphology (40–45(at.%)Al).[16] The average grain or lamellar size is a function of heat treatment time and cooling schedule. Microstructural refinement can be obtained by either rapid solidification or thermomechanical processing. The lamellar structure contains alternating plates of α_2 and γ, which form by eutectoid transformation of primary α during cooling to room temperature. The two-phase lamellar structure has the following orientation relationship: $(0001)_{\alpha2}//(111)_\gamma$ and $[2110]_{\alpha2}//\langle110\rangle_\gamma$. It has been reported that all the α_2 plates in a grain have the same orientation, whereas the γ phase may possess different variants.[159] Studies of microstructures have also shown the presence of twins and antiphase boundaries in the γ plates.[160,161]

Various kinds of duplex microstructures containing mixtures of α_2 and γ phases can be obtained through suitable heat-treatment procedures.[162,163] If an alloy with a given composition is heat treated at a temperature only slightly above the γ-solvus, a small fraction of α forms at grain boundaries or triple points, and thereby retards γ grain growth. The α phase formed in this manner has been found to undergo ordering to form α_2 instead of transforming to $\alpha_2+\gamma$-containing lamellar structure. This structure is referred to as "supersolvus." On the other hand, if heat treatment is carried out at temperatures below the α transus, a structure consisting of large colonies of $\alpha_2+\gamma$ dispersed with tiny islands of undissolved γ is obtained. This structure has been termed *subtransus*. Furthermore, a dual-phase structure containing equiaxed grains of single-phase γ and α_2 can be obtained by aging of alloys containing $\alpha_2+\gamma$ phases below the eutectoid temperature for a relatively long duration. It is also possible to form equiaxed γ structure containing Widmanstatten precipitation of α_2 by heat treating in the γ-nose region (49(at.%)Al at 1125°C as shown in Figure 1.10).[164]

8.4.2.3 Deformation Mechanisms

γ-TiAl (L1$_0$, tP4) has a face-centered tetragonal structure with alternating (002) planes of Ti and Al. In this structure, the c/a ratio is $\approx$1.02, indicating the presence of a small amount of tetragonality, which creates nonequivalence among the $\langle110\rangle$ slip vectors, as depicted in Figure 8.4. As expected, slip occurs on {111} planes in close-packed directions.[165] The dislocation energy is

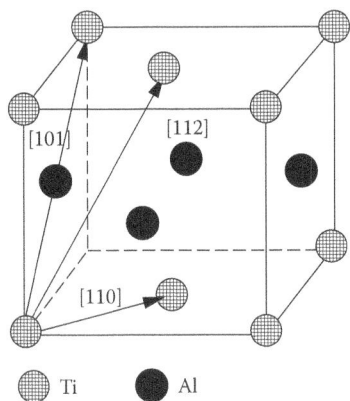

FIGURE 8.4
Schematic illustration of the slip vectors in $L1_0$-structured TiAl.

found to be the minimum for $b = \frac{1}{2}[110]$, which comprises the easy slip mode. The [101] dislocations are found as equivalent to [011], but have higher self-energy than that of $b = \frac{1}{2}[110]$. The Peierls stress for slip systems on the non-close-packed planes such as {110} is obviously higher, and therefore these are activated only at higher temperatures. The APB energy of γ-TiAl is known to be two to four times higher than that of the stacking fault energy,[166,167] and therefore dissociation of dislocations with $b = a/2[\bar{1}10]$ into Shockley partials with $b = a/6[\bar{1}12]$ is preferred. Furthermore, cross-slip of partial dislocations leads to the formation of a sessile configuration, which further reduces the mobility of superdislocations ($b = [\bar{1}01]$).[161]

8.4.2.4 Microstructure–Mechanical Property Relationship

The nature of microstructural constituents and their morphology strongly affect the mechanical behavior of γ-TiAl-based alloys. Alloys containing single-phase γ-TiAl are found to be brittle due to limited dislocation mobility. Studies of postdeformation microstructures of single-phase alloys have shown the presence of superdislocations in undissociated condition, whereas the density of ½[110] easy-slip dislocations is found to be low in the fractured samples.[168] As the Peierls stress is high due to anisotropic chemical bonding, the dislocations are not mobile.

Some of the early studies[162,169–171] have shown that less than 1% plastic strain to fracture has been observed in both fine-grained polycrystalline and single-crystalline, single-phase γ-TiAl. However, a study by Imayev et al. has shown that the tensile ductility of single-phase equiaxed γ-TiAl can be significantly improved by tailoring uniform microstructure with a choice of suitable grain size through appropriate thermomechanical processing.[172] The results are shown in Table 8.4. It may be noted that the highest tensile elongation in the range of 4%–6.9% could be observed for TiAl samples with

TABLE 8.4

Grain Size Corresponding to Maximum
Values of Tensile Elongation Observed for
Single-Phase γ-TiAl (Ti–50.7 at.% Al)

Strain Rate (s^{-1})	Maximum Tensile Elongation (%)	Grain Size (μm)
3.3×10^{-4}	2.7	12
3.3×10^{-2}	6.9	8
8.3×10^{-1}	6	8
8.3×10^{0}	4.5	7
8.3×10^{1}	4.0	8

Source: Nathal, M. V., et al., *Structural Intermetallics 1997*, TMS, Warrendale, PA, 1997, 505–514.

grain size of 8 μm for tests carried out at strain rates between 3.3×10^{-2} and 8.3×10^{-1} s^{-1}. For a compression test carried out at a strain rate of 8.3×10^{-4} s^{-1}, a maximum strain of 35% was recorded for samples with a grain size of 7 μm. The observation of optimum grain size for ductility enhancement has been attributed to stress relaxation from pileups at grain boundaries and deformation twinning. As the number of operative slip systems in γ-TiAl is less than five, twinning improves the strain compatibility between neighboring grains. As the grain size decreases, the nature of deformation changes from planar slip to more homogeneous, but twinning is less preferred. The possibility of twinning is also reduced with decreasing strain rate, as expected, which explains why ductility is also lowered.

The tensile properties of a few selected TiAl-based alloys are depicted in Table 8.5. The Al-lean alloys having a small amount of Ti$_3$Al (α$_2$) have been found to be more ductile than the single-phase γ-based alloys.[157] Generally, the duplex microstructure (mixture of equiaxed γ and lamellar) aids in improving ductility,[175] but lowers both toughness and creep resistance. On the other hand, toughness and creep resistance have been reported to be higher for alloys with lamellar or single-phase microstructures. The plastic fracture strain has been found to be ≤1% for alloys with fully lamellar or single-phase microstructures, whereas it is nearly 3% for those having duplex microstructures and 48 at.% Al content.[155] For this composition (48 at.% Al), heat treatment leads to a microstructure with nearly equal volume fractions of α and γ phases, and therefore significant research has been devoted to further addition of alloying elements to this base alloy. As the duplex microstructure contains two phases after heat treatment, grain growth is inhibited during annealing. Moreover, deformation of the γ phase is made possible by glide of ½[110] dislocations and twinning of 1/6[112] Shockley partials. It is interesting that these dislocations are active only in the alloy with duplex microstructure, but not in the single-phase alloy. It has been proposed that

TABLE 8.5

Tensile Properties of Selected TiAl-Based Alloys

Composition (Microstructural Constituents)	Test Temperature (°C)	Yield Strength (MPa)	Ultimate Tensile Strength (MPa)	Elongation (%)	References
Ti–45Al–8Nb–	RT	840–860	880–920	2.5	Imayev et al.[173]
0.2C ($\alpha_2 + \gamma$)	750	580	600	9	
Ti–43.7Al–	RT	580	620	1	Imayev et al.[173]
3.2(Nb,Cr,Mo)–	750	500	620	3	
0.2B					
Ti–43.5Al–4Nb–	RT	950	975	1.2	Schwaighofer
1Mo–0.1B	800	575	750	10	et al.[174]
Ti–48Al–2Cr–2Nb					Lipsitt[158]
Duplex	RT	480		3.1	
	760	406		50	
Lamellar	RT	455		0.4	
	760	403		2.8	

α_2 absorbs the interstitial oxygen from the γ phase, which in turn modifies the dislocation core structure and alters the stacking fault energy.[176,177] *In situ* straining studies on samples with lamellar structure (α_2/γ and γ/γ) in TEM have shown evidence of mobile interfacial dislocations at the interlamellar interfaces even at ambient temperature.[178]

The fracture toughness of γ-TiAl-based alloys has been found to be strongly influenced by the Al content as well as the nature of the microstructure obtained through different heat treatments.[163,175] Selected results are shown in Table 8.6. If the Ti–48 at.% Al alloy is wrought and heat treated below the α-transus (Figure 1.10) to obtain a duplex microstructure, the fracture toughness is found to be only ≈15 MPa m$^{1/2}$. On the other hand, heat treatment of this alloy above the α-transus leads to the formation of completely lamellar microstructure, and thereby the fracture toughness increases to about 25 MPa m$^{1/2}$. In cast γ-TiAl-based alloys, the fracture toughness values below 45 at.% Al are found to be relatively high (≈20–22 MPa m$^{1/2}$), as a result of equiaxed grain lamellar structure due to formation of β by solidification. Furthermore, the fracture toughness is found to be higher in alloys containing 46%–48% Al with large columnar lamellar grains with wavy boundaries formed during solidification of α. In alloys with >48 at.% Al, fracture toughness decreases with increasing Al content, because the γ phase starts forming at the grain boundaries.[159,180]

It is interesting to note that, in spite of lower ductility, the lamellar structure leads to higher fracture toughness. Here, toughening is caused by interaction of the crack front with the lamellae.[181] Crack arrest or resistance to crack growth is caused by crack-tip blunting due to plastic deformation of the α_2 plates or by crack deflection through interfacial debonding of the lamellae. Therefore, the

TABLE 8.6

Fracture Toughness of TiAl-Based Alloys

	Ti–48Al–2Cr			
Microstructural Constituents	Grain Size (μm)	Lamellar Volume Fraction (%)	K_{IC} (MPa)	References
Near-γ	19		12	Behr et al. (1995)[179]
Duplex	21	20	15	
Fully lamellar	300	95	22	
Ti–48Al–2Cr–2Nb				Huang and
Duplex			14.3 (RT)	Chestnut (1994)[159]
			19.2 (760°C)	
Fully lamellar			28.3 (RT)	
Ti–45Al–2Nb–1.5V–1Mo–0.3Y				Chen et al. (2011)[180]
Fine lamellar+γ+β (along grain boundaries)	100	>90%	23.5	
Ti–45Al–5Nb–0.3Y				
Fully lamellar	110	>95	18	

mechanism of toughening is more effective when the crack front is perpendicular to the edges of α_2 or γ plates.[180] The local plastic strain at the crack tip has been found to be as high as ≈20%, which is substantially higher than the nominal tensile ductility of this material.[182] It has also been shown that deformation of the γ phase by twinning also contributes to toughening by crack-tip blunting.[183] *In situ* tensile straining studies using TEM have shown that the TiAl alloy with noticeable room-temperature ductility shows evidence of crack-tip bridging by γ ligaments between a pair of α_2 lamellae, delamination along the α_2/γ interface, and fracture of β grains.[184] In contrast, specimens with no ductility have shown cracks following the path along the boundary between the colonies of lamellae with a particular orientation, thereby leading to intergranular fracture.

The creep resistance of γ-TiAl-based alloys with lamellar microstructure has been found to be higher than for those with duplex or single-phase (γ) microstructures at higher temperatures, as shown in Figure 8.5.[175,185] Moreover, a study by Zhu et al. on XD™ (exothermic formation of dispersoid, trademark of Martin Marietta Corporation, Baltimore, MD) processed Ti–45Al–2Nb–2Mn (wt.%) alloy dispersed with 0.8 vol.% TiB_2 particles[185] has shown the creep strain rates to be lower and rupture times to be greater for fully lamellar morphology compared with those observed for nearly lamellar microstructure. Reduction of interlamellar spacing leads to increase in the creep resistance of this alloy. Reduction of the creep rates in both primary and secondary stages leads to increase in the time to rupture. The α_2 plates are known to be weaker than γ at elevated temperatures, and therefore do not contribute to creep resistance. It

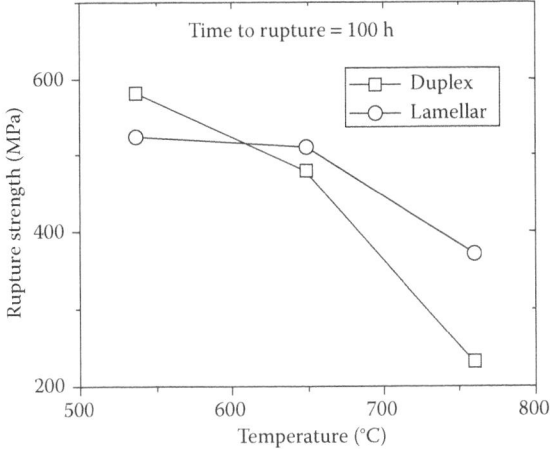

FIGURE 8.5
Plots depicting the variation of 100 h rupture strength with temperature for γ-TiAl-based alloys with duplex and fully lamellar structures. (Data from Huang, S. C. and J. C. Chestnut, *Intermetallic Compounds, Vol. 2, Practice,* Wiley, Chichester, UK, 1994.)

has been shown that coarse grain size (with lamellar morphology having specific orientation in each grain) in alloys with lamellar microstructure makes the greatest contribution to creep resistance.[186] Typically, the grain sizes of duplex, single-phase γ, and fully lamellar structures are ≈25, ≈50, and ≈1000 μm, respectively.[159] Studies on the wrought-processed Ti–46Al–3Nb–2Cr–0.2W based alloy with coarse-grained fully lamellar microstructure have shown the highest creep resistance. Creep resistance of this alloy is further enhanced by increasing the concentration of Nb or W and decrease in Al content,[177] as well as addition of small amounts of C and Si to form fine incoherent particles of silicides and carbides at interlamellar interfaces.[187,188] The creep strength is significantly enhanced by interaction of dislocations with the carbide and silicide precipitates. Presence of metastable α_2 phase in fully lamellar microstructure is known to worsen the creep resistance by promoting degradation of the lamellar microstructure and dynamic recrystallization of the γ phase.[188] Increase in volume during loss of α_2 leads to dynamic recrystallization of γ grains. Significant localized dislocation activity is found in the recrystallized γ phase, which promotes tertiary creep. The presence of small concentrations (ppm levels) of impurities such as C and N as an interstitial solid solution is found to have a beneficial effect on creep strengthening.[189]

Tensile creep studies on γ-TiAl-based alloys with lamellar microstructure have shown absence of a well-defined secondary creep stage because of early formation of cavities at interlamellar or lamellar colony boundaries due to back stress developed during deformation.[190] The tertiary creep regime of γ-TiAl-based alloys has been found to be quite extended. High stresses are developed at colony boundaries because of the difference between the lamellar orientations in neighboring colonies. A study involving *in situ* straining in TEM has shown that sliding and migration of interlamellar (α_2/γ and γ/γ) interfaces occur through cooperative motion of interfacial dislocations, each of which has a step or ledge at the core.[191] As expected, the dislocation velocity and migration rate are strongly affected by the resolved shear stress at the interfaces.

The stress exponent has been found to be ≈ 7.6, which has been attributed to the occurrence of dynamic recrystallization. The postcreep microstructures have shown presence of ½[110] dislocations with no evidence of twinning. On the other hand, creep studies on single-phase γ-TiAl containing a mix of lamellae and equiaxed γ-grains have shown the stress exponent to be ≈ 4.5 and the apparent activation energy to be 300 kJ (g.atom)$^{-1}$, which is close to that for diffusion of Ti in γ-TiAl (≈ 330 kJ mol^{-1}).[192] The creep activation energy has been found to be in the range of 300–380 kJ mol^{-1}, which is suggestive of dislocation climb-controlled creep.[39,193–194] A study on wrought single-phase γ-TiAl and Ti–53Al–1Nb alloys has shown diffusion creep ($n = 1$, $Q_{app} = 192$ kJ mol^{-1}) in the temperature range of 796°C–832°C and dislocation creep ($n = 6$, $Q_{app} = 560$ kJ mol^{-1}) between 832°C and 900°C.[195] The lower apparent activation energy ($Q_{app} = 192$ kJ mol^{-1}) is attributed to the Coble creep mechanism involving grain-boundary diffusion, whereas the cause for the higher value has been ascribed to formation of dislocations. In a similar manner, a study on the creep behavior of γ-TiAl without second phase, as well as γ-TiAl-containing precipitates of α_2-Ti$_3$Al or Ti$_2$AlC in the range of 900°C–1200°C, has shown diffusion creep ($n = 1$) and dislocation-creep ($n = 3$–5) mechanisms at lower and higher stress regimes, respectively.[196] The creep behavior of γ-TiAl-containing precipitates has also exhibited existence of a threshold stress, which decreases with increasing temperature. The existence of threshold stress can be attributed to dislocation particle interactions.

A comparative study of the creep behaviors of TiAl-based alloys with lamellar (Ti–48Al) and duplex (Ti–48Al–2Mn–2Nb) microstructures at 700°C under stresses have shown the stress exponent to be ≈ 19 for both materials at stresses between 310 and 430 MPa.[197] This observation is reported in spite of lower creep resistance exhibited by the alloy with duplex microstructure. However, the stress exponent of the duplex alloy has been found to be ≈ 6 at stresses <300 MPa. Studies of postcreep microstructures have shown evidence of ½[110] dislocations as well as twins in specimens tested under both stress regimes, with the density of these defects being greater at higher stresses. A higher-than-normal value of the stress exponent (≈ 19) has been explained on the basis of back stress at the interlamellar interfaces, which needs to be overcome to generate dislocations for propagation within the γ phase. This back stress is also the cause of existence of threshold stress in both alloys. As expected, the threshold stress has been found to scale with creep resistance, and therefore it is reported to be greater in the alloy with lamellar microstructure. Furthermore, the cast γ-TiAl-based alloy (Ti–48Al–2W–0.5Si (at.%)) subjected to creep in the stress range of 80–325 MPa at temperatures between 700°C and 850°C has shown stress exponents in the range of 4.1–6.6, exhibiting the predominance of dislocation creep.[198]

In TiAl ($\leq$46 at.% Al)-based alloys containing a mixture of γ ($L1_0$, tP4) and β (bcc, cP2) phases with average grain size of 18 μm, superplastic deformation with elongation up to 450% has been observed at 1200°C.[199] Superplasticity has been observed in this alloy with submicrometer grain size at much lower temperatures, such as 850°C.[200] Presence of the B2 phase in the TiAl-based alloys is reported to aid in their superplastic behavior. The following mechanisms have been suggested in the literature regarding the contribution of B2 phase to superplasticity: (i) it acts as a lubricant, and thereby facilitates grain-boundary sliding and grain rotation during superplastic deformation[199]; (ii) it enhances the cohesion between the grain boundaries and therefore delays the onset of cavitation[201]; and (iii) it aids in accommodation of strain at triple junctions caused by grain-boundary sliding.[202] It is possible to stabilize the B2 phase (ordered form of β) at high temperatures by suitable addition of β-stabilizing alloying elements (Cr, Nb, W, and Mo).[203,204] Superplasticity has been observed at 790°C in the Ti–45Al–3Fe–2Mo alloy showing a mixture of three phases (B2+α_2+γ) in its microstructure.[205] In this alloy, addition of Fe and Mo has contributed to grain refinement (average grain size $\approx$ 14 μm). The observation of superplasticity has been ascribed to dynamic recrystallization of β (or B2) grains during high-temperature deformation.

Superplastic behavior has been reported by Lin and Sun in the case of a coarse-grained ($\approx$95 μm average grain size) Ti–47Al–2Mn–2Nb–B alloy containing equiaxed γ phase with a small amount of uniformly distributed α_2 phase in the microstructure, when deformed at temperatures between 1025°C and 1100°C using strain rates in the range of 4×10^{-5}–$1.28 \times 10^{-3}\,s^{-1}$.[206] The strain-rate sensitivity and uniform elongations obtained by analyzing the test results have been reported to be $\geq$0.3% and $\geq$200%, respectively. Studies of postdeformation microstructures and microtexture have shown evidence for formation of much finer grains with sizes in the range of 3–5 μm, as well as subgrain or low angle boundaries, which are transformed to high angle boundaries with misorientations in the range of 15–30°. These observations suggest that dynamic recovery followed by recrystallization promotes superplastic deformation of the originally coarse-grained γ-TiAl.

Alloys with lamellar microstructure have shown anisotropic deformation behavior.[23] It has been shown that both yield strength and ductility are functions of lamellar orientation. The yield stress is found to be high if the lamellae are oriented either parallel or perpendicular to the loading axis. In these orientations, plastic deformation of the γ phase by twinning or slip needs to be transferred to the neighboring α_2 phase across the interphase boundaries. For other orientations, the yield stress is found to be low, as deformation is promoted within the γ phase parallel to the interphase boundaries. The ductility has been found to be highest for orientation of the loading axis at 45° to the lamellae.

8.4.2.5 Effect of Alloying Additions

Studies have shown that alloying γ-TiAl-based alloys with V, Mn, and Cr increases ductility of the alloys with duplex microstructure.[189,207–209] Alloying addition is limited to 1–3 at.% in order to restrict the formation of the B2 phase. The Al concentration is maintained in these ternary alloys in the range of 45–50 at.% in order to ensure duplex microstructure. It should be noted that additions of these alloying elements to single-phase γ alloys or those with lamellar microstructures do not improve the ductility. It has been shown that addition of Mn lowers the stacking fault energy of γ provided α_2 phase is also present,[209] whereas addition of V and Cr leads to substitution of the Al sites, causing reduction in covalency of the Ti–Al bond.[210]

In recent years, significant research has been devoted to the development of disordered β-solidifying γ-TiAl-based alloys containing Nb and Mo (TNM).[211–213] At the ambient temperature, their microstructures contain a mixture of γ-TiAl and α_2-Ti$_3$Al, along with precipitates of β_0-TiAl (ordered phase). The TiAl-based alloys with stabilized β phase are promising due to their impressive high-temperature formability along with creep resistance. Substitutional solid-solution alloying[214,215] with Nb, Mo, Ta, and W, as well as addition of C and N occupying the interstitial sites,[216–218] is known to significantly enhance the creep resistance of the TNM alloys by solid-solution-hardening mechanisms. Alloying with Nb and Mo has been found to reduce the diffusivity in γ-TiAl, which in turn restricts the thermally activated dislocation climb. Furthermore, the formation of Ti$_3$AlC precipitates with cubic perovskite structure contributes to creep strengthening by a precipitation-hardening mechanism.[214,215,217,218] Solubility of C in the octahedral sites (Ti$_2$Al$_4$ and Ti$_4$Al$_2$ type) of γ-TiAl is known to be very low, and it is significantly increased on alloying with Nb due to the formation of antisite defects.[219–221] The Nb atoms occupy the Ti sites, thereby forcing the Ti atoms to occupy the Al sites, and therefore Ti$_6$-type octahedral sites surrounded by Ti(Nb) atoms are formed, which enhance the solubility of C atoms.[220–222]

8.4.3 Ti₃Al

Research on intermetallic alloys based on Ti$_3$Al has been driven by the requirement for materials for use at temperatures beyond the capability of near-α titanium alloys (550°C–600°C). Ti$_3$Al has attracted interest for research and development for aerospace applications because its specific stiffness and stress-rupture properties are comparable to those of nickel-based superalloys.[223] The major shortcoming of this intermetallic is lack of ductility due to its ordered structure and lack of five independent slip systems.

8.4.3.1 Alloying and Phase Composition

In 1978, Blackburn and Smith reported considerable room-temperature ductility in Ti–24Al–11Nb alloy.[224] It was shown that addition of >10 at.% Nb is required to stabilize the high-temperature bcc phase of Ti. The Ti–Al–Nb alloys studied for the last three decades can be classified on the basis of the amount of β-stabilizers (Nb, Mo, Ta, and V)[225] as follows: 10–12 at.% (e.g., Ti–24Al–11Nb,[226] Ti–25Al–8Nb–2Mo–2Ta[222]), 14–17 at.% (e.g., Ti–24Al–10Nb–3V–1Mo,[227] Ti–24Al–17Nb[228]), or 25–30 at.% (e.g., Ti–22Al–27Nb[229]). The alloys with 10–12% Nb contain two-phase microstructures comprising α_2 (Ti$_3$Al, D0$_{19}$, hP8) and β or B2 (disordered or ordered bcc phase), whereas those with 14%–17% Nb contains three phases: α_2, β or B2, and O (a derivative of α_2 with composition based on Ti$_2$AlNb). The alloys with 25–30 at.% β-stabilizer elements are reported to exhibit two-phase microstructure comprising O and β (or B2) phases.[225]

It is possible to obtain various types of microstructures in the Ti–Al–Nb system by quenching from the β phase, continuous cooling transformation, quenching from the β phase followed by aging, and thermomechanical processing.[225] Quenching of the alloys with Nb content <7.5 at.% from the β-phase field causes the formation of α′-martensite with hexagonal structure, which subsequently undergoes ordering to form α_2 with lath morphology, as shown in Figure 8.6a. Through suitable choice of heat treatment parameters during continuous cooling, it is also possible to obtain microstructures exhibiting lamellar or mosaic morphologies. The rate of cooling from the β-phase field affects both size and distribution of the α_2 laths. The microstructures arising from isothermal aging of the quenched-in β contain the plate-like O phase, equiaxed O phase with B2 precipitates, of α_2 and B2 precipitates in the O phase, O + B2 with cellular morphology. A micrograph showing a typical microstructure containing a mixture of equiaxed O-phase grains along with aged B2 is shown in Figure 8.6b.

It has been observed that resistance to both stress rupture and creep is enhanced with increasing concentration of Al and Mo, but beyond optimum levels, these alloying elements adversely affect both toughness and ductility.[225] Interestingly, strength, toughness, and ductility increase with increase in Nb content. In order to understand the effect of alloying on mechanical behavior, the dislocation structures in different types of constituent phases have been thoroughly investigated by various researchers using TEM techniques.

8.4.3.2 Slip Systems

The α_2 phase deforms through operation of the following slip systems: (0001) $\langle 11\bar{2}0 \rangle$, {$10\bar{1}0$}$\langle 11\bar{2}0 \rangle$, and {$11\bar{2}1$}$\langle 11\bar{2}6 \rangle$.[168,223] The primary contribution to plastic deformation of Ti$_3$Al comes from glide of the $\langle 11\bar{2}0 \rangle$ dislocations, which are found to be split into superpartials, 1/6$\langle 11\bar{2}0 \rangle$ on both basal (0001) and prismatic ($10\bar{1}0$) planes.[230,231] The possible slip vectors and superpartials are schematically shown in Figure 8.7a. The CRSS for slip on {$11\bar{2}1$}$\langle 11\bar{2}6 \rangle$ shows anomalous behavior with increase in temperature, quite unlike the trend observed for the other slip systems. As the CRSS for {$11\bar{2}1$}$\langle 11\bar{2}6 \rangle$ slip is higher than that for other systems, it is not found to be operative in polycrystalline

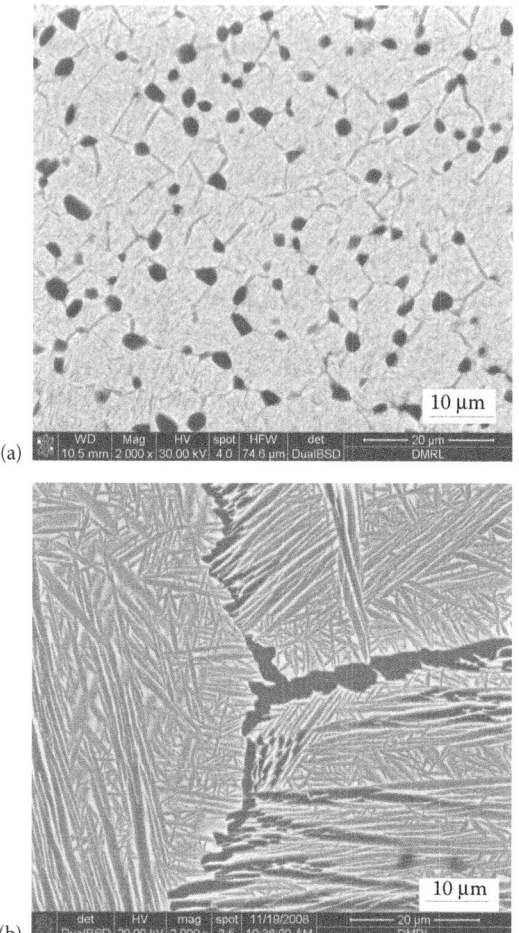

FIGURE 8.6
SEM micrographs of Ti–22–25Al–(22–7)Nb alloy depicting typical microstructures contain-
ing O+B2 phase mixture with the following morphologies: (a) equiaxed O along with aged
B2; (b) transformed B2 showing α_2 laths. (Courtesy of Dr. T. K. Nandy, Defence Metallurgical
Research Laboratory, Hyderabad, India.)

samples. The nature of deformation in basal planes has been found to be
inhomogeneous, with the slip being planar in character, whereas it is more
uniform in the prismatic plane. Whereas superdislocations with screw orientation have been observed in the basal planes, dipoles and dislocation
networks have been observed in the prismatic planes. The brittleness of
polycrystalline α_2 can be attributed to the presence of only four independent
slip systems. Climb of $\langle 11\bar{2}0 \rangle$ dislocations is reported to occur at elevated
temperatures such as 650°C.[168,232,233] Furthermore, dislocations with $b = [0001]$
have been observed to originate during β to α_2 phase transformation.[234,235]

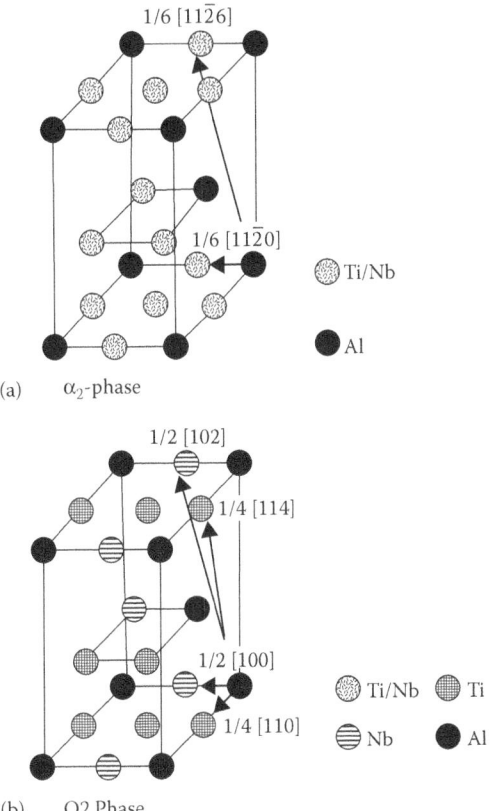

(a) α_2-phase

(b) O2 Phase

FIGURE 8.7
Schematic illustration of slip vectors in (a) the α_2 phase (Ti$_3$Al) and (b) the O2 phase (Ti$_2$AlNb).

Whereas the operation of prismatic slip with movement of $\langle 11\bar{2}0 \rangle$ disloca-tions is more common, addition of Nb promotes basal slip as well as homo-geneous prismatic slip in the α_2 phase.

TEM studies of deformed B2 phase have shown evidence of inhomoge-neous deformation with formation of nonplanar slip bands containing pinned edge dislocations with $\langle 111 \rangle$ Burgers vector.[236] The APB energy for splitting of $\langle 111 \rangle$ dislocations into $\frac{1}{2}\langle 111 \rangle$ superpartials is low enough to per-mit their observation in TEM images. The screw dislocations are less mobile due to their nonplanar character, whereas edge dislocations are pinned by their interaction with surrounding defects. Grain-size refinement helps in increasing the homogeneity of slip.[237] In addition, the [100] dislocations have been observed near α_2/B2 interfaces in a two-phase microstructure, which have originated due to incompatibility stresses.

Studies of the dislocation structure in the O phase have shown evidence for operation of dislocations with $b = [100]$, [102], $\frac{1}{2}[114]$, and $\frac{1}{2}[110]$. The operative slip vectors in the O2 phase are shown schematically in Figure 8.7b. This figure indicates that [100] and ½[110] dislocations are further split into ½[100] and ¼[110] superpartials coupled with APBs, respectively, on basal (001) planes. On the other hand, ¼[114] and [102] dislocations are split into ¼[114] and ½[102] dislocations on prismatic $\{1\bar{1}0\}$ and {010} planes.[225,238] Comparison of slip vectors in α_2 and O2 phases, as shown in Figure 8.5a and b, respectively, indicates that the $\frac{1}{3}\langle11\bar{2}0\rangle$ dislocations in α_2 are equivalent to [100] and ½[110] dislocations in the O2 phase. In a similar manner, the $\frac{1}{3}\langle11\bar{2}6\rangle$ dislocations in α_2 are equivalent to [102] and ½[114] dislocations in the O2 phase. In the low-temperature regime for deformation of the Ti–27Al–25Nb alloy, ½[110], [100], and [102] dislocations have been found to be operative. Studies of the postcreep microstructures of this alloy have shown [001] dislocations in edge orientations, suggesting their involvement in climb, as well as evidence for glide of ½[110] and [100] dislocations on the (110) and (001) planes, respectively.

8.4.3.3 Mechanical Behavior

8.4.3.3.1 Tensile Behavior

The ductility of single-phase α_2-Ti$_3$Al has been found to depend on its grain size.[172] Tension tests on cast and thermomechanically processed stoichiometric α_2-Ti$_3$Al (Ti–25 at.% Al) have shown almost no ductility in coarse-grained samples with grain sizes in the range of 5–10 μm. However, the samples with submicrometer grain sizes, 0.3 and 0.8 μm, have exhibited elongations of 4.8% and 2.5%, respectively. Moreover, the intergranular fraction of the fracture surface increases with decrease in grain size, being 5% and 90% for the samples with grain sizes of 5.5 and 0.3 μm, respectively. The increased ductility observed in the samples with submicrometer grain size has been attributed to increase in homogeneity of slip with decrease in grain size.

As the α_2 phase is brittle, it is necessary to add alloying elements to stabilize the β (or B2) phase in the microstructure. For this purpose, Nb is the preferred alloying element because it not only stabilizes the β phase but also leads to formation of the O phase. Fracture is initiated in these alloys with the onset of cleavage crack nucleation in α_2.[225] It has been reported that the presence of the B2 phase increases the critical strain required for crack nucleation in the α_2 phase, as the stresses due to strain incompatibility at interphase interfaces are accommodated by slip in the former phase. Moreover, the B2 phase has higher strength, and the rule of mixtures can be used to predict the strength of the multiphase alloys.[225] As fracture requires propagation of crack into a more ductile phase, an increase in tensile fracture strain is observed. However, an alloy having single-phase B2 with large grain size shows brittle behavior, because of strain localization caused by inhomogeneous slip. For alloys having microstructure with α_2+O laths formed by decomposition of the β (B2) phase through solution treatment above (α_2+B2)/β (B2) transus temperature followed by cooling at different rates, the yield strength increases with decreasing lath size.[224,237,239] However, maximum ductility is achieved for the alloys having fine basket-weave morphology of $\alpha_2 + O$ laths, without the formation of grain boundary α_2.

8.4.3.3.2 Creep Behavior

Steady-state creep behavior has been investigated in the temperature range of 600°C–800°C for a variety of Ti$_3$Al-based alloys.[240] The activation energy for creep of Ti$_3$Al has been reported as $\approx$206 kJ mol^{-1}. In another study,[241] the stress exponent has been found to be in the range of 4–5 at low and high stresses, whereas it is found to be higher ($\approx$9) at intermediate stresses. On the other hand, the apparent activation energies have been found to be 260, 400, and 370 kJ mol^{-1} at low-, intermediate-, and high-stress regimes, respectively. The rate-controlling mechanism is reported to be dislocation recovery by climb at low stress, and dynamic recrystallization at high stress, which justifies the difference between the apparent activation energies for creep in these two regimes. In the region of intermediate stress, dislocation multiplication is found to be operative. These studies have shown that the creep resistance of the single-phase alloys is superior to those of dual-phase Ti–Al–Nb alloys.[225] Furthermore, the single-phase O has exhibited higher creep strength compared with that of α_2. Although the presence of β (B2) phase contributes to tensile yield strength and ductility, the creep resistance is lowered. It has also been reported that the creep resistance decreases with increasing volume fraction of equiaxed α_2.[232,239] In contrast, the creep resistance is found to be higher for the alloys exhibiting lath morphology. As the α_2 lath size becomes finer with increasing rate of cooling from the transformation temperature, the steady-state creep rate decreases until a critical size is reached. If the α_2 lath size is further reduced, the steady-state creep rates are found to increase. Of course, such a transition in creep behavior related to lath size depends on the applied stress and homologous temperature, which in turn decide whether the operating mechanism is diffusion or dislocation creep.[149] If diffusion creep (stress exponent = 1) is operative, the creep rates are enhanced with increase in the α_2/β interfacial area caused by the reduction in the α_2 lath size. In the dislocation-creep regime (stress exponent = 4–5), dislocation pileups occur at α_2/β interfaces, and therefore the finer size of the α_2 lath lowers the creep rates. Here, the activation energy is found to be same as that for climb-controlled creep (self-diffusion) in Ti. Stress exponents between 1 and 4, as observed by different researchers, indicate transitions between different operating mechanisms.[242–244] It has been also observed that the lath structure offers better resistance against both diffusional and dislocation creep than the equiaxed structure.

Studies on steady-state creep behavior of O-phase-containing Ti–Al–Nb alloys have shown the stress exponents to be in the range of 5–7, which is suggestive of climb-controlled dislocation creep.[227] The climb of [001] dislocations in edge orientation has been found to be operative during creep at high temperatures (650°C–750°C).[245,246] The apparent activation energies for creep of O-phase-containing Ti–Al–Nb-based ternary alloys have been found to be in the range of 294–348 kJ mol^{-1},[245] which are found to agree well with that for interdiffusion in Ti$_3$Al ($\approx$312 kJ mol^{-1}).[247]

8.4.3.3.3 Fatigue and Fracture Behavior

The contribution to room-temperature fracture toughness of Ti–Al–Nb alloys with microstructures having α_2+β phase mixture comes from the role of the β phase in delaying the onset of microcrack nucleation in α_2 and subsequent propagation.[248–250] The presence of β phase delays the onset of microcrack nucleation or blunts the already nucleated cracks in the neighboring α_2 phase. As discussed above, the β phase is able to relax the incompatibility stresses at the α_2/β interfaces. It is desirable to have a continuous network of β phase around α_2 to inhibit crack nucleation. Furthermore, increase in the volume fraction of the β phase in the microstructure enhances the possibility of crack-tip blunting. As the roles of β phase in improving tensile ductility and fracture toughness are identical, a strong correlation is found between these properties. In alloys with microstructures comprising α_2+β phase mixture, α_2 size, and distribution and volume fraction of the β phase, are the parameters affecting the damage-tolerant behavior. The best combination of strength and toughness is found in high-Nb-containing alloys (Ti–22Al–27Nb), with microstructures having O phase as the major constituent, as well as fine α_2 lath size, and a significant volume fraction of β phase. As expected, the fracture toughness is found to increase with temperature due to a rise in resistance to crack propagation.[249] Fatigue fracture is initiated in these alloys by nucleation of microcracks within the α_2 phase, followed by their linkage.[251,252] However, as expected, the growth of fatigue cracks is retarded by the presence of β phase. It should be noted that fatigue-crack growth rates are much faster in Ti$_3$Al-based alloys than those in Ti alloys.[225]

8.5 Iron Aluminides

8.5.1 Basic Information

FeAl is known to have compositions with Al concentrations in the range of 35–50 at.% and ordered bcc (B2, cF2) structure, whereas Fe$_3$Al has ordered cubic D0$_3$ (cF16) crystal structure with $\geq$25 at.% Al. The most commonly used compositions are Fe–28 at.% Al and Fe–40 at.% Al alloys for Fe$_3$Al and FeAl, respectively. These Fe-aluminide-based ordered intermetallic alloys have attracted interest for several decades due to their outstanding resistance to oxidation and sulfidation as well as lower cost.[3,253] Moreover, their densities

(5.4–6.7 g cm^{-3}) are 30% lower than those of similar high-temperature structural materials such as stainless steels and superalloys. In other words, the strength-to-density ratios of Fe$_3$Al- and FeAl-based intermetallics are higher than those of the above materials. Furthermore, it has been shown that both FeAl and Fe$_3$Al are amenable to superplastic deformation,[254–258] which can be used for near-net shaping of the components. It has been shown that, through suitable control of the composition, the ductility of the Fe$_3$Al-based alloys can be enhanced to 15%–20%.[259,260] However, Fe$_3$Al exhibits a sharp decrease in strength above 600°C. The predominant operating slip system in both Fe$_3$Al and FeAl is {110}⟨111⟩, which provides for five independent slip systems,[25] but still these materials exhibit brittle behavior due to environmental embrittlement.[76,261–262]

8.5.2 Environmental Embrittlement

A significant drawback of FeAl and Fe$_3$Al is their susceptibility to degradation in a moist environment[253] by mechanisms similar to that for Ni$_3$Al, as discussed earlier, in Section 8.3.1.2. It was first demonstrated by Liu et al. that Fe$_3$Al and FeAl containing <40 at.% Al are intrinsically ductile, and the cause of poor ductility at the ambient temperature is embrittlement due to moisture.[76,261] Interestingly, even single crystals of FeAl have been reported to be susceptible to environmental embrittlement due to moisture.[1,263] Significantly higher tensile elongation to failure[264,265] and fracture toughness[266] are observed on carrying out mechanical tests in vacuum or dry oxygen atmosphere. The fracture toughness is found to be 78 MPa m$^{1/2}$ in oxygen in Fe–28Al, whereas it is known to vary from 33 MPa m$^{1/2}$ in air to 105 MPa m$^{1/2}$ in oxygen for Fe–35Al. The aluminum-rich Fe–35Al is more susceptible to environmental embrittlement than Fe–28Al.

Atomic hydrogen obtained from moisture attacks Al atoms at crack tips, promoting brittle fracture along the cleavage planes. The bonding mechanism involves Fe-to-H charge transfer.[267] The methods adopted to reduce environmental embrittlement by moisture include control of grain size and shape, and use of alloying elements such as Cr for Fe$_3$Al and B for FeAl.[1] It has been shown that a microstructure having elongated grains with limited transverse boundaries resists hydrogen diffusion in the case of Fe$_3$Al, and therefore ductility is significantly improved.[4] In contrast, an equiaxed grain structure obtained by recrystallization through high-temperature annealing treatments has been found to be susceptible to hydrogen diffusion, resulting in embrittlement. Carrying out stress-relief heat treatment of hot-rolled sheets in the temperature range of 700°C–750°C, which is below the recrystallization temperature, followed by oil quenching retains the elongated grain morphology, and thereby leads to high ductility.[264] It has been shown that this thermomechanical treatment minimizes crack formation at the surfaces, thereby inhibiting the ingress of hydrogen.

8.5.3 Room-Temperature Mechanical Properties

The mechanical properties of Fe-aluminides are known to be strongly dependent on the Al content of the alloy. It has been shown that the yield strength increases until the Al content is increased to 40 at.%, whereas the ambient-temperature ductility decreases. The variation of elongation to failure with Al concentration, as plotted in Figure 8.8, shows the ductility to be greater in vacuum than in air.[264,265] It has been shown that the room-temperature ductility of ordered Fe_3Al can be significantly enhanced either by decreasing the Al concentration to <18 at.% or by increasing it from 25% to 28%,[265,266] which is also obvious from this figure. The increase in ductility of Fe_3Al with reduction of Al content has been attributed to reduction of APB energy, which makes movement of $\langle 111 \rangle$ dislocations by cross-slip easier.[25] Although the ductility is increased by 20% on reducing the Al content, the corrosion resistance is compromised. The yield strength of Fe_3Al decreases on increasing the Al content from 25 to 28 at.%, as precipitation hardening by formation of α phase is eliminated.[266,268]

The poor ductility of FeAl can be attributed to its directional bonding,[25,269] ordered structure, and susceptibility to environmental degradation in a moist environment.[270] FeAl with 36.5 at.% Al concentration has shown a significant increase in tensile elongation to failure (2.2%–17.6%) on change of environment from moist air to dry oxygen, whereas the alloy containing 40 at.% Al has shown poor ductility. The fracture mode of Fe–36.5 at.% Al is found to be transgranular cleavage in air, and intergranular in oxygen,

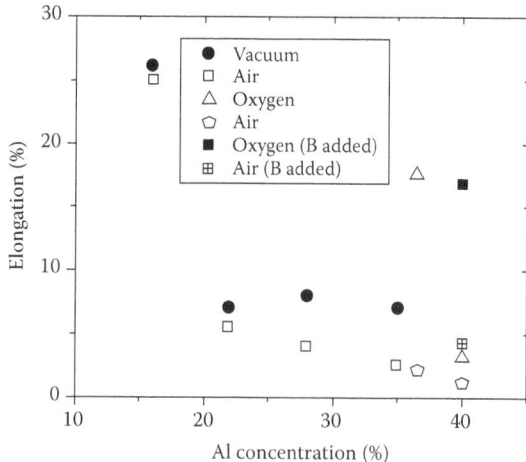

FIGURE 8.8

Plots showing the variation of elongation of binary Fe-aluminide alloy with Al concentration in vacuum, oxygen, and air. (From McKamey, C. G. *Physical Metallurgy and Processing of Intermetallic Compounds*, Chapman & Hall, London, 1996; Sikka, V. K., et al. *Structural Intermetallics*, TMS, Warrendale, PA, 1993.)

whereas brittle grain-boundary failure is observed for Fe–40 Al in both environments.[4] This observation shows that the intrinsic grain-boundary cohesive strength is relatively less in the alloy with the higher Al content. In the presence of oxygen, embrittlement by hydrogen is inhibited in the case of the Fe–36.5 at.% Al alloy. However, as the Al atoms at grain boundaries undergo preferential oxidation, intergranular failure is observed in a dry oxygen environment. Alloying with 300 wppm of B aids in improving the ductility of Fe–40 at.% Al in an oxygen environment, but not in air. This benefit is known to arise from increase in the cohesive strength of the grain boundaries, where B tends to segregate.[4]

As the strength of FeAl is inadequate at elevated temperatures, introduction of point defects through alloying is one of the strategies for mechanical property enhancement. Addition of ternary alloying elements and formation of vacancies impart solid-solution strengthening. In the case of FeAl, solid solubility has been found for 1–5 at.% Cr, Ti, Mn, and Co. High grain-boundary mobility in FeAl is responsible for significant grain growth. Increase in the Al content of this alloy lowers the grain growth rate by reducing the grain-boundary mobility. Of course, presence of oxides in the powder metallurgy-processed FeAl helps in reducing grain growth by pinning of grain boundaries.

Single crystals of both Fe$_3$Al and FeAl are reported to be ductile at room temperature.[25] The single crystals of FeAl are known to exhibit slip along the $\langle 111 \rangle$ direction, with the slip plane being either {211} or {101}.[271] The presence of these dislocations with APBs between the superpartial dislocations having $b = \frac{1}{2}\langle 111 \rangle$ has been observed in B2-structured Fe–35Al alloy using TEM. The APB energy has been found to increase linearly with increasing Al content of FeAl-based intermetallic alloy.[272] The preferred cleavage plane in Fe–40Al single crystals has been found to be {100}.[25] The existence of a preferred cleavage plane leads to anisotropy of fracture properties. High cleavage strength of FeAl has been attributed to directional d-bond formation at the Fe sites.[269]

Strength and ductility have been found to depend on both temperature and composition. It is commonly reported that, with increasing Al content, yield strength and strain hardening rates are found to increase, whereas ductility decreases. For Al content in the range of 25–50 at.%, ductility has been found to be maximal in the case of the Fe–25 at.% Al alloy, which has exhibited 8% elongation to failure for a microstructure with 50 μm grain size.[25,273,274] Furthermore, for the Fe–35 at.% Al alloy, 7% elongation to failure has been reported. A decrease in ductility with increase in grain size has been reported for each of these alloys. On the other hand, 2.5% and 0% tensile elongation has been observed in the case of the Fe–40 at.% Al and Fe–50 at.% Al alloys, respectively. The above literature data indicate that, for FeAl-based alloys, Fe-rich deviations in stoichiometry can lead to appreciable ductility. It is intuitive that the APB energies associated with splitting of $\langle 111 \rangle$ dislocations have a strong influence on their mobility, which in turn affects the tensile ductility. The APB energy can be altered by changes in stoichiometry and suitable alloying additions. Interestingly, the increase in hardness of Fe–Al alloys with Al content has been found to be higher in the samples subjected to annealing followed by air cooling, which indicates that the quenched-in vacancies have a significant role in the process of hardening.[275]

8.5.4 Elevated-Temperature Mechanical Properties

The yield strength of Fe_3Al (25 at.% Al) is known to decrease sharply with increasing temperature above the critical ordering temperature, T_c (814 K or 541°C), as there is a transition from $D0_3$ to B2 phase with relatively simpler structure at higher temperatures.[25] However, it is interesting to note that B2-structured FeAl-based alloys with higher Al content (35–50 at.% Al) retain reasonable strength until 600°C (Figure 8.9).[266,276]

In the temperature range of 400°C–541°C, Fe_3Al shows anomalous behavior characterized by increase in yield strength with temperature, as shown in Figure 8.9.[276] It should be noted that, within this temperature range, the long-range order parameter of Fe_3Al ($D0_3$) decreases from 0.8 to 0. Disordering is accompanied by age hardening through the precipitation of α from the ordered $D0_3$ phase. The anomalous increase in yield strength above 400°C has been attributed to increase in aging kinetics with increase in temperature above 400°C.

As the test temperature is increased in the range of 200°C–730°C, the active slip system in single crystals of FeAl changes from $\{110\}\langle 111\rangle$ to $\{110\}\langle 1\bar{1}0\rangle$, or

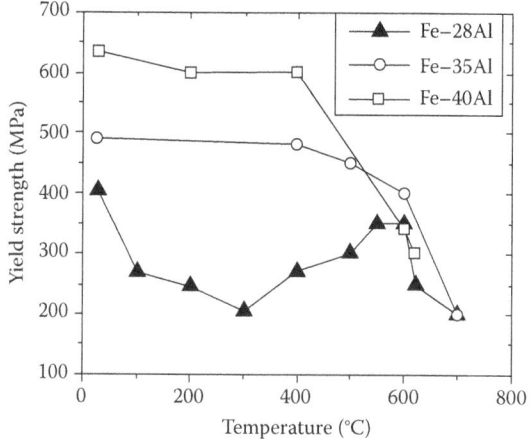

FIGURE 8.9
Plots depicting the variation of yield strength against temperature for Fe-aluminides with different Al concentrations. (From McCamey, C. G., *High Temperature Ordered Intermetallic Alloys II*, MRS, Pittsburgh, 1987; Gaydosh, D. J. and M. V. Nathal., Tensile behavior of Fe-40Al alloys with B and Zr additions. American Institute of Mining, Metallurgical and Petroleum Engineers annual meeting, New Orleans, LA (USA), 2–6 March 1986.)

{100}⟨001⟩. This transition temperature is found to change with deviations from stoichiometry. Similarly to single crystals, polycrystalline binary FeAl exhibits change in slip direction from ⟨111⟩ to ⟨100⟩ as a function of temperature and composition ($\approx$350°C for 50 at.% Al).[277,278] This can be confirmed from the observation of dislocations with both ⟨111⟩ and ⟨100⟩ Burgers vectors in hot-extruded polycrystalline Fe–40 at.% Al and Fe–50 at.% Al alloys.[25] The strength of FeAl decreases with increasing temperature, similarly to the behavior observed for regular metals. However, a drastic decrease in high-temperature strength with sharp increase in ductility has been observed in FeAl above the homologous temperature of 0.4. Such brittle-to-ductile transition in FeAl may be attributed to the onset of ⟨100⟩ slip along with diffusion-assisted process.[279]

Compression creep studies of FeAl in the temperature range of 827°C–1127°C have shown the predominance of two mechanisms with stress exponents (n) of 1 and 5 with nearly similar values of activation energy.[280–283] In the regime of diffusion creep (n $\approx$ 1), the creep rates have been found to increase with decrease in grain size, whereas the opposite trend has been observed in the dislocation-creep regime (n $\approx$ 5). In other words, the dislocation-creep regime is characterized by a Hall–Petch-type mechanism, where creep resistance is derived from grain-boundary strengthening due to fine grain size and alloying with B. It is interesting to note that the grain-boundary strengthening mechanism is operative in the case of FeAl until the homologous temperature of 0.75, which is much higher than that for ordinary metals and alloys.[280,281]

8.5.5 Effect of Alloying

Alloying addition is carried out for Fe_3Al primarily to increase the critical ordering temperature (T_c). Alloying with Ti and Si has been found to raise the T_c as well as the high-temperature tensile strength of Fe_3Al-based alloys significantly.[284] However, these alloys have been found to exhibit brittle behavior at room temperature. It has been possible to increase the ductility of Fe_3Al by ternary alloying with 2–6 at.% Cr. The mechanisms proposed to explain the beneficial effects of Cr on ductility of Fe_3Al are (i) increase in cleavage strength, which restricts cleavage fracture, as well as (ii) reduction of APB energy, which allows easier dissociation of superdislocations.[285]

Ternary alloying of B2 Fe–35Al with 1, 2, or 4 at.% Ti, Cr, V, or Ni, or 1–2 at.% Nb, Mo, Ta, W, or Si, causes significant solid-solution strengthening at 600°C and 700°C as well as room-temperature ductility.[263,286] The compressive creep strength of FeAl has also been found to increase significantly on ternary alloying. FeAl-based alloys containing 1–5 at.% of elements with high solid solubilities, such as Cr, Mn, Ti, or Co, show single-phase microstructures and are strengthened by solid-solution hardening.[25] In contrast, ternary alloys containing 0.8–5 at.% of elements with limited solubility (B, Zr, Ta, Nb, Re, and Hf) show strengthening by forming solid solutions as well as by formation of intermetallic precipitates, which pin dislocations. For example, the coherent precipitates of Fe_2Nb in Fe–25Al–2Nb alloys contribute to strengthening between ambient temperature and 600°C.[273] It has been observed that alloying FeAl with 6 at.% Cr results in improvement of room-temperature ductility from 4% to 8%–10% along with a change in fracture mode from transgranular cleavage to mixed mode.[4,25] This change in behavior can be attributed to reduction in APB energy on alloying with Cr.

Alloying Fe_3Al- and FeAl-based alloys with carbon has been found to be essential for their formation in air. Alloys with <0.06 wt.% C content have exhibited cracking during forging.[277] Moreover, high-carbon-containing Fe_3Al- and FeAl-based alloys have exhibited excellent machinability and could be turned on a lathe.[287] Both these observations have been attributed to the reduced susceptibility of the high-carbon-containing alloys to hydrogen.[288–290] As the Fe_3Al- and FeAl-based alloys have been found to be machinable irrespective of the formation of perovskite carbide, alloy carbides, or graphite, it has been inferred that the interstitial carbon is responsible for reduced susceptibility to hydrogen embrittlement. It has been proposed that the presence of interstitial carbon at the interstitial sites slows the rate of hydrogen diffusion through the iron aluminide lattice. This mechanism in turn can lead to reduced susceptibility to hydrogen embrittlement. It must be emphasized that the alloys remain susceptible to hydrogen embrittlement, as evidenced by increase in tensile ductility, when tested in vacuum.[288] It has been further shown[291–293] that presence of carbon reduces the permeability and diffusivity of hydrogen in both Fe_3Al and FeAl based alloys, and thereby reduces their susceptibility to hydrogen embrittlement.

8.5.6 Effect of Dispersion Strengthening

For strength retention at elevated temperatures, addition of reinforcements has been attempted. A study by Morris et al. on creep of Fe_3Al-based materials has shown significant enhancement in creep resistance on dispersion strengthening.[294] Significant increase in rupture strength has been observed on dispersion strengthening of Fe_3Al and FeAl, as shown in Figure 8.10a and b. Oxide dispersion strengthening with addition of fine nanometric particles of ZrO_2, Al_2O_3, Y_2O_3, and $FeAl_2O_4$ has yielded promising results.[294] These dispersoids contribute to strengthening by inhibiting grain growth by pinning of the grain boundaries. Addition of 0.5 at.% Ti, Hf, and Zr along with 1 at.% B to Fe_3Al leads to *in situ* formation of diborides, which lead to significant improvement in creep resistance. These composites have been processed by both powder metallurgy and ingot metallurgy processing routes. The combination of 2% Mo + 0.1Zr + 0.2B alloy (as in FA-114) has been found to be particularly effective in improving both high-temperature strength and creep resistance.[295] This observation has been attributed to the formation of fine ZrC particles, which pin dislocations.

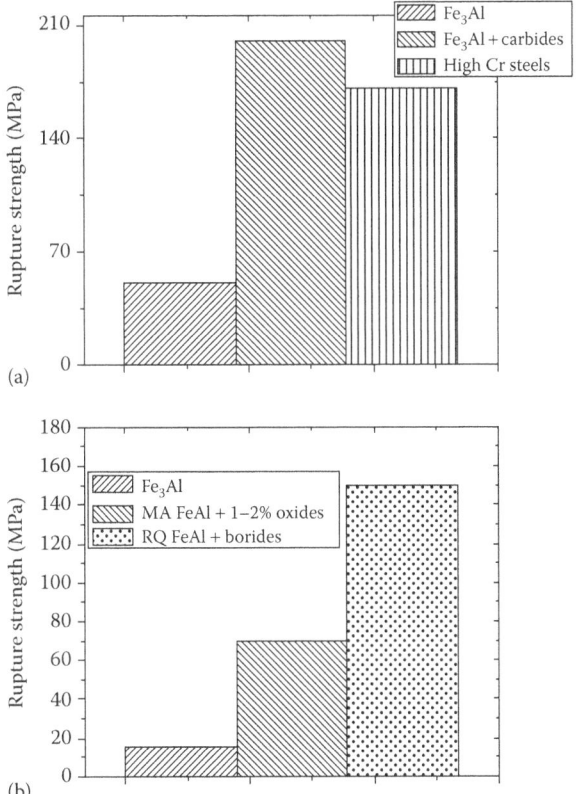

FIGURE 8.10
1000 h rupture strength of Fe_3Al and FeAl with or without dispersion strengthening at (a) 600°C and (b) 700°C.

Carbon additions to Fe-aluminides cause the formation of carbide dispersoids, which lead to significant increase in creep resistance. Solubility of carbon in Fe_3Al alloys is very limited (<0.05 wt.%) and most of the carbon is present as carbide precipitates. If no strong carbide former (such as Ti, Cr, W, or Mo) is added, the carbide present is $Fe_3AlC_{0.5}$ perovskite carbide. The Fe–Al–C system contains a stable $Fe_3AlC_{0.5}$ carbide, which has a perovskite structure.[296] This carbide is stable in alloys with low Al contents (<37 at.% Al), and its dispersion leads to significant strengthening.[297] However, in Fe–Al-based alloys with higher Al concentrations, the graphite is more stable. Therefore, only Fe_3Al-based alloys can be strengthened by the formation of $Fe_3AlC_{0.5}$-type perovskite carbide. On the other hand, carbides such as TiC and WC can be used in FeAl- as well as Fe_3Al-based composites. To prevent graphite formation, the FeAl-based intermetallics may be alloyed with strong carbide formers such as Ti, Cr, or Mo.[298] Formation of carbides by these alloying elements provides significant dispersion strengthening. Recently, carbon nanotubes have also been used to strengthen iron aluminides, but the work is only preliminary and further evaluation is required.[299]

8.6 Summary

The aluminides of Ni (Ni_3Al and NiAl), Ti (Al_3Ti, γ-TiAl, and Ti_3Al), and Fe (Fe_3Al and FeAl) are of interest to the scientific community for various structural applications at ambient and elevated temperatures. The maximum temperature for application of these intermetallic alloys is dependent on their melting points and critical temperatures for ordering, as well as corrosion and oxidation resistance. Among the aforementioned intermetallic alloys, Ni_3Al ($L1_2$), Fe_3Al ($D0_3$), and FeAl (B2) have exhibited ductility in their polycrystalline form at room temperature, whereas Al_3Ti ($D0_{22}$) is found to be brittle at all temperatures. On the other hand, NiAl (B2), γ-TiAl ($L1_0$), and Ti_3Al ($D0_{19}$) have shown BDTT at 400°C, 600°C, and 700°C–800°C, respectively. A sharp decrease in ductility is observed in Ni_3Al, Fe_3Al, and FeAl in a moist environment, where embrittlement occurs due to the interaction of Al with atomic hydrogen. The nature of operative slip systems in the aforementioned aluminides is dependent on crystal structure, tetragonality or deviation from cubic structure, interatomic bond directionality, and APB energy.

Although the close-packed {111} is the preferred slip plane of Ni_3Al, the APB energy of the {010} planes is much lower. Therefore, there is a net driving force for cross-slip of the screw segments of {111}$\langle\bar{1}01]$ dislocations into {010} planes due to the net differences in APB and elastic energies of dislocations. With increase in temperature, cross-slipped segments of screw dislocations in the {010} planes are split into superpartials, which are of sessile nature and lead to the formation of Kear–Wilsdorf locks, thereby causing anomalous increase in yield strength with temperature in the range of 600°C–800°C. Alloying additions to Ni_3Al are carried out for enhancing both mechanical properties and resistance to environmental degradation. Segregation of B to grain boundaries in Ni_3Al is effective in both increasing grain-boundary cohesive strength and inhibiting H embrittlement, thereby increasing the ductility of polycrystalline samples. The elements substituting Al sites in Ni_3Al lead to a greater amount of solid-solution hardening compared with those occupying Ni sites.

B2-structured NiAl is brittle at ambient temperature, showing $\leq 2\%$ elongation due to the directional nature of bonding because of Ni (d)–Al (p) hybridization along the close-packed $\langle 111 \rangle$ direction. This is responsible for formation and movement of dislocations with $b = \langle 100 \rangle$ on {010} and {011} planes. The BDTT for this intermetallic alloy has been reported to be at around 400°C, when the strain incompatibility due to absence of five independent slip systems is accommodated by dislocation climb. It has been possible to enhance the room-temperature ductility of NiAl single crystals by microalloying with Fe, Ga, and Mo. The high-temperature yield and creep strengths of NiAl have been found to be comparable to those of Ni-based superalloys.

Research on Al_3Ti has primarily focused on trying suitable alloying additions in order to stabilize the $L1_2$ structure to achieve ductility at ambient temperatures. In spite of stabilizing the $L1_2$ phase, not much improvement in ductility is observed, except on alloying with Cr and Mn. Among the titanium aluminides, the γ-TiAl-based multiphase alloys have been found to be most promising due to the possibility of tailoring microstructures comprising $\alpha_2 + \gamma$ phase mixture to obtain desirable mechanical properties such as strength, tensile elongation, and fracture toughness. The phases present in the microstructures of the γ-TiAl-based alloys depend on Al content: single-phase γ for 52–54 at.% Al, $\alpha_2 + \gamma$ with lamellar morphology along with equiaxed γ, termed *duplex structure*, for 45–51 at.% Al, and fully lamellar morphology for 40–45 at.% Al. Whereas a duplex structure is desirable for high ductility, a lamellar morphology of phases provides the best combination of strength and fracture toughness.

Although Ti_3Al is known to be brittle, its alloying with Nb leads to an increase in ductility and fracture toughness along with creep resistance. The microstructures of Ti–Al–Nb alloys vary with Nb content: (i) two-phase microstructures comprising $\alpha_2 + \beta$ (or B2) for 10–12 at.% Nb; (ii) three phases, $\alpha_2 + \beta$ (or B2)+O (Ti_2AlNb) for 14%–17% Nb, and (iii) O + β (or B2) for 25–30 at.% Nb. The maximum ductility is achieved for the alloys having a fine basket-weave morphology of $\alpha_2 + O$ laths, without the formation grain boundary α_2.

In iron aluminides, the ductility decreases but strength increases with increasing Al content. A maximum elongation of 15%–20% can be observed through suitable control of composition of Fe_3Al, while its strength decreases sharply above its ordering temperature. These alloys are susceptible to hydrogen embrittlement in the presence of moisture. The presence of C as an interstitial alloying element is known to lower the rate of H diffusion, and thereby enhances the mechanical properties significantly. C-containing alloys are found to be machinable and possess high creep resistance.

References

1. Liu, C. T. 1995. Recent advances in ordered intermetallics. *Mater. Chem. Phys.* 42:77–86.
2. Meier, G. H. and F. S. Pettit. 1992. The oxidation behavior of intermetallic compounds. *Mater. Sci. Eng. A* 153:548–560.
3. Stoloff, N. S., C. T. Liu, and S. C. Deevi. 2000. Emerging applications of intermetallics. *Intermetallics* 8:1313–1320.
4. Deevi, S. C., V. K. Sikka, and C. T. Liu. 1997. Processing, properties, and applications of nickel and iron aluminides. *Prog. Mater. Sci.* 42:177–192.

5. WardClose, C. M., R. Minor, and P. J. Doorbar. 1996. Intermetallic-matrix composites: A review. *Intermetallics* 4:217–229.
6. Cinca, N., C. R. C. Lima, and J. M. Guilemany. 2013. An overview of intermetallics research and application: Status of thermal spray coatings. *J. Mater Res. Technol.* 2:75–86.
7. Liu, C. T. 1992. Environmental embrittlement and grain-boundary fracture in Ni₃Al. *Scr. Metall. Mater.* 27:25–28.
8. Miracle, D. B. 1993. Overview No-104: The physical and mechanical properties of NiA1. *Acta Metall. Mater.* 41:649–684.
9. Risanti, D., J. Deges, L. Falat, S. Kobayashi, J. Konrad, M. Palm, B. Poter, A. Schneider, C. Stallybrass, and F. Stein. 2005. Dependence of the brittle-to-ductile transition temperature (BDTT) on the Al content of Fe-Al alloys. *Intermetallics* 13:1337–1342.
10. Yoshimi, K., S. Hanada, and H. Tokuno. 1994. Effect of frozen-in vacancies on hardness and tensile properties of polycrystalline B2 FeAl. *Mater. Trans. JIM* 35:51–57.
11. Lipsitt, H. A., D. Shechtman, and R. E. Schafrik. 1980. The deformation and fracture of Ti₃Al at elevated temperatures. *Metall. Trans. A* 11:1369–1375.
12. Imayev, V. M., R. M. Imayev, and G. A. Salishchev. 2000. On two stages of brittle-to-ductile transition in TiAl intermetallic. *Intermetallics* 8:1–6.
13. Liu, C. T. and D. P. Pope. 1994. Ni₃Al and its alloys. In *Intermetallic Compounds: Vol. 2, Practice*, eds. J. H. Westbrook and R. L. Fleischer, 17–51. Chichester, UK: Wiley.
14. Shimokawa, T., M. Hosomi, H. Inui, and M. Yamaguchi. 1991. Deformation of Al₃Ti single crystals. In *6th JIM International Symposium on Intermetallic Compounds: Structure and Mechanical Properties*, ed. O. Izumi, 661. Sendai: The Japan Institute of Metals.
15. Darolia, R., D. F. Lahrman, R. D. Field, J. R. Dobbs, K.-M. Chang, E. H. Goldman, and D. G. Konitzer. 1992. Overview of NiAl alloys for high temperature applications. In *Ordered Intermetallics: Physical Metallurgy and Mechanical Behavior*, eds. C. T. Liu, R. W. Cahn, and G. Sauthoff, Vol. 213, 679. NATO ASI Series E: Applied Science. Dordrecht: Kluwer Academic.
16. Huang, S. C. and P. A. Siemers. 1989. Characterization of the high-temperature phase fields near stoichiometric γ-TiAl. *Metall. Trans. A* 20:1899–1906.
17. McCullough, C., J. J. Valencia, C. G. Levi, and R. Mehrabian. 1989. Phase equilibria and solidification in Ti-Al alloys. *Acta Metall.* 37:1321–1336.
18. Nobuki, M. and T. Tsujimoto. 1991. Influence of alloy composition on hot deformation properties of Ti-Al binary intermetallics. *ISIJ Intl.* 31:931–937.
19. Masahashi, N., Y. Mizuhara, M. Matsuo, M. Kimura, M. Hanamura, and H. Fuji. 1991. Ternary alloying of gamma titanium aluminides for hot-workability. In *High-Temperature Ordered Intermetallic Alloys IV*, Materials Research Society Symposium Proceedings, Vol. 213, 795–800. Pittsburgh, PA: MRS.
20. Matsuo, M. 1991. Developments in processing technology of gamma titanium aluminides for potential application to airframe structures. *ISIJ Intl.* 31:1212–1222.
21. Maeda, T., M. Okada, and Y. Shida. 1991. In *Superplasticity in Advanced Materials*, eds. S. Hori, M. Tokizane, and N. Furushiro, 311. Osaka: The Japan Society for Research on Superplasticity.

22. Oliver, B. F. and B. Kad. 1991. A study of αγ peritectic oscillations using containerless liquid directional solidification of γ-TiAl compositions. *J. Less Common Met.* 168(1):81–90.

23. Inui, H., M. H. Oh, A. Nakamura, and M. Yamaguchi. 1992. Room-temperature tensile deformation of polysynthetically twinned (Pst) crystals of TiAl. *Acta Metall. Mater.* 40:3095–3104.

24. Takeyama, M., T. Hirano, and T. Tsujimoto. 1991. In *Intermetallic Compounds: Structure and Mechanical Properties*, ed. O. Izumi, 507. Sendai: The Japan Institute of Metals.

25. Vedula, K. 1994. FeAl and Fe₃Al. In *Intermetallic Compounds, Vol. 2, Practice*, eds. J. H. Westbrook and R.L. Fleischer, 199. Chichester, UK: Wiley.

26. Sikka, V. K. 1991. Mechanical properties of Fe₃Al-based alloys. In *Heat Resistant Materials*, eds. K. Natesan and D. J. Tillack, 141. Materials Park: ASM-International.

27. Sikka, V. K. 1991. Production of Fe₃Al based intermetallic alloys. In *High-Temperature Ordered Intermetallic Alloys IV*, eds. L. A. Johnson, D. P. Pope, and J. O. Stiegler. Materials Research Society Symposium Proceedings, Vol. 213, 907–912. Pittsburgh, PA: MRS.

28. Baligidad, R. G., U. Prakash, V. R. Rao, P. K. Rao, and N. B. Ballal. 1994. Electroslag remelting of an Fe-28at%Al intermetallic alloy. *Ironmak. Steelmak.* 24:321–330.

29. Baligidad, R. G., U. Prakash, V. Ramakrishna Rao, P. K. Rao, and N. B. Ballal. 1996. Processing of Fe₃Al based intermetallic alloys through electroslag remelting. *ISIJ Intl.* 36:1448–1452.

30. Baligidad, R. G., U. Prakash, and A. Radhakrishna. 2010. An Air Induction Melting Process for Preparation of Intermetallic Alloy. Indian Patent No. 242438.

31. Baligidad, R. G., U. Prakash, and A. Radhakrishna. 1998. Processing of high carbon Fe₃Al based intermetallic alloy. *Intermetallics* 6:765–769.

32. Lapin, J., D. Tiberghien, and F. Delannay. 2000. On the parameters affecting the formation of iron aluminides during pressure-assisted infiltration of aluminium into a preform of steel fibres. *Intermetallics* 8:1429–1438.

33. Tiberghien, D., J. Lapin, S. Ryelandt, and F. Delannay. 2002. On the control of the residual porosity in iron aluminides processed by reactive squeeze-infiltration of aluminium into a preform of steel fibres. *Mater. Sci. Eng. A* 323:427–435.

34. Liu, C. T., V. K. Sikka, J. A. Horton, and E. H. Lee. 1988. Alloy development and mechanical properties of nickel aluminide (Ni₃Al) alloys. ORNL-6483. Oak Ridge, TN: Oak Ridge National Laboratory.

35. Sikka, V. K. 1989. Nickel aluminides: New advanced alloys. *Mater. Manuf. Proc.* 4:1–24.

36. German, R. M., A. Bose, and N. S. Stoloff. 1989. Powder processing of high temperature aluminides. In *High Temperature Ordered Intermetallic Alloys III*, eds. C. T. Liu, A. I. Taub, N. S. Stoloff, and C. C. Koch, Materials Research Society Symposium Proceedings, Vol. 133, 403–414. Warrendale, PA: MRS.

37. Sims, D. M., A. Bose, and R. M. German. 1987. Reactive sintering of nickel aluminide. *Prog. Powder Metall.* 43:575–596.

38. Bose, A., B. H. Rabin, and R. M. German. 1988. Reactive sintering nickel-aluminide to near full density. *Powder Metall. Int.* 20(3):25–30.

39. Martin, P. L., M. G. Mendiratta, and H. A. Lipsitt. 1983. Creep deformation of TiAl and TiAl + W alloys. *Metall. Trans. A* 14:2170–2174.

40. Huang, S. C. and E. L. Hall. 1991. The effects of Cr additions to binary TiAl-base alloys. *Metall. Trans. A* 22:2619–2627.
41. Sikka, V. K., R. H. N. Baldwin, J. H. Reinshagen, J. R. Knibloe, and R. N. Wright. 1991. Powder processing of Fe_3Al based iron aluminide alloys. In *High-Temperature Ordered Intermetallic Alloys IV*, eds. L. A. Johnson, D. P. Pope, and J. O. Stiegler, Materials Research Society Symposium Proceedings, Vol. 213, 901–906. Pittsburgh, PA: MRS.
42. Sikka, V. K., B. G. Gleseke, and R. H. Baldwin. 1991. Mechanical properties of Fe_3Al-based alloys. In *Heat Resistant Materials*, eds. K. Natesan and D. J. Tillack, 363–371. Materials Park, OH: ASM International.
43. Gaydosh, D. J. and M. A. Crimp. 1985. Rapidly solidified NiAl and FeAl. In *High-Temperature Ordered Intermetallic Alloys I*, eds. C. C. Koch, C. T. Liu, and N. S. Stoloff, Materials Research Society Symposium Proceedings, Vol. 39, 429–436. Pittsburgh, PA: MRS.
44. Liang, X., H. K. Kim, J. C. Earthman, and E. J. Lavernia. 1992. Microstructure and elevated temperature behavior of a spray-atomized and co-deposited $Ni_3Al/SiC/TiB_2$ intermetallic matrix composite. *Mater. Sci. Eng. A* 153:646–653.
45. Liang, X. and E. J. Lavernia. 1992. Interfacial behavior in a Ni_3Al/TiB_2 intermetallic matrix composite. *Mater. Sci. Eng. A* 153:654–661.
46. Strum, M. J. and G. A. Henshall. 1993. Fracture behavior of vanadium/vanadium silicide *in situ* composites. In *High-Temperature Ordered Intermetallic Alloys V*, eds. I. Baker, R. Darolia, J. D. Whittenberger, and M. H. Yoo, Materials Research Society Symposium Proceedings, Vol. 288, 1093. Warrendale, PA: MRS.
47. Mason, D. P. and D. C. Vanaken. 1993. The effect of microstructural scale on hardness of $MoSi_2$-Mo_5Si_3 eutectics. *Scr. Metall. Mater.* 28:185–189.
48. Wright, R. N., M. T. Anderson, and J. K. Wright. 1998. Microstructure and properties of an oxide dispersion-strengthened iron aluminide. *Mater. Sci. Eng. A* 258:285–290.
49. Wolski, K., F. Thevenot, and J. LeCoze. 1996. Effect of nanometric oxide dispersion on creep resistance of ODS-FeAl prepared by mechanical alloying. *Intermetallics* 4:299–307.
50. Nourbakhsh, S., W. H. Rhee, O. Sahin, and H. Margolin. 1992. Mechanical behavior of a fiber reinforced Ni_3Al matrix composite. *Mater. Sci. Eng. A* 153:619–627.
51. Glushko, V. I., S. T. Mileiko, and N. V. Kondrashova. 1993. Fabrication and some properties of Ni_3Al matrix $Al_2O_3+ZrO_2+Y_2O_3$-fiber composites. *J. Mater. Sci. Lett.* 12:915–917.
52. Newkirk, J. and D. Dixon. 1992. Interaction of tantalum with reinforcements in γ TiAl. *Mater. Sci. Eng. A* 153:662–667.
53. MacKay, R. A., P. K. Brindley, and F. H. Froes. 1991. Continuous fiber-reinforced titanium aluminide composites. *JOM* 43:23–29.
54. Kesapradist, J., K. Ono, and K. Fukaura. 1992. Producing Ni_3Al matrix composite material by vacuum hot pressing and heat treatment of nickel-plated aluminum sheets with Al_2O_3 fiber. *Mater. Sci. Eng. A* 153:641–645.
55. Westbrook, J. H. 1957. Temperature-dependence of the hardness of secondary phases common in turbine bucket alloys. *Trans. AIME* 209:898–904.
56. Flinn, P. A. 1960. Theory of deformation in superlattices. *Trans. Metall. Soc. AIME* 218:145–154.
57. Davies, R. G. and N. S. Stoloff. 1965. On the yield stress of aged Ni-Al alloys. *Trans AIME* 233:714–719.

58. Copley, S. M. and B. H. Kear. 1967. Temperature and orientation dependence of the flow stress in off-stoichiometric Ni_3Al (γ' phase). *Trans. Metall. Soc. AIME* 239:977–984.

59. Statonbevan, A. E. and R. D. Rawlings. 1975. Dislocation-structures in deformed single-crystal $Ni_3(Al, Ti)$. *Philos. Mag.* 32:787–800.

60. Umakoshi, Y., D. P. Pope, and V. Vitek. 1984. The asymmetry of the flow-stress in $Ni_3(Al,Ta)$ single-crystals. *Acta Metall.* 32:449–456.

61. Kear, B. H. and H. G. F. Wilsdorf. 1962. Dislocation configurations in plastically deformed Cu_3Au alloys. *Trans. Metall. Soc. AIME* 224:382–386.

62. Marcinkowski, M. J., N. Brown, and R. M. Fisher. 1961. Dislocation configurations in $AuCu_3$ and AuCu type superlattices. *Acta Metall.* 9:129–137.

63. Kear, B. H., A. F. Giamei, J. M. Silcock, and R. K. Ham. 1968. Slip and climb processes in gamma' precipitation hardened nickel-base alloys. *Scr. Metall.* 2:287–293.

64. Douin, J., P. Veyssiere, and P. Beauchamp. 1986. Dislocation line stability in Ni_3Al. *Philos. Mag. A* 54:375–393.

65. Dimiduk, D. M. 1991. Dislocation structures and anomalous flow in $L1_2$ compounds. *J. de Physique III* 1:1025–1053.

66. Bonneville, J., T. Kruml, J.-L. Martin, B. Matterstoc, B. Viguier, and P. Spätig. 1997. Characteristics of plastic flow in $Ni_3(Al,Hf)$ single crystals. *Mater. Sci. Eng. A* 234–236:770–773.

67. Masahashi, N., T. Takasugi, and O. Izumi. 1988. Mechanical properties of Ni_3Al containing C, B and Be. *Acta Metall.* 36(7):1823–1836.

68. Mishima, Y., S. Ochai, and Y. M. Yodogawa. 1986. Mechanical properties of Ni_3Al with ternary addition of transition metal elements. *Trans. Jpn Inst. Mater.* 27:41–50.

69. Kear, B. H. and M. F. Hornbecker. 1966. Cross-slip, antiphase defects, and work hardening in ordered Cu_3Au. *Acta Metall.* 14(5):659–677.

70. Thornton, P. H., R. G. Davies, and T. L. Johnston. 1970. Temperature dependence of flow stress of γ phase based upon Ni_3Al. *Metall. Trans. AIME* 1:207–218.

71. Takeuchi, S. and E. Kuramoto. 1973. Temperature and orientation dependence of yield stress in Ni_3Ga single-crystals. *Acta Metall.* 21:415–425.

72. Paidar, V., D. P. Pope, and V. Vitek. 1984. A theory of the anomalous yield behavior in L12 ordered alloys. *Acta Metall.* 32:435–448.

73. Yoo, M. H. 1987. Effects of elastic anisotropy on the anomalous yield behavior of cubic ordered alloys. In *High Temperature Ordered Intermetallic Alloys II*, eds. N. S. Stoloff, C. C. Koch, C. T. Liu, and O. Izumi, Materials Research Society Symposium Proceedings, Vol. 81, 207–212. Pittsburgh, PA: MRS.

74. Liu, C. T. 1993. Ni_3Al aluminide alloys. In *Structural Intermetallics*, eds. R. Darolia, J. J. Lewandowski, C. T. Liu, P. L. Martin, D. B. Miracle, and M. V. Nathal, 365–378. Warrendale, PA: TMS.

75. Liu, C. T. and W. C. Oliver. 1991. Environmental embrittlement and grain-boundary fracture in Ni_3Si. *Scr. Metall. Mater.* 25:1933–1937.

76. Liu, C. T., E. H. Lee, and C. G. McKamey. 1989. An environmental effect as the major cause for room-temperature embrittlement in FeAl. *Scr. Metall.* 23:875–880.

77. Liu, C. T., C. G. McKamey, and E. H. Lee. 1990. Environmental effects on room-temperature ductility and fracture in Fe_3Al. *Scr. Metall. Mater.* 24:385–389.

78. Liu, C. T. and E. P. George. 1990. Environmental embrittlement in boron-free and boron-doped FeAl (40 at-percent Al) alloys. *Scr. Metall. Mater.* 24:1285–1290.

79. George, E. P., C. T. Liu, and D. P. Pope. 1992. Environmental embrittlement: The major cause of room-temperature brittleness in polycrystalline Ni₃Al. *Scr. Metall. Mater.* 27:365–370.

80. Aoki, K. and O. Izumi. 1979. Improvement in room temperature ductility of the intermetallic compound Ni₃Al by ternary trace element addition. *Nippon Kinzoku Gakkaishi* [*J. Jpn. Inst. Met.*] 43:1190.

81. Taub, A. I., S. C. Huang, and K. M. Chang. 1984. Improved strength and ductility of Ni₃Al by boron modification and rapid solidification. *Metall. Trans. A* 15:399–402.

82. Liu, C. T. and C. L. White. 1985. Design of ductile polycrystalline Ni₃Al alloys. In *High Temperature Ordered Intermetallic Alloys I*, eds. C. C. Koch, C. T. Liu, and N. S. Stoloff, Materials Research Society Symposium Proceedings, Vol. 39, 365–380. Pittsburgh, PA: MRS.

83. George, E. P. and C. T. Liu. 1997. Mechanical properties of Ni₃Al and FeAl: Recent developments. In *Structural Intermetallics 1997*, eds. M. V. Nathal, R. Darolia, C. T. Liu, P. L. Martin, D. B. Miracle, R. Wagner, and M. Yamaguchi. 703–712. Warrendale, PA: TMS.

84. White, C. L., R. A. Padgett, C. T. Liu, and S. M. Yalisove. 1984. Surface and grain-boundary segregation in relation to intergranular fracture: Boron and sulfur in Ni₃Al. *Scr. Metall.* 18:1417–1420.

85. Liu, C. T., C. L. White, and J. A. Horton. 1985. Effect of boron on grain-boundaries in Ni₃Al. *Acta Metall.* 33:213–229.

86. Chen, S. P., A. F. Voter, R. C. Albers, A. M. Boring, and P. J. Hay. 1989. Theoretical studies of grain-boundaries in Ni₃Al with boron or sulfur. *Scr. Metall.* 23:217–222.

87. Bond, G. M., I. M. Robertson, and H. K. Birnbaum. 1987. Effect of boron on the mechanism of strain transfer across grain boundaries in Ni₃Al. *J. Mater. Res.* 2:436–440.

88. Schulson, E. M., T. P. Weihs, D. V. Viens, and I. Baker. 1985. The effect of grain size on the yield strength of Ni₃Al. *Acta Metall.* 33:1587–1591.

89. Khadkikar, P. S., K. Vedula, and B. S. Shabel. 1987. The role of boron in ductilizing Ni₃Al. *Metall. Trans. A* 18:425–428.

90. Baker, I., E. M. Schulson, and J. A. Horton. 1987. *In situ* straining of Ni₃Al in a transmission electron microscope. *Acta Metall.* 35:1533–1541.

91. Chiba, A., S. Hanada, S. Watanabe, T. Abe, and T. Obane. 1994. Relation between ductility and grain boundary character distributions in Ni₃Al. *Acta Metall. Mater.* 42(5):1733–1738.

92. Lee, K. H. and C. L. White. 1995. Intrinsic ductility of Ni₃Al with and without boron at 77 K. *Scr. Metall. Mater.* 32(11): 1871–1875.

93. Ochiai, S., Y. Oya, and T. Suzuki. 1984. Alloying behavior of Ni₃Al, Ni₃Ga, Ni₃Si and Ni₃Ge. *Acta Metall.* 32:289–298.

94. Guard, R. W. and J. H. Westbrook. 1959. Alloying behavior of Ni₃Al (γ′ phase). *Trans. Metall. Soc.* 215:807–814.

95. Rawlings, R. D. and A. E. Statonbevan. 1975. Alloying behavior and mechanical properties of polycrystalline Ni₃Al (gamma′ phase) with ternary additions. *J. Mater. Sci.* 10:505–514.

96. Dimiduk, D. M., T. A. Parthasarathy, S. Rao, and C. Woodward. 1992. In *Ordered Intermetallics: Physical Metallurgy and Mechanical Behavior*, eds. C. T. Liu, R. W. Cahn, and G. Sauthoff. NATO ASI Series E: Applied Sciences, Vol. 213, 237. Boston, MA: Kluwer Academic.

97. Huang, S. C., A. I. Taub, and K. M. Chang. 1984. Boron extended solubility and strengthening potency in rapidly solidified Ni₃Al. *Acta Metall.* 32:1703–1707.

98. Fleischer, R. L. 1963. Substitutional solution hardening. *Acta Metall.* 11:203–209.

99. Hazzledine, P. M. and J. H. Schneibel. 1989. Inverse creep in Ni₃Al. *Scr. Metall.* 23:1887–1892.

100. Schneibel, J. H. and J. A. Horton. 1988. Evolution of dislocation structure during inverse creep of a nickel aluminide: Ni–23.5 Al–0.5 Hf–0.2B (at.%). *J. Mater. Res.* 3:651–655.

101. Hemker, K. J., M. J. Mills, and W. D. Nix. 1991. An investigation of mechanisms that control intermediate temperature creep of Ni₃Al. *Acta Metall. Mater.* 39(8):1901–1913.

102. Rong, T. S., I. P. Jones, and R. E. Smallman. 1995. Dislocation mechanisms in creep of Ni₃Al at intermediate temperature. *Acta Metall. Mater.* 43(4):1385–1393.

103. Zhu, W. H., D. Fort, I. P. Jones, and R. E. Smallman. 1998. Orientation dependence of creep of Ni₃Al at intermediate temperature. *Acta Mater.* 46(11):3873–3881.

104. Nicholls, J. R. and R. D. Rawlings. 1977. Steady-state creep of an alloy based on the intermetallic compound Ni₃Al (γ′). *J. Mater. Sci.* 12:2456–2464.

105. Rong, T. S., I. P. Jones, and R. E. Smallman. 1995. A TEM study of ⟨110⟩{111} slip in crept polycrystalline Ni₃Al. *Acta Metall. Mater.* 43(8):3085–3092.

106. Klotz, U. E., R. P. Mason, E. Göhring, and E. Arzt. 1997. High-temperature creep in a coarse-grained oxide-dispersion strengthened Ni₃Al alloy. *Mater. Sci. Eng. A.* 231:198–205.

107. Lapin, J. 1999. High temperature creep of precipitation-strengthened Ni₃Al-based alloy. *Intermetallics* 7:599–609.

108. Shah, D. M. and D. N. Duhl. 1987. Evaluation of multicomponent nickel base L1₂ and other intermetallic alloys as high temperature structural materials. In *High Temperature Ordered Intermetallics II*, eds. N. S. Stoloff, C. C. Koch, C. T. Liu, and O. Izumi, Materials Research Society Symposium Proceedings, Vol. 81, 411–417. Pittsburgh, PA: MRS.

109. Han, Y. F., S. H. Li, S. Ma, and Y. N. Tan. 1992. A DS casting Ni₃Al base superalloy for gas turbine blades and vanes. Paper presented at the first Pacific Rim International conference on advanced materials and processing, Hongzhou, China, 23–27 June.

110. Han, Y. F., Xing, Z. P., and Chaturvedi, M. C. 1997. Development and engineering application of a DS cast Ni₃Al alloy IC6. In *Structural Intermetallics 1997*, eds. M. V. Nathal, R. Darolia, C. T. Liu, P. L. Martin, D. B. Miracle, R. Wagner, and M. Yamaguchi, 713–719. Warrendale, PA: TMS.

111. Miracle, D. B. and R. Darolia. 1994. NiAl and its alloys. In *Intermetallic Compounds: Vol. 2, Practice*, eds. J. H. Westbrook and R. L. Fleischer, 53. Chichester, UK: Wiley.

112. Wasilewski, R. J. 1966. Elastic constants and Young's modulus of NiAl. *Trans. Metall. Soc. AIME* 236:455–457.

113. Darolia, R. 1991. NiAl alloys for high-temperature structural applications. *J. Met.* 43:44–49.

114. Wasilewski, R. J., S. R. Butler, and J. E. Hanlon. 1967. Plastic deformation of single-crystal NiAl. *Trans. Metall. Soc. AIME* 239:1357–1364.

115. Ball, A. and R. E. Smallman. 1966. The deformation properties and electron microscopy studies of the intermetallic compound NiAl. *Acta Metall.* 14:1349–1355.

116. Loretto, M. H. and Wasilewski, R. J. 1971. Slip systems in NiAl single crystals at 300 K and 77 K. *Philos. Mag.* 23:1311–1328.

117. Company, R. G., M. H. Loretto, and R. E. Smallman. 1973. The determination of the ½⟨111⟩ {110} antiphase boundary energy of NiAl. *J. Microsc.* 98:174–179.

118. Veyssière, P. and R. Noebe. 1992. Weak-beam study of ⟨111⟩ superlattice dislocations in NiAl. *Philos. Mag. A* 65:1–13.

119. Pascoe, R. T. and C. W. A. Newey. 1968. The mechanical behaviour of the intermediate phase NiAl. *Metal Sci.* 2:138–143.

120. Clapp, P. C., M. J. Rubins, S. Charpenay, J. A. Rifkin, and Z. Z. Yu. 1989. Some thermodynamic properties of NiAl calculated by molecular dynamics solution. In *High Temperature Ordered Intermetallic Alloys III*, eds. C. T. Liu, A. I. Taub, N. S. Stoloff, and C. C. Koch, Materials Research Society Symposium Proceedings, Vol. 133, 29–36. Pittsburgh, PA: MRS.

121. Rao, S. I., C. Woodward, and T. A. Parthasarathy. 1991. Empirical interatomic potentials for $L1_0$ TiAl and B2 NiAl. In *High Temperature Ordered Intermetallics IV*, eds. L. A. Johnson, J. O. Stiegler, and D. P. Pope, Materials Research Society Symposium Proceedings, Vol. 213, 125–130. Pittsburgh, PA: MRS.

122. Fu, C. L. and M. H. Yoo. 1991. First principles investigation of mechanical behavior of B2 type aluminides: FeAl and NiAl. In *High Temperature Ordered Intermetallics IV*, eds. L. A. Johnson, J. O. Stiegler, and D. P. Pope, Materials Research Society Symposium Proceedings, Vol. 213:667–672. Pittsburgh, PA: MRS.

123. Hong, T. and A. J. Freeman. 1991. Effect of antiphase boundaries on the electronic structure and bonding character of intermetallic systems: NiAl. *Phys. Rev. B* 43:6446–6458.

124. Hahn, K. H. and K. Vedula. 1989. Room-temperature tensile ductility in polycrystalline B2 NiAl. *Scr. Metall.* 23:7–12.

125. Vedula, K., K. H. Hahn, and B. Boulogne. 1989. Room temperature tensile ductility of polycrystalline B2 NiAl. In *High Temperature Ordered Intermetallics III*, eds. C. T. Liu, A. I. Taub, N. S. Stoloff, and C. C. Koch, Materials Research Society Symposium Proceedings, Vol. 133, 299–304. Pittsburgh, PA: MRS.

126. Schulson, E. M. and D. R. Barker. 1983. A brittle to ductile transition in NiAl of a critical grain size. *Scr. Metall.* 17:519–522.

127. Lahrman, D F., R. D. Field, and R. Darolia. 1991. The effect of strain rate on the mechanical properties of single crystal NiAl. In *High Temperature Ordered Intermetallics IV*, eds. L. A. Johnson, J. O. Stiegler, and D. P. Pope, Materials Research Society Symposium Proceedings, Vol. 213, 603–608. Pittsburgh, PA: MRS.

128. Miracle, D. B., S. Russell, and C. C. Law. 1989. Slip system modification in NiAl. In *High Temperature Ordered Intermetallics III*, eds. C. T. Liu, A. I. Taub, N. S. Stoloff, and C. C. Koch, Materials Research Society Symposium Proceedings, Vol. 133, 225–230. Pittsburgh, PA: MRS.

129. Field, R. D., D. F. Lahrman, and R. Darolia. 1991. The effect of alloying on slip systems in (001) oriented NiA1 single-crystals. *Acta Metall. Mater.* 39:2961–2969.

130. Darolia, R., D. Lahrman, and R. Field. 1992. The effect of iron, gallium and molybdenum on the room-temperature tensile ductility of NiAl. *Scr. Metall. Mater.* 26:1007–1012.

131. Field, R. D., D. F. Lahrman, and R. Darolia. 1991. Slip systems in (001) oriented NiA1 single-crystals. *Acta Metall. Mater.* 39:2951–2959.

132. Vandervoort, R. R., A. K. Mukherjee, and J. E. Dorn. 1966. Elevated temperature deformation mechanisms in β-NiAl. *Trans. ASM* 59:930–944.

133. Rudy, M. and G. Sauthoff. 1986. Dislocation creep in the ordered intermetallic (Fe, Ni)Al phase. *Mater. Sci. Eng.* 81:525–530.

134. Sauthoff, G. 1989. Intermetallic phases: Materials development and prospects. *Z. Metallk.* 80(5):337–344.

135. Polvani, R. S., W. S. Tzeng, and P. R. Strutt. 1976. High-temperature creep in a semi-coherent NiAl-Ni$_2$AlTi alloy. *Metall. Trans. A* 7:33–40.

136. Whittenberger, J. D., E. Arzt, and M. J. Luton. 1990. Preliminary investigation of a NiAl composite prepared by cryomilling. *J. Mater. Res.* 5:271–277.

137. Jha, S. C., R. Ray, and D. J. Gaydosh. 1989. Dispersoids in rapidly solidified B2 nickel aluminides. *Scr. Metall.* 23:805–810.

138. Yaney, D. L. and W. D. Nix. 1988. Mechanisms of elevated-temperature deformation in the B2 aluminides NiAl and CoAl. *J. Mater. Sci.* 23:3088–3098.

139. Hancock, G. F. and B. R. McDonell. 1971. Diffusion in the intermetallic compound NiAl. *Phys. Stat. Sol. A* 4:143–150.

140. Yamaguchi, M. and H. Inui. 1994. Al$_3$Ti and its L1$_2$ variations. In *Intermetallic Compounds, Vol. 2, Practice*, eds. J. H. Westbrook and R. L. Fleischer, 147–173. Chichester, UK: Wiley.

141. Vasudevan, V. K., R. Wheeler, and H. L. Fraser. 1989. The mechanisms of plastic deformation of rapidly solidified Al$_3$Ti and Al$_{67}$Ni$_8$Ti$_{25}$ intermetallic compounds. In *High Temperature Ordered Intermetallics III*, eds. C. T. Liu, A. I. Taub, N. S. Stoloff, and C. C. Koch, Materials Research Society Symposium Proceedings, Vol. 133:705–710. Pittsburgh, PA: MRS.

142. Hug, G., J. Douin, and P. Veyssiere. 1989. Surface energies of planar defects and dislocation processes in Al$_3$Ti and TiAl. In *High Temperature Ordered Intermetallics III*, eds. C. T. Liu, A. I. Taub, N. S. Stoloff, and C. C. Koch, Materials Research Society Symposium Proceedings, Vol. 133, 125–130. Pittsburgh, PA: MRS.

143. Morris, D. G. and R. Lerf. 1991. Plastic deformation of the intermetallic Al$_3$Ti. *Philos. Mag. A* 63:1195–1206.

144. Umakoshi, Y., M. Yamaguchi, T. Yamane, and T. Hirano. 1988. Deformation and improvement of ductility of the intermetallic compounds Al$_3$V and Al$_3$(V,Ti). *Philos. Mag. A* 58(4):651–665.

145. Khantha, M., V. Vitek, and D. P. Pope. 1990. Miniaturized disk bend testing and microstructure of 3.8 MeV Zr^{3+} irradiated Zr$_3$Al. In *High Temperature Ordered Intermetallics IV*, eds. L. A. Johnson, J. O. Stiegler, and D. P. Pope, Materials Research Society Symposium Proceedings, Vol. 213:763–768. Pittsburgh, PA: MRS.

146. Khantha, M., V. Vitek, and D. P. Pope. 1992. An atomistic study of dislocations and their mobility in a model D0$_{22}$ alloy. *Mater. Sci. Eng. A* 152:89–94.

147. Yamaguchi, M. and Y. Umakoshi. 1990. The deformation-behavior of intermetallic superlattice compounds. *Prog. Mater. Sci.* 34:1–148.

148. Wheeler, R., V. K. Vasudevan, and H. L. Fraser. 1990. Factors influencing the deformation mechanisms in the intermetallic compounds Al$_3$Ti and Al$_3$V. *Philos. Mag. Lett.* 62:143–151.

149. Seibold, A. 1981. Phase-equilibria in the ternary-system Ti-Fe-O and Ti-Al-Fe. *Z. Metallk.* 72:712–719.

150. Mazdiyasni, S., D. B. Miracle, D. M. Dimiduk, M. G. Mendiratta, and P. R. Subramanian. 1989. High-temperature phase-equilibria of the L1$_2$ composition in the Al-Ti-Ni, Al-Ti-Fe, and Al-Ti-Cu systems. *Scr. Metall.* 23:327–331.

151. Mabuchi, H., K. I. Hirukawa, and Y. Nakayama. 1989. Formation of structural L12 compounds in TiAl$_3$-base alloys containing Mn. *Scr. Metall.* 23:1761–1765.

152. Sparks, C. G., W. D. Porter, J. H. Schneibel, W. C. Oliver, and C. G. Golec. 1991. In *Alloy Phase Stability and Design*, eds. G. M. Stocks, D. P. Pope, and A. F. Giamei, Materials Research Society Symposium Proceedings, Vol. 186, 175. Pittsburgh, PA: MRS.

153. Powers, W. O. and J. A. Wert. 1990. Microstructure, deformation, and fracture characteristics of an $Al_{67}Pd_8Ti_{25}$ intermetallic alloy. *Metall. Trans. A* 21:145–151.

154. Kumar, K. S. and S. A. Brown. 1992. Compression behavior of quaternary and higher-order solid-solution L12 trialuminides. *Scr. Metall. Mater.* 26:197–202.

155. Srinivasan, S., S. R. Chen, and R. B. Schwarz. 1992. Synthesis of Al/Al_3Ti two-phase alloys by mechanical alloying. *Mater. Sci. Eng. A* 153:691–695.

156. Dipietro, M. S., K. S. Kumar, and J. D. Whittenberger. 1991. Compression behavior of TiB_2-particulate-reinforced composites of $Al_{22}Fe_3Ti_8$. *J. Mater. Res.* 6:530–538.

157. Whittenberger, J. D., K. S. Kumar, S. Brown, M. S. DiPietro, and S. C. Farmer. In *Light Weight Alloys for Aerospace Applications II*, eds. E.-W. Lee and N.-J. Kim, 327. Warrendale, PA: TMS.

158. Lipsitt, H. A. 1985. Titanium aluminides: An overview. In *High Temperature Ordered Intermetallics I*, eds. C. C. Koch, C. T. Liu, and N. S. Stoloff, Materials Research Society Symposium Proceedings, Vol. 39, 351–364. Pittsburgh, PA: MRS.

159. Huang, S. C. and J. C. Chestnut. 1994. Gamma TiAl and its alloys. In *Intermetallic Compounds, Vol. 2, Practice*, eds. J. H. Westbrook and R. L. Fleischer, 73–90. Chichester, UK: Wiley.

160. Feng, C. R., D. J. Michel, and C. R. Crowe. 1990. Antiphase domains and (011) twins in TiAl. *Philos. Mag. Lett.* 61:95–100.

161. Inui, H., A. Nakamura, M. H. Oh, and M. Yamaguchi. 1991. High-resolution electron-microscope study of lamellar boundaries in Ti-rich TiAl polysynthetically twinned crystals. *Ultramicroscopy* 39:268–278.

162. Schectman, D., M. J. Blackburn, and H. A. Lipsitt. 1974. The plastic deformation of TiAl. *Metall. Trans. A* 5:1373–1381.

163. Huang, S. C. and D. S. Shih. 1991. Microstructure-property correlation in TiAl-base alloys. In *Microstructure/Property Relationships in Titanium Aluminides and Alloys*, eds. Y. W. Kim and R. R. Boyer, 105. Warrendale, PA: TMS.

164. Takeyama, M. 1992. Microstructural evolution and tensile properties of titanium-rich TiAl alloy. *Mater. Sci. Eng. A* 152:269–276.

165. Marcinkowski, M. J., N. Brown, and R. M. Fisher. 1961. Dislocation configurations in $AuCu_3$ and AuCu type superlattices. *Acta Metall.* 9:129–137.

166. Greenberg, B. A. 1970. Splitting of superdislocations in ordered CuAu-type structure. *Phys. Stat. Sol.* 42(1):459–468.

167. Greenberg, B. A. 1973. New types of superdislocations in ordered $L1_0$ structure. *Phys. Stat. Sol.* 55(1):59–67.

168. Court, S. A., V. K. Vasudevan, and H. L. Fraser. 1990. Deformation mechanisms in the intermetallic compound TiAl. *Philos. Mag. A* 61:141–158.

169. McAndrew, J. B. and H. D. Kessler. 1956. Ti-36%A1 as base for high temperature alloys. *J. Met.* 8:1348–1353.

170. Huang, S. C. and E. L. Hall. 1991. Plastic-deformation and fracture of binary TiAl-base alloys. *Metall. Trans. A* 22:427–439.

171. Kawabata, T., T. Kanai, and O. Izumi. 1985. Positive temperature dependence of the yield stress in TiAl $L1_0$ type superlattice intermetallic compound single crystals at 293–1273 K. *Acta Metall.* 33:1355–1366.

172. Imayev, V. M., G. A. Salishchev, R. M. Imayev, M. R. Shagiev, N. K. Gabdullin, and A. V. Kuznetsov. 1997. An approach to ductility improvement of TiAl and titanium aluminides based on microstructure control. In *Structural Intermetallics 1997*, eds. M. V. Nathal, R. Darolia, C. T. Liu, P. L. Martin, D. B. Miracle, R. Wagner, and M. Yamaguchi, 505–514. Warrendale, PA: TMS.
173. Imayev, V. M., T. Oleneva, R. Imayev, H.-J. Christ, and H.-J. Fecht. 2012. Microstructure and mechanical properties of low and heavy alloyed γ-TiAl + α_2-Ti$_3$Al based alloys subjected to different treatments. *Intermetallics* 26:91–97.
174. Schwaighofer, E., H. Clemens, S. Mayer, J. Lindemann, J. Klose, W. Smarsly, and V. Güther. 2014. Microstructural design and mechanical properties of a cast and heat treated intermetallic multi-phase γ-TiAl based alloy. *Intermetallics* 44:128–140.
175. Huang, S. C. 1992. Microstructures and property tradeoffs in wrought TiAl-base alloys. *Metall. Trans. A* 23:375–377.
176. Hall, E. L. and S. C. Huang. 1989. Stoichiometry effects on the deformation of binary TiAl alloys. *J. Mater. Res.* 4:595–602.
177. Vasudevan, V. K., M. A. Stucke, S. A. Court, and H. L. Fraser. 1989. The influence of second phase Ti$_3$Al on the deformation mechanisms in TiAl. *Philos. Mag. Lett.* 59:299–307.
178. Hsiung, H. M., A. J. Schwartz, and T. G. Nieh. 2004. *In situ* TEM observations of interface sliding and migration in a refined lamellar TiAl alloy. *Intermetallics* 12:727–732.
179. Behr, R., A. Wanner, H. Clemens, and W. Glatz. 1995. Fatigue crack growth characteristics and fracture toughness of γ-TiAl base alloy sheet material. In *High Temperature Ordered Intermetallic Alloys VI*, eds. I. Baker, S. Hanada, J. Horton, R. D. Noebe, and D. S. Schwartz, Materials Research Society Symposium Proceedings, Vol. 364, 781–786. Pittsburgh, PA: MRS.
180. Chen, Y., H. Niu, F. Kong, and S. Xiao. 2011. Microstructure and fracture toughness of a β phase containing TiAl alloy. *Intermetallics* 2011:1405–1410.
181. Chan, K. S. 1992. Understanding fracture toughness in gamma TiAl. *J. Met.* 44(5):30–38.
182. Kim, S.-W., Y.-S. Na, J.-T. Yeom, S. E. Kim, and Y. S. Choi. 2014. An *in situ* transmission electron microscopy study on room temperature ductility of TiAl alloys with fully lamellar microstructure. *Mater. Sci. Eng. A* 589:140–145.
183. Deve, H. E., A. G. Evans, and D. S. Shih. 1992. A high toughness γ-titanium aluminide. *Acta Metall. Mater.* 40:1259–1265.
184. Mitao, S., S. Tsuyama, and K. Minakawa. 1991. Effects of microstructure on the mechanical properties and fracture of γ-base titanium aluminides. *Mater. Sci. Eng. A* 143:51–62.
185. Zhu, H., D. Y. Seo, K. Maruyama, and P. Au. 2008. Effect of microstructure on creep deformation of 45XD TiAl alloy at low and high stresses. *Mater. Sci. Eng. A.* 483–484:533–536.
186. Takahashi, T., H. Nagai, and H. Oikawa. 1990. Effects of grain size on creep behaviour of Ti-50 mol.% Al intermetallic compound at 1100 K. *Mater. Sci. Eng. A* 128:195–200.
187. Kim, Y.-W. and S.-L. Kim. 2014. Effects of microstructure and C and Si additions on elevated temperature creep and fatigue of gamma TiAl alloys. *Intermetallics* 53:92–101.

188. Karthikeyan, S. and M. J. Mills. 2005. The role of microstructural stability on compression creep of fully lamellar γ-TiAl alloys. *Intermetallics* 13:985–992.
189. Blackburn, M. J. and M. P. Smith. 1981. Titanium alloys of the TiAl types. US Patent 4294615.
190. Es-Souni, M., A. Bartels, and R. Wagner. 1995. Creep behavior of a fully transformed near γ-TiAl alloy Ti-48Al-2Cr. *Acta Metall. Mater.* 43(1):153–161.
191. Hsiung, L. M., A. J. Schwartz, and T. G. Nieh. 2004. *In situ* TEM observations of interface sliding and migration in a refined lamellar TiAl alloy. *Intermetallics* 12: 727–732.
192. Kroll, S., H. Mehrer, N. Stolwijk, C. Herzig, R. Rosenkranz, and G. Frommeyer. 1992. Titanium self-diffusion in the intermetallic compound γ-TiAl. *Z. Metallk.* 83(8): 591–595.
193. Kampe, S. L., J. D. Bryant, and L. Christodoulou. 1991. Creep deformation of TiB₂ reinforced near-γ titanium aluminides. *Metall. Trans. A* 22:447–454.
194. Hayes, R. W. and B. London. 1992. On the creep deformation of a cast near gamma TiAl alloy Ti48Al1Nb. *Acta Metall. Mater.* 40:2167–2175.
195. Hayes, R. W. and P. L. Martin. 1995. Tension creep of wrought single phase γ-TiAl. *Acta Metall. Mater.* 43(7):2761–2772.
196. Gorzel, A. and G. Sauthoff. 1999. Diffusion creep of intermetallic TiAl alloys. *Intermetallics* 7:371–380.
197. Morris, M. A. and T. Lipe. 1997. Creep deformation of duplex and lamellar TiAl alloys. *Intermetallics* 5:329–337.
198. Lundström, D. and B. Karlsson. 1997. Creep behavior of a cast γ-TiAl based alloy for gas turbine applications. In *Structural Intermetallics 1997*, eds. M. V. Nathal, R. Darolia, C. T. Liu, P. L. Martin, D. B. Miracle, R. Wagner, and M. Yamaguchi, 461–468. Warrendale, PA: TMS.
199. Masahashi, N. J., Y. Mizuhara, M. Matsuo, T. Hanamura, M. Kimura, and K. Hashimoto. 1991. High temperature deformation behavior of titanium aluminide based gamma plus beta micro-duplex alloy. *ISIJ Intl.* 31:728–737.
200. Imayev, R. M. and V. M. Imayev. 1991. Mechanical behavior of TiAl submicrocrystalline intermetallic compound at elevated temperatures. *Scr. Metall. Mater.* 25:2041–2046.
201. Vanderschueren, D., M. Nobuki, and M. Nakamura. 1993. Superplasticity in a vanadium alloyed gamma plus beta phased Ti-Al intermetallic. *Scr. Metall. Mater.* 28:605–610.
202. Nieh, T. G., L. M. Hsiung, and J. Wadsworth. 1999. Superplastic behavior of a powder metallurgy TiAl alloy with a metastable microstructure. *Intermetallics* 7:163–170.
203. Herzig, C, T. Przeorski, M. Friesel, F. Hisker, S. Divinski, and Y.-W. Kim. 2001. Tracer solute diffusion of Nb, Zr, Cr, Fe, and Ni in γ-TiAl: Effect of preferential site occupation. *Intermetallics* 9:461–472.
204. Kainuma, R., Y. Fujita, H. Mitsui, I. Ohnuma, and K. Ishida. 2000. Phase equilibria among α (hcp), β (bcc) and γ (L1₀) phases in Ti-Al base ternary alloys. *Intermetallics* 8:855–867.
205. Qiu, C., Y. Liu, W. Zhang, B. Liu, and X. Liang. 2012. Development of a Nb-free TiAl-based intermetallics with a low-temperature superplasticity. *Intermetallics* 27:46–51.
206. Lin, D. and F. Sun. 2004. Superplasticity in large-grained TiAl alloy. *Intermetallics* 12:875–883.

207. Tsujimoto, T. and K. Hashimoto. 1989. Structure and properties of TiAl-base alloys containing Mn. In *High Temperature Ordered Intermetallic Alloys III*, eds. C. C. Koch, C. T. Liu, N. S. Stoloff, and A. I. Taub, Materials Research Society Symposium Proceedings, Vol. 133, 391–396. Pittsburgh, PA: MRS.

208. Huang, S. C. and E. L. Hall. 1991. The effects of Cr additions to binary TiAl-base alloys. *Metall. Trans. A* 22:2619–2627.

209. Veyssière, P. and G. Hug. 1989. The use of weak-beam dark-field TEM in the analysis of defects in intermetallics. In *Proceedings of International Symposium on Electron Microscopy in Plasticity and Fracture Research of Materials*, Dresden, October 1989.

210. Morinaga, M., J. Saito, N. Yukawa, and H. Adachi. 1990. Electronic effect on the ductility of alloyed TiA1 compound. *Acta Metall. Mater.* 38:25–29.

211. Chladil, H., H. Clemens, A. Otto, V. Güther, S. Kremmer, A. Bartels, and R. Gerling. 2006. Charakterisierung einer β-erstarrenden γ-TiAl-Basislegierung. *BHM Berg- und Hüttenmännische Monatshefte* 151:356–361.

212. Clemens, H., W. Wallgram, S. Kremmer, V. Güther, A. Otto, and A. Bartels. 2008. Design of novel β-solidifying TiAl alloys with adjustable β/B2-phase fraction and excellent hot workability. *Adv. Eng. Mater.* 10:707–713.

213. Clemens, H. and S. Mayer. 2013. Design, processing, microstructure, properties, and applications of advanced intermetallic TiAl alloys. *Adv. Eng. Mater.* 15:191–215.

214. Appel, F., J. D. H. Paul, and M. Oehring. 2011. *Gamma Titanium Aluminide Alloys: Science and Technology*. Weinheim, Germany: Wiley-VCH.

215. Appel, F., M. Oehring, and R. Wagner. 2000. Novel design concepts for gamma-base titanium aluminide alloys. *Intermetallics* 8:1283–1312.

216. Kawabata, T., M. Tadano, and O. Izumi. 1991. Effect of carbon and nitrogen on mechanical properties of TiAl alloys. *ISIJ Intl.* 31:1161–1167.

217. Tian, W. H. and M. Nemoto. 1997. Effect of carbon addition on the microstructures and mechanical properties of γ-TiAl alloys. *Intermetallics* 5:237–244.

218. Christoph, U., F. Appel, and R. Wagner. 1997. Dislocation dynamics in carbon-doped titanium aluminide alloys. *Mater. Sci. Eng. A* 239–240:39–45.

219. Denquin, A., S. Naka, A. Huguet, and A. Menand. 1993. Atom-probe investigation of the partitioning of interstitial elements in two-phase $\gamma + \alpha_2$ TiAl-based alloys. *Scr. Metall. Mater.* 28:1131–1136.

220. Menand, A., A. Huguet, and A. Nérac-Partaix. 1996. Interstitial solubility in γ and α_2 phases of TiAl-based alloys. *Acta Mater.* 44:4729–4737.

221. Menand, A., H. Zapolsky-Tatarenko, and A. Nérac-Partaix. 1998. Atom-probe investigations of TiAl alloys. *Mater. Sci. Eng. A* 250:55–64.

222. Scheu, C., E. Stergar, M. Schober, L. Cha, H. Clemens, and A. Bartels. 2009. High carbon solubility in a γ-TiAl-based Ti-45Al-5Nb-0.5C alloy and its effect on hardening. *Acta Mater.* 57:1504–1511.

223. Lipsitt, H. A. 1985. Titanium aluminides: An overview. In *High Temperature Ordered Intermetallic Alloys I*, eds. C. C. Koch, C. T. Liu, and N. S. Stoloff, Materials Research Society Symposium Proceedings, Vol.39, 351–364. Pittsburgh, PA: MRS.

224. Blackburn, M. J. and M. P. Smith. 1978. Research to conduct an exploratory and analytical investigation of alloys. AFML-TR-78-18, 1978. Wright-Patterson AFB, OH: Air Force Materials Laboratory.

225. Banerjee, D. 1994. Ti_3Al and its alloys. In *Intermetallic Compounds, Vol. 2, Practice*, eds. J. H. Westbrook and R. L. Fleischer, 91. Chichester, UK: Wiley.

226. Marquardt, B. J., G. K. Scarr, J. C. Chestnut, C. G. Rhodes, and H. L. Fraser. 1989. Research and development for improved toughness aluminides. WRDC-TR-89-4133. Wright Research and Development Center. Dayton, OH: Wright-Patterson Air Force Base.

227. Blackburn, M. J. and M. P. Smith. 1982. Research and development on composition and processing of titanium aluminide alloys for turbine engines. AFWAL-TR-82-4086. Air-Force-Wright-Aeronautical-Laboratories. Dayton, OH: Wright-Patterson Air Force Base.

228. Blackburn, M. J. and M. P. Smith. 1989. Improved toughness alloys based on aluminides. WRDC-TR-89-4095. Wright Research and Development Center. Dayton, OH: Wright-Patterson Air Force Base.

229. Rowe, R. G., D. G. Konitzer, A. P. Woodfield, and J. C. Chestnut. 1991. Tensile and creep behavior of ordered orthorhombic Ti_2AlNb-based alloys. In *High Temperature Ordered Intermetallic Alloys IV*, eds. L. A. Johnson, D. P. Pope, and J. O. Stiegler, Materials Research Society Symposium Proceedings, Vol. 213, 703. Pittsburgh, PA: MRS.

230. Minonishi, Y. 1990. Fourfold dissociation of 1/3(112bar0) superlattice dislocations in Ti_3Al. *Philos. Mag. Lett.* 62:153–158.

231. Minonishi, Y. 1991. Plastic-deformation of single-crystals of Ti_3Al with Do19 structure. *Philos. Mag. A* 63:1085–1093.

232. Mishra, R. S. and D. Banerjee. 1990. Microstructure and steady state creep in Ti-24Al-11Nb. *Mater. Sci. Eng. A* 130:151–164.

233. Thomas, M., A. Vassel, and P. Veyssière. 1988. In *Proceedings of 6th world conference on titanium*, eds. P. Lacombe, R. Tricot, and G. Beranger, *J. de Physique*, Vol. II, 1085. Les Ulis, France.

234. Thomas, M., A. Vassel, and P. Veyssière. 1989. C-slip in Ti_3Al. *Philos. Mag.* 59:1013–1026.

235. Lofvander, J. P. A., S. A. Court, M. H. Loretto, and H. L. Fraser. 1989. Mobility of C dislocations in Ti_3Al. *Philos. Mag. Lett.* 60:111–116.

236. Banerjee, D., A. K. Gogia, and T. K. Nandy. 1990. Deformation structure in a Ti-24Al-11Nb alloy. *Metall. Trans. A* 21:627–639.

237. Gogia, A. K., D. Banerjee, and T. K. Nandy. 1990. Structure, tensile deformation, and fracture of a Ti_3Al-Nb alloy. *Metall. Trans. A* 21:609–625.

238. Banerjee, D., R. G. Rowe, and E. L. Hall. 1991. Deformation of the orthorhombic phase in Ti-Al-Nb alloys. In *High Temperature Ordered Intermetallic Alloys IV*, eds. L. A. Johnson, D. P. Pope and J. O. Stiegler, Materials Research Society Symposium Proceedings, Vol. 213, 285–290. Pittsburgh, PA: MRS.

239. Gogia, A. K., T. K. Nandy, K. Muraleedharan, and D. Banerjee. 1992. The effect of heat treatment and niobium content on the room temperature tensile properties and microstructure of Ti_3Al-Nb alloys. *Mater. Sci. Eng. A* 159:73–86.

240. Mendiratta, M. G. and H. A. Lipsitt. 1980. Steady-state creep-behavior of Ti_3Al-base intermetallics. *J. Mater. Sci.* 15:2985–2990.

241. Nagae, Y., K. Maruyama, and H. Oikawa. 1996. General creep characteristics of single-phase polycrystalline intermetallics. *Mater. Sci. Eng. A* 213:32–36.

242. Cho, W., A. W. Thompson, and J. C. Williams. 1990. Creep-behavior of Ti-25al-10Nb-3V-1Mo. *Metall. Trans. A* 21:641–657.

243. Hayes, R. W. 1991. On the creep-behavior of the Ti_3Al titanium aluminide Ti-25Al-10Nb-3V-1mo. *Acta Metall. Mater.* 39:569–577.

244. Soboyejo, W. O., R. J. Ledrich, and D. S. Schwartz. 1991. In *Microstructure/Property Relationships in Titanium Aluminides and Alloys*, eds. Y. W. Kim and R. R. Boyers, 407. Warrendale, PA: TMS.

245. Nandy, T. K. and D. Banerjee. 2000. Deformation mechanisms in the O phase. *Intermetallics* 8:1269–1282.

246. Nandy, T. K. and D. Banerjee. 1997. The mechanical behavior of the intermetallic Ti$_2$AlNb. In *Structural Intermetallics 1997*, eds. M. V. Nathal, R. Darolia, C. T. Liu, P. L. Martin, D. B. Miracle, R. Wagner, and M. Yamaguchi, 777–786. Warrendale, PA: TMS.

247. Rusing, J., and C. Herzig. 1996. Concentration and temperature dependence of titanium self-diffusion and interdiffusion in the intermetallic phase Ti$_3$Al. *Intermetallics* 4:647–657.

248. Chan, K. S. 1990. Fracture and toughening mechanisms in an alpha-2 titanium aluminide alloy. *Metall. Trans. A* 21:2687–2699.

249. Chan, K. S. 1992. Influence of microstructure on intrinsic and extrinsic toughening in an alpha-2 titanium aluminide alloy. *Metall. Trans. A* 23:183–199.

250. Chan, K. S. 1993. Toughening mechanisms in titanium aluminides. *Metall. Trans. A* 24:569–583.

251. Davidson, D. A. 1991. Titanium alloys: Fatigue crack growth mechanisms and crack tip micromechanics. In *Microstructure/Property Relationships in Titanium Aluminides and Alloys*, eds. Y. W. Kim and R. R. Boyers, 447. Warrendale, PA: TMS.

252. Davidson, D. L., J. B. Campbell, and R. A. Page. 1991. The initiation and growth of fatigue cracks in a titanium aluminide alloy. *Metall. Trans. A* 22:377–391.

253. Stoloff, N. S. 1998. Iron aluminides: Present status and future prospects. *Mater. Sci. Eng. A* 258:1–14.

254. Lin, D. L., A. D. Shan, and D. Q. Li. 1994. Superplasticity in Fe$_3$Al-Ti alloy with large grains. *Scr. Metall. Mater.* 31:1455–1460.

255. Li, D. Q., A. D. Shan, Y. Liu, and D. L. Lin. 1995. Study of superplastic deformation in an FeAl based alloy with large grains. *Scr. Metall. Mater.* 33:681–685.

256. Lin, D., T. L. Lin, A. Shan, and M. Chen. 1996. Superplasticity in large-grained Fe$_3$Al alloys. *Intermetallics* 4:489–496.

257. Lin, D. L., D. Q. Li, and Y. Liu. 1998. Superplasticity in large-grained FeAl-based intermetallic alloys. *Intermetallics* 6:243–256.

258. Lin, D. and Y. Liu. 1999. Microstructural evolution and mechanisms of superplasticity in large-grained iron aluminides. *Mater. Sci. Eng. A* 268:83–89.

259. Sikka, V. K. 1991. Processing and fabrication of Fe$_3$Al-based alloys. In *Heat Resistant Materials*, eds. K. Natesan and D. J. Tillak, 149–158. Materials Park, OH: ASM.

260. McKamey, C. G. and C. T. Liu. 1990. Chromium addition and environmental embrittlement in Fe$_3$Al. *Scr. Metall. Mater.* 24(11):2119–2122.

261. Liu, C. T., C. G. McKamey, and E. H. Lee. 1990. Environmental effects on room-temperature ductility and fracture in Fe$_3$Al. *Scr. Metall. Mater.* 24:385–389.

262. Liu, C. T. and E. P. George. 1991. Effect of aluminum concentration and boron dopant on environmental embrittlement in FeAl aluminides. In *High Temperature Ordered Intermetallics IV*, eds. L. A. Johnson, D. P. Pope, and J. O. Stiegler, Materials Research Society Symposium Proceedings, Vol. 213, 527–532. Pittsburgh, PA: MRS.

263. Stoloff, N. S. and C. T. Liu. 1994. Environmental embrittlement of iron aluminides. *Intermetallics* 2:75–87.
264. McKamey, C. G. 1996. Iron aluminides. In *Physical Metallurgy and Processing of Intermetallic Compounds*, eds. N. S. Stoloff and V. K. Sikka, 351–391. London: Chapman & Hall.
265. Sikka, V. K., S. Viswanathan, and C. G. McKamey. 1993. In *Structural Intermetallics*, eds. R. Darolia, J. J. Lewandowski, C. T. Liu, P. L. Martin, D. B. Miracle, and M. V. Nathal, 483–491. Warrendale, PA: TMS.
266. McCamey, C. G., J. A. Horton, and C. T. Liu. 1987. Effect of aluminum addition on ductility and yield strength of Fe_3Al alloys with 0.5 wt.% TiB_2. In *High Temperature Ordered Intermetallic Alloys II*, eds. N. S. Stoloff, C. C. Koch, C. T. Liu, and O. Izumi, Materials Research Society Symposium Proceedings, Vol. 81, 321–328. Pittsburgh, PA: MRS.
267. Fu, C. L. and G. S. Painter. 1991. 1st principles investigation of hydrogen embrittlement in FeAl. *J. Mater. Res.* 6:719–723.
268. Inouye, H. 1985. Effects of $D0_3$ transitions on the yield behavior of Fe-Al alloys. In *High-Temperature Ordered Intermetallics—I*, eds. C. C. Koch, C. T. Liu, and N. S. Stoloff, Materials Research Society Symposium Proceedings, Vol. 39, 255–262. Pittsburgh, PA: MRS.
269. Fu, C. L. and M. H. Yoo. 1992. Deformation-behavior of B2 type aluminides—FeAl and NiAl. *Acta Metall. Mater.* 40:703–711.
270. Ko, S. H., R. Gnanamoorthy, and S. Hanada. 1997. The effect of environment on tensile ductility and fracture toughness of iron aluminides. *Mater. Sci. Eng. A* 222:133–139.
271. Yamagata, T. and H. Yoshida. 1973. Deformation behavior of FeAl single crystals. *Mater. Sci. Eng.* 12:95–100.
272. Ray, I. L. F., R. C. Crawford, and D. J. Cockayne. 1970. Weak-beam technique applied to superlattice dislocations in an iron-aluminium alloy. *Philos. Mag.* 21:1027–1032.
273. Mendiratta, M. G., S. Ehlers, D. Dimiduk, W. Kerr, S. Mazdiyasini, and H. A. Lipsitt. 1987. A review of recent developments in iron aluminides. In *High Temperature Ordered Intermetallic Alloys II*, eds. N. S. Stoloff, C. C. Koch, C. T. Liu, and O. Izumi, Materials Research Society Symposium Proceedings, Vol. 81, 393–404. Pittsburgh, PA: MRS.
274. Chatterjee, D. K. and M. G. Mendiratta. 1981. *J. Met.* 33(12):5.
275. Schmidt, B., P. Nagpal, and I. Baker. 1989. Annealing studies of B2 FeAl. In *High Temperature Ordered Intermetallic Alloys III*, eds. C. C. Koch, C. T. Liu, N. S. Stoloff, and A. I. Taub, Materials Research Society Symposium Proceedings, Vol. 133, 755–760. Pittsburgh, PA: MRS.
276. Gaydosh, D. J. and M. V. Nathal. 1986. Tensile behavior of Fe-40Al alloys with B and Zr additions. In Annual Meeting of TMS-AIME, New Orleans, LA, 2–6 March, 86N25453, NASA-TM-87290, E-2997, CONF-860340-7. Available from NTIS, PC A03/MF A01.
277. Mendiratta, M. G., H. M. Kim, and H. A. Lipsitt. 1984. Slip directions in B2 Fe-Al alloys. *Metall. Trans. A* 15:395–399.
278. Mendiratta, M. G. and H. A. Lipsitt. 1985. $D0_3$-domain structures in Fe_3Al-X alloys. In *High Temperature Ordered Intermetallic Alloys I*, eds. C. C. Koch, C. T. Liu, and N. S. Stoloff, Materials Research Society Symposium Proceedings, Vol. 39, 155–162. Pittsburgh, PA: MRS.

279. Baker, I. and D. J. Gaydosh. 1987. The microstructure and tensile properties of extruded melt-spun ribbons of iron-rich B2 FeAl. In *High Temperature Ordered Intermetallic Alloys II*, eds. N. S. Stoloff, C. C. Koch, C. T. Liu, and O. Izumi, Materials Research Society Symposium Proceedings, Vol. 81, 315–320.

280. Jung, I., M. Rudy, and G. Sauthoff. 1987. Creep in ternary B2 aluminides and other intermetallic phases. In *High Temperature Ordered Intermetallic Alloys II*, eds. N. S. Stoloff, C. C. Koch, C. T. Liu, and O. Izumi, Materials Research Society Symposium Proceedings, Vol. 81, 263–274. Pittsburgh, PA: MRS.

281. Whittenberger, J. D. 1983. Influence of thermomechanical processing on elevated temperature slow plastic flow properties of B2 aluminide Fe-39.8 at.% Al. *Mater. Sci. Eng.* 57:77–85.

282. Whittenberger, J. D. 1986. The influence of grain size and composition on slow plastic flow in FeAl between 1100 and 1400 K. *Mater. Sci. Eng.* 77:103–113.

283. Vedula, K. and J. R. Stephens. 1987. B2 aluminides for high temperature applications. In *High Temperature Ordered Intermetallic Alloys II*, eds. N. S. Stoloff, C. C. Koch, C. T. Liu, and O. Izumi, Materials Research Society Symposium Proceedings, Vol. 81, 381–392. Pittsburgh, PA: MRS.

284. Morris, D. G., M. A. Munoz-Morris, and C. Baudin. 2004. The high-temperature strength of some Fe_3Al alloys. *Acta Mater.* 52:2827–2836.

285. McKamey, C. G., J. A. Horton, and C. T. Liu. 1989. Effect of chromium on properties of Fe_3Al. *J. Mater. Res.* 4:1156–1163.

286. Titran, R. H., K. Vedula, and G. Anderson. 1985. The effect of surface oxide films on the mechanical behavior of NiAl. In *High Temperature Ordered Intermetallic Alloys I*, eds. C. C. Koch, C. T. Liu, and N. S. Stoloff, Materials Research Society Symposium Proceedings, Vol. 39, 319–326. Pittsburgh, PA: MRS.

287. Baligidad, R. G., U. Prakash, V. R. Rao, P. K. Rao, and N. B. Ballal. 1996. Effect of hot working on room temperature mechanical properties and stress rupture behaviour of ESR processed Fe-16 wt% Al intermetallic alloys. *ISIJ Intl.* 36:1215–1221.

288. Prakash, U. and G. Sauthoff. 2001. Machinable iron aluminides containing carbon. *Scr. Mater.* 44:73–78.

289. Baligidad, R. G., U. Prakash, V. R. Rao, P. K. Rao, and N. B. Ballal. 1996. Processing of Fe_3Al based intermetallic alloys through electroslag remelting. *ISIJ Intl.* 36:1448–1452.

290. Baligidad, R. G., U. Prakash, V. R. Rao, P. K. Rao, and N. B. Ballal. 1996. Effect of carbon content on mechanical properties of electroslag remelted Fe_3Al based intermetallic alloys. *ISIJ Intl.* 36:1453–1458.

291. Baligidad, R. G., U. Prakash, and A. Radhakrishna. 1997. High temperature tensile and creep properties of a cast aim and ESR intermetallic alloy based on Fe_3Al. *Mater. Sci. Eng. A* 231:206–210.

292. Parvathavarthini, N., U. Prakash, and R. K. Dayal. 2002. Effect of carbon addition on hydrogen permeation in an Fe_3Al-based intermetallic alloy. *Intermetallics* 10:329–332.

293. Prakash, U., N. Parvathavarthini, and R. K. Dayal. 2007. Effect of composition on hydrogen permeation in Fe-Al alloys. *Intermetallics* 15:17–19.

294. Morris, D. G. and S. Gunther. 1996. Strength and ductility of Fe-40Al alloy prepared by mechanical alloying. *Mater. Sci. Eng. A* 208:7–19.

295. Schneibel, J. H., P. Grahle, and J. Rösler. 1992. Mechanical alloying of FeAl with Y_2O_3. *Mater. Sci. Eng. A* 153:684–690.

296. Stoloff, N. S., D. A. Alven, and C. G. McKamey. 1997. An overview of Fe₃Al alloy development with emphasis on creep and fatigue. In *International Symposium on Nickel and Iron Aluminides*, eds. S. C. Deevi, V. K. Sikka, P. J. Maziasz, and R. W. Cahn, 65. Materials Park, OH: ASM International.

297. Palm, M. and G. Inden. 1995. Experimental determination of phase equilibria in the Fe-Al-C system. *Intermetallics* 3:443–454.

298. Pang, L. and K. S. Kumar. 1998. Mechanical behavior of an Fe-40Al-0.6C alloy. *Acta Mater.* 46:4017–4028.

299. Pang, L. X., D. J. Xing, A. Q. Zhang, J. Xu, J. S. Zhang, and R. H. Fan. 2010. Carbon nanotube reinforced intermetallic. *Adv. Comp. Mater.* 19:261–267.

9

Aluminides: Oxidation Behavior

9.1 Introduction

Protection against oxidation in aluminides and aluminide matrix composites is expected to be due to the formation of alumina scale. Of course, the thermodynamic stability of the oxide scale, along with the strength of its adhesion to the underlying alloy substrate, needs to be considered. The difficulty in formation of protective Al_2O_3 scale arises because of the competitive oxidation of other alloy constituents, which leads to formation of multiple oxide phases. As a result, internal growth stresses are generated. The scale continuity is interrupted due to spallation or loss of volatile oxidation products. The products of oxidation are also found to vary with temperature. The mechanisms of oxide-scale formation and spallation, as well as the factors influencing such phenomena and also the issues of scale adherence to the alloy underneath, are discussed in this chapter.

9.2 Nickel Aluminides

In Ni-aluminide-based intermetallic alloys, the Al_2O_3 scale grows by inward diffusion of oxygen anions through the grain boundaries and outward diffusion of Al^{3+} through the grain body, as schematically depicted in Figure 9.1a. If the activity of Al is not sufficiently high, oxides of Ni (NiO) as well as other reactive constituents also form simultaneously, and their growth rate is controlled by the cationic diffusion through the oxide scale. Compressive stresses are generated by volume expansion due to formation of different oxides within the scale as well as mismatch between the coefficients of thermal expansion (CTEs) of the oxide scale and the alloy substrate underneath. Lateral growth of oxide scale takes place as oxides are formed inside the scale due to diffusion of O^{2-} through the grain boundaries. The compressive stresses lead to buckling of the oxide scale, made of brittle

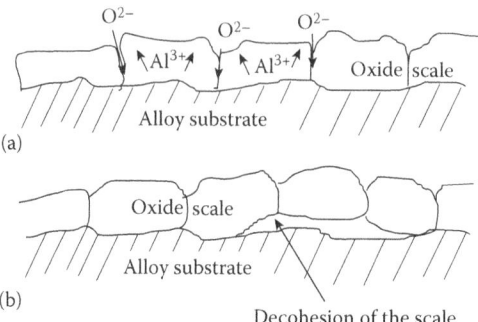

FIGURE 9.1
Schematic illustration of oxide scale formed on an aluminide-based intermetallic alloy showing: (a) diffusion paths of metallic cations and oxygen anions; (b) decohesion and buckling of the oxide scale due to formation of voids at the alloy–oxide interface.

constituents, followed by decohesion at the alloy–oxide interface or cracking and spallation, as depicted schematically in Figure 9.1b.

9.2.1 Ni₃Al

Oxygen embrittlement is considered to be the main reason for low ductility and brittle fracture in Ni_3Al at elevated temperatures. Prior oxidation to form a protective oxide scale on the surface has been found to be partially beneficial for improving the resistance against embrittlement.[1,2] However, this scale cracks on application of a very small amount of strain. The oxidation resistance of Ni_3Al is not considered satisfactory because of the unprotective nature of the oxide scale formed. The variation of mass gain with time for this material follows a parabolic rate law, exhibiting a significant increase in the rate of oxidation between 1100°C and 1200°C.[3] At 900°C, the oxide scale formed on Ni_3Al contains an outer layer of NiO, an intermediate layer of $NiAl_2O_4$, and an inner layer of Al_2O_3. In contrast, Al_2O_3 is the only oxidation product formed at 1200°C. These oxide scales are reported to undergo spallation on being subjected to thermal cycles. The poor adhesion of oxide scale is due to the formation of voids at the alloy–oxide interface due to selective oxidation of Al in Ni_3Al.[3] The selective oxidation causes a net flux of vacancies from the matrix to the alloy–oxide interface. These vacancies coalesce to form microvoids at the alloy–oxide interface, and as a result, spallation of oxide scale is observed.

Alloying Ni_3Al with Zr or Hf is reported to enhance the adherence of the oxide scale to the alloy substrate when subjected to thermal cycles between 127°C and 1027°C.[4,5] In addition, alloying Ni_3Al-based alloy with Cr has been attempted for further enhancement of oxidation resistance.[6,7] Comparative study of the oxidation behavior of $Ni_{78}Al_{21}Zr_1B_{0.2}$ and $Ni_{73}Al_{18}Cr_8Zr_1B_{0.2}$ in the temperature range of 560°C–1300°C has shown the Cr addition to be beneficial up to 1150°C. The oxide scale formed on the Cr-alloyed Ni_3Al exposed at temperatures $\leq 1150°C$ has been found to comprise three layers, as mentioned above, with Cr atoms substituting for Ni or Al sites. The Al-rich inner oxide layer is reported to be thin and dense. In contrast, grain boundaries in the $Ni_{78}Al_{21}Zr_1B_{0.2}$ alloy are preferentially oxidized with formation of the oxides of Al and Zr. At temperatures higher than 1150°C, blisters are found to form at alloy grain boundaries due to selective oxidation of Al, which causes the formation of Kirkendall vacancies at the alloy–oxide interfaces. Microvoids are formed in the oxide scale of the $Ni_{73}Al_{18}Cr_8Zr_1B_{0.2}$ alloy by vaporization of CrO_3. With increase in stress, the blister walls are cracked. Hence, alloying with Cr worsens the oxidation resistance of the Ni_3Al-based alloy at temperatures >1150°C.

Diffusion barrier coatings have been developed for improving the oxidation resistance of Ni_3Al-based superalloys such as IC6 ($Ni_{76}Al_{15}Mo_{7.5}$ $B_{0.09}C<0.3$). In spite of its excellent mechanical properties from ambient temperature to 1100°C, the suitability of this alloy for high-temperature application is limited by its inferior resistance to oxidation and corrosion. Coatings of NiCrAlY[8–10] or NiCrAlYSi[11] have been attempted, but the outcome is not satisfactory due to interdiffusion of Mo, Cr, and Al between the coating and the alloy substrate. Recently, a coating containing an outer Al reservoir layer and duplex Re–Cr–Ni–Mo-based diffusion barrier layer has shown promising results.[12] To obtain this coating, the alloy was first subjected to electroplating with Re(Ni)/Ni films, followed by Cr-pack cementation. Subsequently, electroplating was carried out with a Ni film, followed by Al-pack cementation at 1000°C, as a result of which an Al reservoir layer is formed along with layers of Ni_2Al_3 and γ-Ni(Cr,Mo,Al). After oxidation at 1150°C in air, the Al reservoir layer is changed to a γ-$Ni_{79}Cr_4Mo_5Al_{12}$ (all in at%) layer, on which a protective α-Al_2O_3 scale is formed.

9.2.2 NiAl

The high-temperature oxidation resistance of NiAl is superior to that of Ni_3Al because of higher activity of Al in the former alloy. Therefore, NiAl has been used as a bond coat in thermal barrier coatings of Ni-based superalloys. This intermetallic alloy comprises the basis of coatings used for oxidation and corrosion resistance of Ni-based and Co-based superalloys used in gas turbine engines for several decades.[13,14] A continuous and protective layer of Al_2O_3 forms over a range of temperatures and compositions in a single-phase field of NiAl.[15] The growth rate of Al_2O_3 is enhanced due to a low parabolic rate constant. However, the α-Al_2O_3 scale formed on NiAl is susceptible to spallation on cyclic exposure at elevated temperatures due to formation of voids at oxide–alloy interfaces. Addition of minor amounts of noble metallic elements such as Pt, Pd, and Ru[16–18] or reactive elements such as Hf, Zr, La, and Dy[19–23] has been found to improve the cyclic oxidation resistance of NiAl. Alloying with Hf helps by suppressing the growth of such voids. As La and Dy have lower solubility in NiAl than that of Hf or Zr, these elements form precipitates, which subsequently oxidize. Oxides of Hf, Zr, La, or Dy aid in increasing the scale adhesion by acting as pegs at the oxide scale–alloy interface.[23] It has been reported that, on exposure of the NiAl–Dy–Cr alloy at 1200°C, nucleation of α-Cr occurs at the alloy–oxide interfaces.[24] Subsequent oxidation of α-Cr to form CrO_3 (g) followed by escape of the latter in gaseous form leads to the formation of voids at the alloy–oxide interfaces. Formation of voids reduces the net area of interfacial contact, which in turn promotes spallation. In contrast, an alloy containing both Cr and Si (NiAlDyCr–3Si) has shown excellent resistance to cyclic oxidation and scale spallation. Cr and Si are precipitated together at the grain boundaries, and outward diffusion and subsequent oxidation of these elements are restricted. However, the higher concentration of Si (>5 at.%) enhances internal oxidation and lowers oxidation resistance.

Formation of metastable phases (θ, δ, or γ) and their phase transformations (θ→α and γ→α) in alumina scale within the intermediate temperature range of 800°C–1200°C leads to generation of internal stresses and cracking.[25] The stresses are generated as phase transition from cubic to hexagonal trigonal crystal structure in Al_2O_3 leads to 13% reduction in volume. Transverse cracks generated in the oxide scale due to the internal stresses lead to accelerated oxidation, which in turn is responsible for the formation of surface ridges. It has also been reported that θ-Al_2O_3 is less dense and grows faster than α-Al_2O_3. Addition of Zr and Y has been found to reduce the rate of θ→α transformation, besides improving the scale adhesion. The size of ions controls the way the kinetics of θ→α transformation is affected by alloying additions. Whereas larger ions of Zr, Y, La, and Hf are reported to reduce the rate of such phase transformation, Ti ions with smaller size have the opposite effect.

At temperatures ≥1200°C, α-Al_2O_3 has been found to form quickly on oxide dispersion-strengthened NiAl, whereas metastable Al-oxides (θ and γ-Al_2O_3) form first at temperatures ≤1100°C.[26] Addition of Y_2O_3 dispersoids to NiAl increases the duration of the transient stage by delaying the formation of α-Al_2O_3. Therefore, preoxidation of NiAl-based alloys at 1200°C for 3 h enhances the oxidation resistance at lower temperatures, as the oxide scale is uniformly thick and contains no voids. At temperatures of 1050°C and 1100°C, cracks are reported to form due to thermal stresses.

9.3 Titanium Aluminides

9.3.1 Al₃Ti

Al_3Ti is the only intermetallic in the Al–Ti binary system, which forms a protective scale of Al_2O_3 in air. The cast ingots of Al_3Ti have a small volume fraction of metallic Al as second phase on account of peritectic solidification, and a rapid rate of oxidation is observed for short durations of exposure at temperatures above the melting point (660°C) of Al. However, a study by Smialek and Humphrey on isothermal oxidation behavior of Al_3Ti has shown that parabolic rate law is followed at temperatures above 1000°C, as a protective scale of Al_2O_3 is formed.[27] In contrast to the nature of scale formed on isothermal exposure, a mixture of Al_2O_3 and TiO_2 is reported to form on cyclic oxidation.[28] The thermal shock caused by cyclic exposure causes spallation of the oxide scale. Preferential oxidation from Al at the top surface results in enrichment of Ti, which leads to the formation of TiAl. The presence of a two-phase mixture comprising $Al_3Ti + TiAl$ at the top surface is responsible for the formation of TiO_2 in addition to that of Al_2O_3 in the oxide scale. The formation of nonprotective scale of $Al_2O_3 + TiO_2$ on the top of the TiAl phase leads to an accelerated rate of environmental degradation.

As the single-phase Al_3Ti possesses excellent oxidation resistance, attempts have been made to use it as a protective coating against environmental degradation for Ti_3Al and TiAl-based alloys, which are susceptible to oxidation. Such coatings have been processed by the pack aluminizing process.[29-31] It has also been reported that the best oxidation resistance is obtained for an optimum range of coating thickness ($\approx$8–16 mg cm^{-2}),[30] whereas a higher rate of degradation has been found for relatively thicker coatings, which undergo cracking. Cracks are formed due to brittleness of the Al_3Ti coating as well as mismatch of its CTE (15×10^{-6}°C^{-1}) with that of Ti_3Al (CTE $\approx 10 \times 10^{-6}$°C^{-1}).

The oxidation behavior of ternary Al_3Ti–X (X = Ag, Ni, Fe, Mn, or Cr) has been investigated under both isothermal[32,33] and cyclic[28] conditions. For these alloys, the oxidation resistance resulting from ternary alloying with Cr has been found to be the best, whereas the worst results have been obtained for Ag. Interestingly, formation of TiO_2 has been observed in the ternary alloy containing Ag. As the ternary alloys with X = Mn and Cr have shown impressive compressive ductility along with limited tensile elongation to failure, their oxidation behavior has attracted significant interest. Cyclic exposure tests at 1200°C have shown the oxidation resistance from alloying with Cr to be far superior to that from alloying with Mn.[28] A continuous scale of Al_2O_3 is formed in the Al_3Ti–Cr alloy, whereas a mixture of Al_2O_3 and TiO_2 is formed in the Al_3Ti–Mn alloy. It has been further shown that doping of the $Al_{67}Cr_8Ti_{25}$ alloy with oxygen-active elements would reduce the spallation.

9.3.2 TiAl

Insufficient oxidation resistance of TiAl at elevated temperatures (temperatures exceeding 700°C) is a major shortcoming of this intermetallic alloy.[34,35] The oxide scale formed on TiAl is made of a mixture of Al_2O_3 and TiO_2, and therefore it is not protective.[36] However, it has been reported that, on addition of suitable alloying elements such as Nb, W, and Mo, it is possible to improve the oxidation resistance significantly by forming a continuous scale containing only Al_2O_3.[37-40] The results depicted in Figure 9.2 show that alloying of

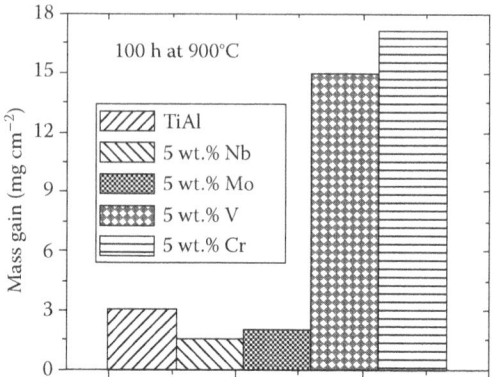

FIGURE 9.2
Bar charts showing mass gain by TiAl and TiAl–5 wt.% X alloys (X=Nb, Mo, Cr, and V) on exposure at 900°C for 100 h. (Data from Kim, B., et al., *Scr. Metall. Mater.*, 33, 1117–1125, 1995.)

TiAl with 5 wt.% Nb or Mo is beneficial for oxidation resistance, whereas addition of a similar amount of Cr or V has an adverse effect.[38] The oxidation resistance is found to improve almost linearly with increase in concentration of Nb, W, or Mo, such that the γ-alloys can be used at temperatures up to 900°C without the application of any coating. The positive effect of Nb or Mo or a combination of both alloying elements on oxidation resistance of TiAl is presented using bar charts in Figure 9.3.[36,39,40] It is also known that Nb additions up to 10 at.% improve the oxidation resistance of TiAl by reducing TiO_2 growth and promoting the development of a more protective layer.[40] It has been shown that a layered oxide scale comprising $TiO_2/Al_2O_3/Nb_2O_5$ is more protective at high temperatures than TiO_2.[38] On the basis of Pt marker-based experiments, it has been concluded that oxidation of TiAl progresses

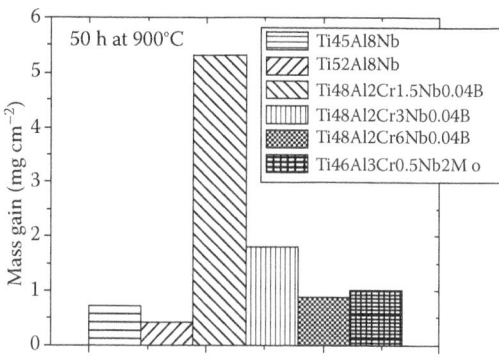

FIGURE 9.3
Bar charts showing mass gain by TiAl alloyed with Nb, Mo, or both along with other elements, drawn using data from literature. (From Pilone, D., et al., *Intermetallics*, 43, 131–137, 2013; Pilone, D. and F. Felli, *Intermetallics*, 26, 36–39, 2012; Lin, J. P., et al., *Intermetallics*, 19, 131–136, 2011.)

by simultaneous outward and inward diffusion of Ti^{4+} and O^{2-}, respectively. In contrast, oxide-scale growth in TiAl alloyed with Nb occurs by inward diffusion of O^{2-} ions.

It has been shown that addition of <4 at.% Cr causes deterioration of the oxidation resistance, whereas alloying with >8 at.% Cr leads to the formation of a protective oxide scale.[41,42] It has been shown that surface alloying of TiAl with Cr significantly enhances oxidation resistance on isothermal exposure at 900°C.[43] In the initial stages of oxidation (up to 30 h), the oxide scale is dominated by the presence of Cr_2O_3 and TiO_2, in spite of the fact that free energies of formation increase in the following order: $Al_2O_3 < TiO_2 < Cr_2O_3$. However, Al_2O_3 and TiO_2 are found to be the main constituents after about 100 h of exposure. The observation of Al_2O_3 only after extended exposure is attributed to slow growth rate of Al_2O_3 due to its low parabolic rate constant. Moreover, the diffusion coefficient of Ti is reported to be higher than that of Al up to a temperature of around 1350°C in air.[44] The disappearance of Cr_2O_3 from the oxide scale is attributed to its solubility in Al_2O_3. The oxide scale of the Cr-alloyed TiAl has been found to be dense and protective, in contrast to the loose and porous scale observed in the case of Cr-free TiAl. This improvement has been attributed to increased oxide-scale adhesion and preferential oxidation of Al caused by alloying TiAl with Cr. Perkins et al. have shown that external Al_2O_3 scales can be formed on TiAl alloys at 1100°C in oxygen and at 1400°C in air by increasing the diffusivity of Al atoms through stabilization of the β phase and simultaneously decreasing the solubility of oxygen in the matrix through alloying with Cr and V.[41] Although addition of Ta is reported to reduce the oxidation rate,[45] the oxide scale is found to spall if subjected to rapid thermal cycling.[42] The cyclic oxidation resistance of TiAl is reported to be enhanced on alloying with Cr, W, Nb,[46] and Y.[47]

Alloying elements such as Nb, W, Mo, and Cr substitute for Ti sites in the TiAl lattice. Therefore, addition of these elements raises the activity of Al in comparison with that of Ti, which in turn makes the formation of a continuous scale of Al_2O_3 thermodynamically feasible.[48,49] Furthermore, it has been shown that addition of the aforementioned alloying elements increases the diffusivity of Al and thereby enhances its available amount at the surface so as to promote the formation of Al_2O_3. Another possible mechanism is reduction in oxygen diffusivity by decrease in its solubility through lowering of the electron/atom ratio.[48] Substitution of the Ti sites by transition-metal alloying elements with valence of +5 or +6, which is higher than that of Ti, leads to reduction in anion-vacancy concentration, which in turn lowers oxygen diffusivity through the oxide scale. It has been observed that alloying of TiAl with only 0.1 at.% P lowers the oxidation rate by reducing the diffusion rate of oxygen through lowering its vacancy concentration.[50] It has also been observed that addition of a sufficient amount of Si leads to the formation of protective SiO_2.

Suitable quaternary alloying addition is also found to enhance the oxidation resistance of some of the γ-TiAl-based ternary alloys. It has been observed that addition of 2 at.% Nb significantly improves the oxidation resistance of the otherwise ductile Ti–48 at.% Al–2 at.% Cr alloy, but not that of the Ti–48 Al–2 V alloy.[51]

9.3.3 Ti$_3$Al

Exposure of Ti$_3$Al-based alloys at elevated temperatures ($\geq$700°C) leads to their degradation through formation of surface cracks in the form of annular rings. This type of environmental damage is expected to lower the creep/stress rupture life at elevated temperatures. The poor oxidation resistance of Ti$_3$Al-based alloys is attributed to formation of nonprotective oxide scale comprising a mixture of TiO$_2$ and Al$_2$O$_3$, with the volume fraction of the former oxide increasing on approaching the alloy substrate, as expected on the basis of thermodynamic considerations. The parabolic rate constant for oxidation of Ti$_3$Al is found to be between the corresponding values for formation of TiO$_2$ and Al$_2$O$_3$.[52] The initial oxidation of Ti$_3$Al results in formation of TiO$_2$ with specific orientations, which evolve into a layered structure comprising an inner TiO$_2$-rich region, an intermediate Al$_2$O$_3$-rich region, and an outer TiO$_2$ layer.[53–55] Furthermore, the interface between the substrate and the oxide is made of an interstitial solid solution of oxygen in the alloy.

A study comparing mass gains at 700°C has shown that the oxidation resistance of Ti$_3$Al is superior to those of other Ti-based alloys for high-temperature applications, such as Ti–6Al–4V, Ti-1100, and IMI 834.[56] However, the mass gain of γ-TiAl has been found to be lower than that of Ti$_3$Al because of higher activity of Al in the former intermetallic alloy. The addition of Nb to Ti$_3$Al up to a specific amount (for example, Nb content $\approx$13 at.%[57]) increases resistance to oxidation by forming a relatively dense oxide scale comprising TiO$_2$.Nb$_2$O$_5$.[57–59] The high density of the oxide layer is attributed to finer grain size of the oxidation products. The oxidation kinetics is found to follow the parabolic rate law. For the Ti–24Al–14Nb–3V–0.5Mo–0.3Si (molar fraction, %) alloy, the isothermal oxidation resistance is reported to be good, with the kinetics being parabolic until 700°C, whereas it is found to be poor at 900°C. Moreover, the oxide scale is found to crack during cooling after being exposed at 1000°C due to thermal shock.[60]

9.4 Iron Aluminides

9.4.1 Oxidizing Environment

Oxidation and corrosion behavior of the Fe-aluminides is controlled by the nature and composition of the oxide scale formed. The minimum concentration of Al required to form a continuous, stable, and protective scale of Al$_2$O$_3$ is reported to be 15 at.% Al in a typical binary Fe–Al alloy.[61] This minimum Al content has been found to depend on the method of processing and composition. In a cast binary Fe–Al alloy, approximately 16–19 at.% Al is reportedly necessary to inhibit internal oxidation and growth of iron oxides at temperatures in the range of 800°C–900°C.[62,63] In the oxide dispersion-strengthened Fe–Al alloys, the minimum concentration of Al required for the formation of protective Al_2O_3 scale is reduced by 3–4 at.% due to much finer grain size.[64] For lower concentrations of Al, the oxide scale comprises Fe_2O_3 and Fe_3O_4. At intermediate Al contents, a transient Al_2O_3 layer with Fe nodules is formed. Formation of Fe nodules can be avoided by increasing the Al content. The Al concentration required to inhibit the formation of Fe nodules can be lowered by additions of small amounts of Cr,[63] whereas it is increased on alloying with Ni.[62] Both Fe_3Al and FeAl have Al concentrations exceeding the minimum amount required for the formation of the protective Al_2O_3 scale. Alloying with Cr is known to accelerate the formation of the protective Al_2O_3 scale.[65]

In the temperature range of 600°C–800°C, the oxide scale formed on Fe-aluminides is known to contain γ-Al_2O_3,[66] and formation of θ-Al_2O_3 has been also reported.[63] Oxidation experiments carried out on Fe_3Al at temperatures from 950°C to 1150°C have shown that formation of α-Al_2O_3 and θ-Al_2O_3 in the oxide scale is promoted at higher and lower temperatures, respectively.[67] Higher oxidation kinetics at lower temperatures has been attributed to the formation of θ-Al_2O_3 with cubic structure containing cation vacancies. Formation of θ-Al_2O_3 has also been reported during transient stages at intermediate temperatures.

The oxidation resistance of B2-structured FeAl (35–50 at.% Al) is known to be superior to that of 304 stainless steel at 1200°C.[68] As expected, the oxidation resistance of FeAl at this temperature is found to scale with concentration of Al. The oxide scale of the FeAl comprises primarily α-Al_2O_3.[69] The mismatch between the CTEs of FeAl and α-Al_2O_3 scale is large, and therefore cyclic exposure leads to spallation of the scale. Pest oxidation has also been reported in the case of FeAl.[70] At lower temperatures, the oxide scale is reported to contain θ-Al_2O_3.[71] The isothermal oxidation behavior of Fe–40 at.% Al alloy in the form of a 200 μm sheet has been investigated in the temperature range of 800°C–1000°C by Lang et al.[72] It has been reported that at temperatures ≤950°C the oxide scale formed in the early stages contains a mixture of θ- and α-Al_2O_3. The formation of θ-Al_2O_3 is responsible for enhanced kinetics of oxidation. The oxidation rate is slower at later stages, as growth of α-Al_2O_3 dominates. Between 1000°C and 1050°C, the oxide scale comprises α-Al_2O_3, and therefore the oxidation kinetics is found to follow the parabolic rate law, except for the initial hour, which probably involves formation of θ-Al_2O_3. At temperatures ≥1150°C, the parabolic rate constant is found to be lower, with the oxide scale being made of columnar grains of α-Al_2O_3, which are three to four times coarser than those formed at lower temperatures. The decrease in parabolic rate constant has been attributed to a change in oxidation mechanism to being controlled by diffusion of oxygen through the grain boundaries of the columnar α-Al_2O_3 grains. Alloying FeAl with small amounts of

Zr ($\approx$0.07 at.%) enhances oxidation resistance by improving the adhesion of the oxide scale to the alloy substrate.[72–79] It has been observed that presence of sulfur adversely affects the adherence of the Al_2O_3 scale. Whereas alloying with IIIB and IVB elements forming sulfides more stable than Al_2S_3 increases the scale adherence, addition of VB and VIB elements whose sulfides are less stable than Al_2S_3 is known to have an adverse effect.[77] Moreover, IIIB elements have been found to be more effective at lower concentrations than IVB elements as the former elements react preferentially with sulfur. It has also been observed that addition of rare-earth alloying elements to Fe–Cr–Al alloys leads to significant improvement in oxidation resistance by lowering the growth stresses and spallation of the oxide scales.[79]

Efforts have been made to improve the oxidation resistance of Fe_3Al by addition of different alloying elements. Alloying Fe_3Al with Ti, Mo, and V has been reported to significantly reduce oxidation resistance.[80] Addition of oxygen-active elements such as Y and Zr has been found to enhance the oxidation resistance of Cr-alloyed Fe_3Al at temperatures exceeding 900°C.[81] Studies on Fe_3Al–2.5 at.% Cr alloys containing minor concentrations of oxygen-active elements have shown their long-duration oxidation behavior to be comparable to that of FeCrAlY alloys and NiAl in the range of 1000°C–1100°C, but inferior at higher temperatures.[82] Furthermore, Fe_3Al-based alloys not containing oxygen-active elements exhibit worse oxidation resistance in air than in oxygen.[64] This observation has been attributed to internal nitridation occurring in air due to the presence of defective oxide scale.

9.4.2 Corrosive Environment (Other than Oxidizing Atmosphere)

The Fe-aluminides are of interest because of their potential for use in coal gasifiers with low oxygen ($<10^{-19}$ atm) and high sulfur activities ($>10^{-8}$ atm). The behavior of Fe-aluminides in different types of oxidative and corrosive environments at elevated temperatures has been extensively reviewed by Tortorelli and Natesan.[64] In sulfidizing atmospheres, the response of Fe-aluminides depends on the nature of the sulfur-containing environment, sulfur activity, and partial pressure of oxygen. A protective scale of Al_2O_3 is known to form even at low partial pressures of oxygen and high sulfur activity.[83] In pure sulfur vapor or an atmosphere containing a mixture of H_2S and H_2, the scale of Al_2O_3 does not form. Fe-aluminides containing >18 at.% Al have shown impressive resistance to degradation until 750°C.[84–88] In the range of 750°C–900°C, the Fe–28 at.% Al alloy forms a scale of Al_2S_3 at a pressure of 10^{-3} Pa, but degrades rapidly at pressure >133 Pa.

To evaluate corrosion resistance under sulfidizing conditions, Fe-aluminide-based alloys are usually exposed in an environment comprising an H_2S–H_2–H_2O mixture at 700°C–800°C. Fe_3Al-based alloys containing 0–10 at.% Cr and 1–2 at.% Nb, Mo, V, Zr, or Y were exposed in the product gas stream of an operating coal gasifier.[89] A majority of these alloys are reported to resist oxidation and sulfidation in gas mixtures with oxygen partial pressures

$<10^{-21}$ atm. The Fe$_3$Al-based alloy containing 28 at.% Al and 2 at.% Cr has shown excellent resistance to corrosion in a sulfidizing atmosphere, whereas higher Cr content lowers the resistance to sulfur attack.[82] In other words, the corrosion resistance of this alloy can be compared to that of 304L stainless steel. At a given Cr level, addition of Mo leads to further improvement in the resistance of the Fe$_3$Al–Cr alloy to localized corrosion induced by chloride-containing solutions.[90]

In the presence of carbon-bearing gases, resistance to environmental degradation requires the formation of protective Al$_2$O$_3$ scale, which needs to be permitted under the available partial pressure of oxygen and fugacity of carbon. If the carbon fugacity is sufficiently high, a mixture of Fe and Al carbides is expected to form, causing embrittlement of the surface. This leads to "metal dusting" of the Fe-aluminides, which in turn enhances the ingress of carbon, leading to supersaturation in terms of carburization, formation of M$_3$C (such as Fe$_3$C) carbides, and deposition of carbon.[91] It has been shown that substantial degradation of Fe$_3$Al occurs in environments with high carbon activity. However, addition of 2–4 at.% Cr has been found to reduce the amount of degradation significantly.[91] FeAl-based alloys exhibit superior carburization resistance compared with that of conventional heat-resistant alloys, because of the formation of protective scale of Al$_2$O$_3$ in the case of the former type of alloy.[92]

The resistance of Fe$_3$Al to chlorine-bearing gases (2% HCl) at elevated temperatures has been found to be impressive.[93] Under such atmospheres, the alloys containing 5 at.% Cr are found to be superior to those with 2 at.% Cr. Fe$_3$Al–5 at.% Cr (28 at.% Al) alloys have shown excellent resistance to degradation in an environment of H$_2$S–H$_2$–H$_2$O–HCl gases in the range of 450°C–550°C, as reported by Saunders et al.[94] On the other hand, the Fe–16Al–5Cr alloy has shown significant susceptibility to degradation in the same environment, indicating that a minimum activity of Al is required to ensure the formation of protective Al$_2$O$_3$ scale.

9.5 Summary

The oxidation behavior of aluminides of Ni, Ti, and Fe is controlled by the growth kinetics of the protective Al$_2$O$_3$ scale relative to the rate of formation of oxides of other alloy constituents. As expected, the oxidation resistance of the aluminides containing a higher amount of Al, such as NiAl, Al$_3$Ti, and FeAl is excellent, and superior to that of Ni$_3$Al, γ-TiAl, and Fe$_3$Al, respectively, which have lower concentrations of Al. The oxidation resistance of the latter group of Al-lean aluminides can be significantly increased through addition of suitable alloying elements. Oxidation resistance of Ni$_3$Al is improved by alloying with Cr, Zr, and Hf, whereas that of γ-TiAl is enhanced by Nb, W,

Mo, and Cr. Addition of oxygen-active elements such as Zr and Y to Fe_3Al-Cr alloys is found to increase its oxidation resistance. These alloying elements either improve the adhesion of the Al_2O_3 scale or lower diffusivity of oxygen ions by reducing vacancy concentration. Lowering the activity of Ni, Ti, or Fe in Ni_3Al, γ-TiAl, and Fe_3Al, respectively, by addition of suitable alloying elements also helps in enhancing the driving force for Al_2O_3 scale formation. In the Al_2O_3 scale, internal stresses are generated due to phase transformations at temperatures below 1000°C. The Fe-aluminides also exhibit excellent resistance to hot corrosion under sulfidizing, chloride-rich, and carburizing environments.

References

1. Liu, C. T. and D. P. Pope. 1994. Ni_3Al and its alloys. In *Intermetallic Compounds, Vol. 2, Practice*, eds. J. H. Westbrook and R. L. Fleischer, 17–51. Chichester, UK: Wiley.
2. Liu, C. T. and C. L. White. 1987. Dynamic embrittlement of boron-doped Ni_3Al alloys at 600°C. *Acta Metall.* 35:643–649.
3. Kuenzly, J. D. and D. L. Douglass. 1974. The oxidation mechanism of Ni_3Al containing yttrium. *Oxid. Met.* 8:139–178.
4. Taniguchi, S. and T. Shibata. 1986. Cyclic oxidation behavior of Ni_3Al −0.1b base alloys containing a Ti, Zr, or Hf addition. *Oxid. Met.* 25:201–216.
5. Taniguchi, S., T. Shibata, and H. Tsuruoka. 1986. Isothermal oxidation behavior of Ni_3Al −0.1b base alloys containing Ti, Zr, or Hf additions. *Oxid. Met.* 26:1–17.
6. Taniguchi, S. and T. Shibata. 1987. Oxidation behavior of Ni_3Al-0.1B containing 2Cr. *Oxid. Met.* 28:155–163.
7. Pan, Y. C., T. H. Chuang, and Y. D. Yao. 1991. Long-term oxidation behavior of Ni_3Al alloys with and without chromium additions. *J. Mater. Sci.* 26:6097–6103.
8. Lang, F. and T. Narita. 2007. Improvement in oxidation resistance of a Ni_3Al-based superalloy IC6 by rhenium-based diffusion barrier coatings. *Intermetallics* 15:599–606.
9. Wang, B., R. F. Huang, G. H. Song, J. Gong, C. Sun, L. S. Wen, and Y. F. Han. 2001. Interdiffusion behavior of Ni-Cr-Al-Y coatings deposited by arc-ion plating. *Oxid. Met.* 56:1–13.
10. Han, Y. F., Z. P. Xing, M. C. Chaturvedi, and Q. Xu. 1997. Oxidation resistance and microstructure of Ni–Cr–Al–Y–Si coating on Ni_3Al based alloy. *Mater. Sci. Eng. A* 239–240:871–876.
11. Zhang, D. B., S. K. Gong, H. B. Xu, and Y. F. Han. 2002. Influence of Mo diffusion on the thermal behavior of TBCs on Ni3Al based alloy IC-6. *Acta Metall. Sin. (Eng. Lett.)* 15:15.
12. Huo, X., J. S. Zhang, B. L. Wang, F. J. Wu, and Y. F. Han. 1999. Evaluation of NiCrAlYSi overlay coating on Ni_3Al based alloy IC-6 after an engine test. *Surf. Coat. Technol.* 114:174–180.
13. Goward, G. W. 1970. Current research on the surface protection of superalloys for gas. Turbine engines. *JOM* 22(10):31–39.

14. Boone, D. H. and G. Goward. 1970. The use of nickel-aluminum intermetallic systems as coatings for high temperature nickel base alloys. In *Ordered Alloys— Structural Applications and Physical Metallurgy. Proceedings of 3rd Bolton Landing Conference*, eds. B. Kear, C. T. Sims, N. S. Stoloff, and J. H Westbrook, 545. Baton Rouge: Claitor's Publication Division.

15. Hutchings, R. and M. H. Loretto. 1978. Compositional dependence of oxidation rates of NiAl and CoAl. *Met. Sci.* 12:503.

16. Hou, P. Y. and K. F. McCarty. 2006. Surface and interface segregation in beta-NiAl with and without Pt addition. *Scr. Mater.* 54:937–941.

17. Leyens, C., B. A. Pint, and I. G. Wright. 2000. Effect of composition on the oxidation and hot corrosion resistance of NiAl doped with precious metals. *Surf. Coat. Technol.* 133–134:15–22.

18. Swadźba, R., M. Hetmańczyk, M. Soźanska, B. Witala, and L. Swadźba. 2011. Structure and cyclic oxidation resistance of Pt, Pt/Pd-modified and simple aluminide coatings on CMSX-4 superalloy. *Surf. Coat. Technol.* 206:1538–1544.

19. Guo, H. B., L. D. Sun, H. F. Li, and S. K. Gong. 2008. High temperature oxidation behavior of hafnium modified NiAl bond coat in EB-PVD thermal barrier coating system. *Thin Solid Films* 516:5732–5735.

20. Barrett, C. A. 1988. Effect of 0.1 at-percent zirconium on the cyclic oxidation resistance of beta-NiAl. *Oxid. Met.* 30:361–390.

21. Jedliński, J. and S. Mrowec. 1987. The influence of implanted yttrium on the oxidation behaviour of β-NiAl. *Mater. Sci. Eng.* 87:281–287.

22. Pint, B. A. 1998. The oxidation behavior of oxide-dispersed beta-NiAl: I. Short-term performance at 1200°C. *Oxid. Met.* 49:531–559.

23. Li, D. Q., H. B. Guo, D. Wang, T. Zhang, S. K. Gong, and H. B. Xu. 2013. Cyclic oxidation of beta-NiAl with various reactive element dopants at 1200°C. *Corros. Sci.* 66:125–135.

24. He, J., Y. Luan, H. B. Guo, H. Peng, Y. L. Zhang, T. Zhang, and S. K. Gong. 2013. The role of Cr and Si in affecting high-temperature oxidation behaviour of minor Dy doped NiAl alloys. *Corros. Sci.* 77:322–333.

25. Pint, B. A., M. Treska, and L. W. Hobbs. 1997. The effect of various oxide dispersions on the phase composition and morphology of Al_2O_3 scales grown on beta-NiAl. *Oxid. Met.* 47:1–20.

26. González-Carrasco, J. L., P. Pérez Zubiaur, P. Adeva, and J. Chao. 1999. Oxidation behaviour of an ODS NiAl-based intermetallic alloy. *Intermetallics* 7:69–78.

27. Smialek, J. L. and D. L. Humphrey. 1992. Oxidation-kinetics of cast $TiAl_3$. *Scr. Metall. Mater.* 26:1763–1768.

28. Parfitt, L. J., J. L. Smialek, J. P. Nic, and D. E. Mikkola. 1991. Oxidation behavior of cubic phases formed by alloying Al_3Ti with Cr and Mn. *Scr. Metall. Mater.* 25:727–731.

29. Kabbaj, M., A. Galerie, and M. Caillet. 1985. Composite $TiAl_3$-tin coatings on titanium. *J. Less Common Met.* 108:1–21.

30. Smialek, J. L., M. A. Gedwill, and P. K. Brindley. 1990. Cyclic oxidation of aluminide coatings on $Ti_3Al + Nb$. *Scr. Metall. Mater.* 24:1291–1296.

31. Mabuchi, H., K. I. Hirukawa, and Y. Nakayama. 1989. Formation of structural L12 compounds in $TiAl_3$-base alloys containing Mn. *Scr. Metall.* 23:1761–1765.

32. Hirukawa, K., H. Mabuchi, and Y. Nakayama. 1991. Oxidation behavior of $TiAl_3$-based alloys with the L12 structure. *Scr. Metall. Mater.* 25:1211–1216.

33. Chen, S. P., W. Zhang, Y. H. Zhang, G. X. Hu, T. F. Li, J. N. Shen, and L. J. Zhou. 1992. Oxidation behavior of Al$_3$Ti-based Ll2-type intermetallic alloys. *Scr. Metall. Mater.* 27:455–460.

34. Becker, S., A. Rahmel, M. Schorr, and M. Schutze. 1992. Mechanism of isothermal oxidation of the intermetallic TiAl and of TiAl alloys. *Oxid. Met.* 38:425–464.

35. Yoshihara, M. and Y. W. Kim. 2005. Oxidation behavior of gamma alloys designed for high temperature applications. *Intermetallics* 13:952–958.

36. Pilone, D., F. Felli, and A. Brotzu. 2013. High temperature oxidation behaviour of TiAl-Cr-Nb-Mo alloys. *Intermetallics* 43:131–137.

37. Ye, H. Q. 1999. Recent developments in Ti$_3$Al and TiAl intermetallics research in China. *Mater. Sci. Eng. A* 263:289–295.

38. Kim, B. G., G. M. Kim, and C. J. Kim. 1995. Oxidation behavior of TiAl-X (X = Cr, V, Si, Mo or Nb) intermetallics at elevated-temperature. *Scr. Metall. Mater.* 33:1117–1125.

39. Pilone, D. and F. Felli. 2012. Isothermal oxidation behaviour of TiAl-Cr-Nb-B alloys produced by induction melting. *Intermetallics* 26:36–39.

40. Lin, J. P., L. L. Zhao, G. Y. Li, L. Q. Zhang, X. P. Song, F. Ye, and G. L. Chen. 2011. Effect of Nb on oxidation behavior of high Nb containing TiAl alloys. *Intermetallics* 19:131–136.

41. Perkins, R. A., K. T. Chiang, and G. H. Meier. 1987. Formation of alumina on Ti-Al alloys. *Scr. Metall.* 21:1505–1510.

42. Mckee, D. W. and S. C. Huang. 1992. The oxidation behavior of gamma-titanium aluminide alloys under thermal cycling conditions. *Corros. Sci.* 33:1899–1914.

43. He, Z. Y., Z. X. Wang, F. Zhang, Z. Y. Wang, and X. P. Liu. 2013. Oxidation behavior of TiAl alloy treated by plasma surface chromizing process. *Surf. Coat. Technol.* 228:S287–S291.

44. Mishin, Y. and C. Herzig. 2000. Diffusion in the Ti-Al system. *Acta Mater.* 48:589–623.

45. McAndrew, J. B. and H. D. Kessler. 1956. Ti-36%A1 as base for high temperature alloys. *J. Met.* 8:1348–1353.

46. McKee, D. W. and S. C. Huang. 1991. Oxidation behavior of gamma-titanium aluminide alloys. In *High Temperature Ordered Intermetallics IV*, eds. L. A. Johnson, D. P. Pope, and J. O. Stiegler, Materials Research Society Symposium Proceedings, Vol. 213, 939. Pittsburgh, PA: MRS.

47. Barinov, S. M., T. T. Nartov, Z. I. Kornilova, Yu. L. Krasulin, and E. M. Lazarev. 1987. Oxidability of some cast and sintered alloys based on the intermetallic compounds TiAl, NiAl, and Ni$_3$Al. *Soviet Powder Metallurgy and Metal Ceramics* 26(12):1002–1006.

48. Meier, G. H., D. Appalonia, R. A. Perkins, and K. T. Chiang. 1989. Oxidation of titanium based alloys. In *Oxidation of High Temperature Intermetallics*, eds. T. Grobstein and J. Doychak, 185. Warrendale, PA: TMS.

49. Kasahara, K., K. Hashimoto, H. Doi, and T. Tsujimoto. 1989. Oxidation behavior of intermetallic compounds TiAl at high-temperatures. *J. Jpn. Inst. Met.* 53:58–62.

50. Ikematsu, Y., T. Hanamura, H. Morikawa, M. Tanino, and J. Takamura. 1991. Oxidation resistance of phosphorus-doped titanium aluminides. In *Intermetallic Compounds—Structure and Mechanical Properties*, ed. O. Izumi, 191. Sendai: The Japan Institute of Metals.

51. Huang, S. C., D. W. McKee, D. S. Shih, and J. C. Chestnutt. 1991. TiAl alloys containing chromium and niobium. In *Intermetallic Compounds—Structure and Mechanical Properties*, ed. O. Izumi, 363. Sendai: The Japan Institute of Metals.

52. Khobaib, M. and F. W. Vahldiek. 1988. High temperature oxidation behavior of Ti₃Al alloys. In *Space Age Metal Technology, Proceedings of 2nd International SAMPE Metals and Processing Metals Conference* II, 262. Covina, CA: SAMPE.
53. Welsch, G., S. L. Friedman, and A. I. Kahveci. 1991. *In situ* investigation of the oxidation of titanium aluminide alloys. In *Microscopy of Oxidation*, eds. M. J. Bennett and G. W. Lorimar, 193. London: Institute of Metals.
54. Perkins, R. A., K. T. Chiang, and G. H. Meier. 1987. Formation of alumina on TiAl alloys. *Scr. Metall. Mater.* 21:1505–1510.
55. Shanabarger, M. R. 1992. The initial oxidation of α2(Ti₃Al) and γ(TiAl) titanium aluminide alloys. *Mater. Sci. Eng. A* 153:608–612.
56. Peters, M., Y. T. Lee, K. -J. Grundhoff, H. Schurmann, and G. Welsch. 1991. Influence of processing on microstructure and mechanical properties of Ti-1100 and IMI 834. In *Titanium 90: Products and Applications*, Proceedings of the Symposium on Microstructure/Property Relationships in Titanium Aluminides and Alloys presented at the 1990 TMS Fall Meeting; Detroit, MI, USA, 1:199. Dayton, OH: Titanium Development Association.
57. Fang, W., S.-H. Ko, H. Hashimoto, T. Abe, and Y.-H. Park. 2002. High temperature oxidation behavior of Ti₃Al–Nb alloys prepared by pulse discharge sintering. *Mater. Sci. Eng. A* 329–331:708–712.
58. Welsch, G. and A. I. Kahveci. 1988. Oxidation behavior of titanium aluminide alloys. In *Oxidation of High Temperature Intermetallics*, eds. T. Grobstein and J. Doychak, 207. Warrendale, PA: TMS.
59. Wiedmann, K. E., S. N. Sankaran, R. K. Clark, and T. A. Wallace. 1988. Static and dynamic oxidation of Ti-14A1-21Nb. In *Oxidation of High Temperature Intermetallics*, eds. T. Grobstein and J. Doychak, 195. Warrendale, PA: TMS.
60. Yu-Hai, Q., L. Mei-Shuan, and L. Bin. 2009. Isothermal oxidation behavior of Ti₃Al-based alloy at 700–1000°C in air. *Trans. Nonferr. Met. Soc. China* 19:525–529.
61. Vedula, K. 1994. FeAl and Fe₃Al. In *Intermetallic Compounds, Vol. 2, Practice*, eds. J. H. Westbrook and R. L. Fleischer, 199. Chichester, UK: Wiley.
62. Tomaszewicz, P. and G. R. Wallwork. 1983. Observations of nodule growth during the oxidation of pure binary iron aluminum-alloys. *Oxid. Met.* 19:165–185.
63. Devan, J. H. and P. F. Tortorelli. 1993. The oxidation sulfidation behavior of iron-alloys containing 16–40 AT% aluminum. *Corros. Sci.* 35:1065–1071.
64. Tortorelli, P. F. and K. Natesan. 1998. Critical factors affecting the high-temperature corrosion performance of iron aluminides. *Mater. Sci. Eng. A* 258:115–125.
65. Tortorelli, P. F. and J. H. DeVan. 1992. Behavior of iron aluminides in oxidizing and oxidizing/sulfidizing environments. *Mater. Sci. Eng. A* 153:573–577.
66. Sakiyama, M., P. Tomaszewicz, and G. R. Wallwork. 1979. Oxidation of iron-nickel aluminum-alloys in oxygen at 600–800°C. *Oxid. Met.* 13:311–330.
67. Babu, N., R. Balasubramaniam, and A. Ghosh. 2001. High-temperature oxidation of Fe₃Al-based iron aluminides in oxygen. *Corros. Sci.* 43:2239–2254.
68. Schmidt, B., P. Nagpal, and I. Baker. 1989. Annealing studies of B2 FeAl. In *High-Temperature Ordered Intermetallic Alloys III*, eds. C. C. Koch, C. T. Liu, N. S. Stoloff, and A. I. Taub, Materials Research Society Symposium Proceedings, Vol. 133, 755–760. Pittsburgh, PA: MRS.
69. Stephens, J. R. 1985. The B2 aluminides as alternative materials. In *High-Temperature Ordered Intermetallic Alloys I*, eds. C.C. Koch, C.T. Liu, N.S. Stoloff, Materials Research Society Symposium Proceedings, Vol. 39, 381–396. Pittsburgh, PA: MRS.

70. Sawada, H., A. Masuda, Y. Atobe, and T. Atoki. 1960. Disintegration of Fe-Al alloy. *Journal of the Japan Institute of Metals and Materials* 11:732–735.
71. Smialek, J. L., J. Doychak, and D. J. Gaydosh. 1988. Oxidation behavior of Fe-Al+Zr,Hf,B. *Oxidation of High Temperature Intermetallics*, 83. Warrendale, PA: TMS.
72. Lang, F. Q., Z. M. Yul, S. Gedevanishvili, S. C. Deevi, and T. Narita. 2003. Isothermal oxidation behavior of a sheet alloy of Fe-40 at.% Al at temperatures between 1073 and 1473 K. *Intermetallics* 11:697–705.
73. Lang, F. Q., Z. M. Yu, S. Gedevanishvili, S. C. Deevi, and T. Narita. 2003. Corrosion behavior of Fe-40Al sheet in N_2–$11.2O_2$–$7.5CO_2$ atmospheres with various SO_2 contents at 1273 K. *Intermetallics* 11:135–141.
74. Barrett, C. A., A. S. Khan, and C. E. Lowell. 1981. The effect of zirconium on the cyclic oxidation of NiCrAl alloys. *J. Electrochem. Soc.* 128:25–32.
75. Prescott, R., D. F. Mitchell, M. J. Graham, and J. Doychak. 1995. Oxidation mechanisms of β-NiAl + Zr determined by SIMS. *Corros. Sci.* 37(9):1341–1364.
76. Pint, B. A., J. R. Martin, and L. W. Hobbs. 1993. O/SIMS characterization of the growth mechanism of doped and undoped α-Al_2O_3. *Oxid. Met.* 39(3/4):167–195.
77. Sigler, D. R. 1989. Aluminum oxide adherence on Fe-Cr-Al alloys modified with group IIIB, IVB, VB, and VIB elements. *Oxid. Met.* 32(5/6):337–355.
78. Hou, P. Y. 1989. Beyond the sulfur effect. *Oxid. Met.* 52(3/4):337–351.
79. Forest, C. and J. H. Davidson. 1995. Some observations on the effects of sulfur and active elements on the oxidation of Fe–Cr–Al alloys. *Oxid. Met.* 43(5/6):479–490.
80. DeVan, J. H. 1989. Oxidation behavior of Fe3Al and derivative alloys. In *Oxidation of High Temperature Intermetallics*, eds. T. Grobstein and J. Doychak, 107. Warrendale, PA: TMS.
81. Tortorelli, P. F. and J. H. DeVan. 1994. Compositional influences on the high-temperature corrosion resistance of iron aluminides. In *Processing, Properties, and Applications of Iron Aluminides*, eds. J. H. Schneibel, and M. A. Crimp, 257. Warrendale, PA: TMS.
82. Pint, B. A., P. F. Tortorelli, and I. G. Wright. 1996. The oxidation behavior of ODS iron aluminides. *Mater. Corros.* 47:663–674.
83. Natesan, K. 1997. Corrosion performance of alumina scales in coal gasification environments. *Mater. High Temp.* 14(2):137–145.
84. Strafford, K. N. and R. Manifold. 1972. Effects of aluminum alloying additions on sulfidation behavior of iron. *Oxid. Met.* 5:85–112.
85. Patnaik, P. C. and W. W. Smeltzer. 1985. Sulfidation properties of Fe-Al alloys at 1173 K in H_2S-H_2 atmospheres. *Oxid. Met.* 23:53–75.
86. Patnaik, P. C. and W. W. Smeltzer. 1985. Sulfidation properties of Fe-Al alloys (6–28 a/o Al) at 1173 K in sulfur vapor at $P_{s2} = 1.45 \times 10^{-3}$Pa. *J. Electrochem. Soc.* 132:1226–1232.
87. Smith, P. J. and W. W. Smeltzer. 1987. A method for long-term sulfidation of metal at low sulfur pressures and its application to sulfidation of an Fe-20 at% Al alloy at 1023 K. *Oxid. Met.* 28:291–308.
88. Mrowec, S. 1995. The problem of sulfur in high-temperature corrosion. *Oxid. Met.* 44:177–209.
89. DeVan, J. H. 1991. Corrosion performance of iron aluminide (Fe_3Al) in coal conversion process environments. In *Heat Resistant Materials*, eds. K. Natesan and D. J. Tillack, 235. Materials Park, OH: ASM International.

90. Buchanan, R. and J. G. Kim. 1992. Fe3Al-type iron aluminides: Aqueous corrosion properties in a range of electrolytes and slow-strain-rate ductilities during aqueous corrosion. ORNL report, Subcontract 88-0768CT92/02, Springfield, VA: National Technical Information Service.

91. Grabke, H. J. 1995. Metal dusting of low- and high-alloy steels. *Corrosion* 51:711–720.

92. Strauss, S., R. Krajak, M. Palm, and H. J. Grabke. 1996. Metal dusting of Fe_3Al and $(Fe,Ni)_3Al$. *Mater. Corros.* 47:701–702.

93. Natesan, K. and R. N. Johnson. 1995. Corrosion performance of Fe-Cr-Al and Fe aluminide alloys in complex gas environments. In *Heat-Resistant Materials II*, eds. K. Natesan, P. Ganesan, and G. Lai, 591–599. Materials Park, OH: ASM International.

94. Saunders, S. R. J., D. D. Gohil, J. P. Banks, M. U. Sheriff, P. F. Tortorelli, J. H. De Van, and I. G. Wright. 1997. Behavior of fecralloy and iron aluminides alloys in coal gasification atmospheres containing HCl. *Mater. Sci. Forum* 251–254:583–590.

Index